Measuring the Universe
The Cosmological Distance Ladder

Springer
London
Berlin
Heidelberg
New York
Barcelona
Hong Kong
Milan
Paris

Stephen Webb

Measuring the Universe

The Cosmological Distance Ladder

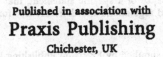

Published in association with
Praxis Publishing
Chichester, UK

Dr Stephen Webb
Department of Mathematical Sciences
Loughborough University
Loughborough
Leicestershire, UK

SPRINGER-PRAXIS SERIES IN ASTRONOMY AND ASTROPHYSICS
SERIES EDITOR: John Mason B.Sc., Ph.D.

ISBN 1-85233-106-2 Springer-Verlag Berlin Heidelberg New York

British Library Cataloguing in Publication Data
Webb, Stephen
    Measuring the universe : the cosmological distance ladder.
    – (Springer-Praxis series in astronomy & astrophysics)
    1. Cosmological distances
    I. Title
    523.1

    ISBN 1852331062

Library of Congress Cataloging-in-Publication Data
Webb, Stephen, 1963–
    Measuring the universe : the cosmological distance ladder /
Stephen Webb.
        p.   cm. – (Springer-Praxis series in astronomy and
astrophysics)
    "Published in association with Praxis Publishing, Chichester, UK."
    Includes bibliographical references and index.
    ISBN 1-85233-106-2 (alk. paper)
    1. Cosmological distances–Measurement.   I. Title.   II. Series.
QB991.C66W43   1999
523.1–dc21   98-43764   CIP

© Praxis Publishing Ltd, Chichester, UK, 1999
Printed by MPG Books Ltd, Bodmin, Cornwall, UK
Reprinted 2001

Cover design: Jim Wilkie

Printed on acid-free paper supplied by Precision Publishing Papers Ltd, UK

To my parents, Ron and Ronnie

Table of contents

Author's preface

How big is the observable universe? This has been perhaps one of the most hotly disputed questions in astronomy over the last three decades. For the most part the dispute has been conducted in the technical literature, which means that the beginning student of astronomy has been sheltered from the controversy. When the dispute is discussed in lectures, it is usually dealt with perfunctorily; more often it is left to an advanced course. This is a pity because the essence of the dispute is easily stated. Simply put, astronomers disagree about the distance to quite nearby objects. The uncertainty in the size and age of the observable universe is effectively just a reflection of our uncertainty in the distance to nearby stars and galaxies.

It is not surprising that there should be this uncertainty. All that astronomers have to work with are the photons that their telescopes collect; their observations are subject to all sorts of biases; and whichever cosmological model they prefer, one thing is certain: stars and galaxies are *very* far away. Determining the distance to an object is thus probably the hardest task that an astronomer faces. The wonder is not that astronomers disagree about the cosmological distance scale; rather, it is that they can say anything definite about distances at all.

So how *do* astronomers deduce the distance to a star, or a galaxy, or a quasar? The answer is that they make use of a 'ladder', which has taken many years of painstaking effort to construct. By understanding the universe and the objects it contains on one rung of the ladder, astronomers get clues that let them reach out and understand objects on the next rung of the ladder. In this way, step by step, they can determine distances in the universe. This book tells the story of how astronomers have calibrated the various rungs of the distance ladder, beginning with distances on the Earth and ending with the limits of the observable universe.

The book is aimed primarily at students taking a first or second course in astronomy or astrophysics, but it is not intended to be a stand-alone course textbook. I hope instead that it may prove useful as a companion to a more conventional textbook. I have tried to write the book in such a way that it will also be accessible to amateur astronomers and to students of physics at A-level; some mathematics is unavoidable, but I have attempted to keep it to a minimum.

The organisation of the book is as follows. Each chapter — eleven in all — describes the historical development of our understanding of one particular rung of the ladder. (I take this approach because it fits nicely with the subject matter, but also because the field is so fast-moving. Readers can use the Web addresses given in the 'further reading' sections to keep up to date with the latest discoveries. However, the book is not intended to be a formal history of the subject, which is a task that would have been quite beyond me.)

After a summary, and a short section that contains a mix of problems requiring both essay-type answers as well as numerical solutions, every chapter ends with a 'further reading' section. These 'further reading' sections contain the sources I have drawn upon most heavily during the writing of this book, some general references at the level of *Scientific American*, some of the classic papers that have helped build the distance ladder and, for the advanced student, references to the recent research literature.

The story of how astronomers came to build the cosmological distance ladder is a fascinating one. Equally fascinating are the astronomers themselves, and I have tried to give biographical sketches of some of the important characters in the story.

Acknowledgements

I would first like to express my thanks to the dozens of astronomers around the world who answered my questions with great patience.

For their help in obtaining references I am grateful to the library staff at the universities of Sheffield, Leicester and Loughborough, and in particular to Mrs Maureen Webster of the Hicks physics library at Sheffield University. David Hanon, Richard Dreiser, Richard French, Bonnie Lundt, Dorothy Schaumberg and Ikuyo Tagawa Garber were of great help in my attempts to find suitable illustrations.

I am indebted to Dr Catherine Molloy of Cardiff University, who provided the original idea for the book; to Bob Marriott, who did an excellent job of copy editing it; and to Clive Horwood of Praxis, who has been a constant source of guidance throughout its preparation.

I would particularly like to thank Dr Nial Tanvir of Cambridge University and Professor Jerry Griffiths of Loughborough University, who made many valuable comments on an early draft, and who caught many errors. I am only too aware that errors will persist; these are, of course, my sole responsibility and I would be most grateful to have them brought to my attention.

Finally, I would like to thank my friends and family — particularly Peter, Jackie and Emily — for tolerating me while I was writing.

Figure credits

I gratefully acknowledge the following sources of illustrations reproduced in this book, and thank the authors, archivists and publishers who have granted their permission.

The portraits of Brahe, Copernicus, Curtis, Herschel, Huggins, Kepler and Pickering are courtesy of the Mary Lea Shane Archives of the Lick Observatory, University of California-Santa Cruz. The portraits of de Sitter, Draper, Galileo, Newton and Russell are courtesy of the Yerkes Observatory. The portrait of Cannon is courtesy of the Whitin Observatory, Wellesley College. The portraits of Baade, Einstein, Hubble, Humason and Kapteyn are courtesy of the Mount Wilson and Las Campanas Observatories. The portraits of Leavitt, Zel'dovich and Penzias and Wilson are courtesy of the American Institute of Physics, Emilio Segré Visual Archives. (The photograph of Zel'dovich shows him talking to Andrei Sakharov at an international conference on quantum gravity held in Moscow in 1987.) The portrait of Hertzsprung is courtesy of Dorritt Hoffleit. The portrait of Shapley is courtesy of Owen Gingerich and Helen Sawyer Hogg. The portrait of Eddington is courtesy of the Astronomical Society of the Pacific. The portrait of Slipher is courtesy of the Lowell Observatory. The portrait of Guth is courtesy of Donna Coveny/MIT. The portrait of Sandage is courtesy of the Archives, California Institute of Technology.

Figure 1.1 is copyright © NASA/STScI/AURA. Figure 7.3 is from *Pulsating Stars and Cosmic Distances* by Robert P. Kraft; courtesy of Ikuyo Tagawa Garber, copyright © estate of Bunji Tagawa. Figure 7.4 is courtesy of David Hanon. Figures 8.1, 8.2 and 9.1 are copyright © Anglo-Australian Observatory; photography by David Malin. Figures 9.2 and 9.3 were taken by R. Sahai and J. Trauger using the Wide Field Planetary Camera 2; copyright © NASA/STScI/AURA. Figures 9.4 and 9.9 are adapted from Jacoby *et al.* (see p. 234 for the reference) copyright © Astronomical Society of the Pacific. The Wide Field Planetary Camera 2 was used to obtain figure 9.14 (W. N. Colley, E. Turner and J. A. Tyson), figure 9.15 (M. Franx and G. Illingworth) and figure 9.16 (C. S. J. Pun and R. Kirshner); all these images are copyright © NASA/STScI/AURA. Figures 10.3 and 10.4 are adapted from Ned Wright's cosmology tutorial Web site. Figure 10.6 is adapted from *Gravity and Spacetime* by J. A. Wheeler (1990). Figure 11.4 is courtesy of David Hanon. Figure 11.5 is adapted from *Introduction to Cosmology* by J. V. Narlikar (1983). Figure 11.6 is adapted from a diagram by Robin Ciardullo. Figure 11.7 is courtesy of Arjun Dey and copyright © Johns Hopkins University. Figure 11.8 is copyright © NASA/STScI/AURA.

1

Introduction

In October 1963, at the height of the Cold War, a treaty came into force that banned the testing of nuclear weapons in space. Signing a treaty and abiding by its terms are two different things, of course. The United States wanted proof that their Cold War enemies were sticking to the test ban. Therefore, before most countries had even signed the treaty, the United States Air Force launched the first of the *VELA* spy satellites. The name of the satellites came from the Spanish verb 'velar': to watch. The *VELA* satellites, hunting in pairs, watched for signs of a nuclear explosion.

A nuclear explosion emits lots of neutrons and γ-rays. The detection of these particles would, according to the intelligence agencies, be evidence of a clandestine nuclear test. The *VELA* satellites therefore carried an array of neutron and γ-ray detectors. The original detectors were rather poor, but with each launch their efficiency increased. In 1969, the American astronomer Ray William Klebesadel (1932–) and his colleagues were developing the detectors for *VELA 5*. As part of this work, Klebesadel combed through data from the two *VELA 4* satellites. He found that the detectors recorded a γ-ray event on 2 July 1967. He checked further, and found that this event also triggered the still operational *VELA 3* satellites. Had the γ-rays come from a bomb?

Later *VELA* satellites found more γ-rays. Lots of them: about one burst every fortnight. At first the US authorities were worried, but they soon realised that neither China nor the Soviet Union could test nuclear weapons every other week. Neither could they deliberately jam the *VELA* detectors as a means of hiding the tell-tale sign of a weapons test. Satellite malfunction was suggested and ruled out. Eventually the intelligence community lost interest, and passed the puzzle over to astronomers.

News of γ-ray bursts became public in 1973. Improvements in detector technology since then mean that we now see γ-ray bursts at the rate of about one a day. In total, over 2000 bursts have been studied. They occur at random and each burst is different. Some last for an hour; others for less than a second. Some emit their energy smoothly; others do so in spasms. About the only thing they have in common is that they emit most of their energy as γ-rays, along with a few X-rays. Unfortunately, it is difficult to make sharp images at γ-ray wavelengths so it is difficult to know precisely which part of the sky a particular γ-ray burst comes from. This in turn makes it difficult to deduce the distance scale associated with γ-ray bursts. Do they come from the outskirts of our solar system, from the halo of our Galaxy, from remote galaxies — or perhaps none of these places?

Without knowing the distance to γ-ray bursts it is impossible to begin to understand their origin. In the absence of hard data, theorists were free to speculate. Some explained the bursts as the result of colliding comets. Others suggested that magnetic flares on the surface of stars held the clue. Still others resorted to those mysterious black holes to solve the problem. Scientists proposed more than 100 different mechanisms to explain the origin of the bursts. (At one time this meant there were fewer observed bursts than there were theories to explain them!) It is little wonder that, since 1973, γ-ray bursts have been top of the list of astronomical enigmas.

The situation changed in the early hours of 28 February 1997. The Italian–Dutch X-ray satellite *BeppoSAX* recorded the γ-ray burst GRB 970228 with an X-ray camera. Since X-rays are easier to pinpoint on the sky than γ-rays, astronomers for the first time had the chance to train the world's best telescopes on the precise location of a γ-ray burster. Within 20 hours of the *BeppoSAX* discovery a team of Dutch and English astronomers, using the Isaac Newton Group of Telescopes at La Palma in the Canary Islands, found the first visible light counterpart of a γ-ray burster. Other telescopes soon joined in, and watched the source gradually fade from view. (See figure 1.1.) Since that first discovery, *BeppoSAX* has pinpointed several more bursts. At last, astronomers can study γ-ray bursts in detail. These first thorough observations bear out what most astronomers were starting to believe anyway: the bursts come from the depths of the universe.

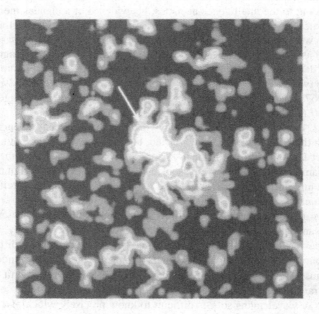

FIGURE 1.1: The visible fireball that accompanied GRB 970228. The host galaxy may be the object below and to the right of the fireball.

For γ-ray bursts to shine so brightly at such large distances they must be *incredibly* energetic events. One possible explanation for them is the *fireball model*, developed by the Hungarian–American astronomer Peter Istvan Mészáros (1943–) and the Astronomer Royal, Martin John Rees (1942–). In this picture a cataclysmic event — perhaps the merging of two neutron stars in a distant galaxy, or maybe the explosion of a very massive star that is still within the cloud of dust and gas from which it formed — releases a vast amount of energy in a small region of space. The resulting fireball moves at close to the speed of light, and smashes a blast wave into the surrounding gas. The interaction of the blast wave with the gas creates the most violent explosions known to mankind. These events emit more energy in a few seconds than our Sun will generate in its entire lifetime. They make the nuclear explosions that *VELA* searched for seem as powerful as a damp safety match.

As I hope this tale demonstrates, attempting to estimate the distance to astronomical objects is painfully difficult. It took three decades of intensive work by hundreds of astronomers around the world to deduce the distance scale of γ-ray bursts. (Even now, there is an outside chance that the source of the bursts is much closer and less energetic than generally believed.) There are many other celestial objects whose distance remains uncertain. This should not be surprising. After all, what have astronomers got to work with? A brief flash of radiation, in the case of γ-ray bursters. A gradual drip-drip-drip of photons, in the case of faint galaxies. The wonder is that astronomers can determine the distance to *any* celestial object.

This book tells the story of how astronomers deduce distances. We begin, as the ancient astronomers did, by measuring the size of the Earth. This information helps us set the scale of the solar system. In turn, we can use this to deduce the distance to nearby stars. A study of the nearby stars gives us information that helps us uncover the size of the Galaxy. And in this way, step by careful step, we can build a ladder that takes us to remote galaxies. We can press out to even greater distances. Eventually we can even answer one of the most difficult questions of all: how big is the universe?

Before we begin, though, it is worth reviewing how we measure distances in everyday life.

1.1 DISTANCES IN EVERYDAY LIFE

Most of us are so highly adept at judging distance that we take the skill for granted. We *know* how far away something is just by looking at it. This is just as well, since otherwise we would need a tape measure to handle routine tasks, and crossing a busy road would be too hazardous to contemplate. Although the methods we use to judge distance may seem so obvious as to require no mention, it *is* worth considering how they work. We can then apply the same methods to the sky, where our everyday intuitions may not apply.

The first and most obvious clue to judging distance comes about when one object is partly in front of another and so partly blocks our view of it. The object at the front is closer than the object that is obscured.

Perspective is another powerful clue for judging distance. Parallel lines, such as street kerbs or rows of houses, seem to converge as they head off into the distance.

After a few years of childhood experience, we know the true size of nearly all the objects we are ever likely to encounter. They become *standard rods*. If we know the true size, s, of an object we can judge its distance, d, from its apparent angular size, θ. For small angles the relation between these quantities is just

$$d \approx \frac{s}{\theta} \tag{1.1}$$

with θ measured in radians. The brain solves this equation automatically; figure 1.2 illustrates the idea.

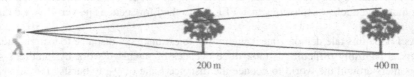

FIGURE 1.2: When the angles involved are small, the angle made by the distant tree is half that made by the nearby tree. We know from experience that mature trees have more or less the same height, so immediately we know that the distant tree is twice as far as the nearby tree. From the apparent angular size of trees we can usually quite accurately estimate their distance.

This method is one of the commonest ways we judge distance. It works so well that we hardly ever question it. But this means we can be fooled. If our standard rod is not standard after all, so that we have an incorrect value of s in (1.1), our distance estimate will be in error. Several years ago, a highly effective* advertisement appeared on British commercial television. It showed a thirsty man crawling through the sun-baked desert towards an ice-cold bottle of beer. The bottle was huge, so clearly it was in the foreground of the scene. The man was a small figure in the background. It seemed that he had many miles to crawl before reaching the beer. Then, in one of those trick moments beloved of advertising directors, the man bumped his nose on the bottle. When he got to his feet we saw how we had been tricked: the bottle was as tall as the man. It filled the screen not because it was in the foreground but because it was twenty times bigger than any 'normal' bottle. Television tricks apart, we are unlikely to be fooled in this way. In everyday life there are usually enough visual clues for us to distinguish correctly between a nearby bonsai and a distant tree. However, it serves as a warning when we try to use this method in astronomy, where we have no prior knowledge of the real size of objects.

Just as we know the true size of everyday objects, we know how fast they can move. This gives us another distance clue. For example, a nearby train zooms past us; a train on the horizon seems almost still.

Perhaps the most important way we estimate distance comes from *parallax*. Parallax occurs when we look at the same object from different vantage points. Our view changes, and the amount by which it changes tells us the object's distance. For example, imagine taking ten paces along a street. As you walk, nearby objects seem to move quickly. After ten paces a street lamp that was in front of you has moved behind you, and out of your field of view. Objects in the middle distance appear to move more slowly. Those same

---

*Perhaps it was not so effective: I cannot remember the brand name being advertised.

ten paces only slightly change your view of houses on the other side of the street. Objects in the far distance do not seem to move at all.

You do not have to move bodily for parallax effects to work. A slight tilt of your head is enough to produce relative motions of near and far objects that your brain can recognise and turn into distance information. Indeed, you need not move at all. Since your left and right eyes look out on the world from slightly different vantage points, your brain receives enough parallax information for you to estimate the distance of nearby objects.

To understand the parallax effect, do the experiment shown in figure 1.3. Close your left eye. Now hold a pencil vertically at arm's length and align it, using just your right eye, with some distant object — perhaps a door. Now close your right eye and open your left eye. You will see that the pencil appears to have moved relative to the door. The closer the pencil is to your eyes, the bigger the parallax shift. (If you want further proof of the importance of stereoscopic vision for estimating distance, try playing football or riding a bicycle with one eye closed.)

FIGURE 1.3: Using just your left eye, line up a pencil with some distant object. Now close your left eye and look at the pencil with your right eye. Note how the pencil shifts relative to the distant object. The closer you hold the pencil, the greater is the shift in position.

None of these tricks work in the darkness of night, of course. Nevertheless, if we can see lights at night we can easily judge their distance. We know the true luminosity of street lights, car headlamps, house lights, and so on. They are *standard candles*. By comparing their apparent brightness with their true brightness we get their distance: the dimmer they appear, the farther they must be.

When we observe the world around us, our brain takes in all the available information on perspective, relative sizes and speeds, parallax, and much else besides, and effortlessly produces a consistent picture of distances. Our brain does less well when we look up at the night sky. There seems to be no way to estimate the distance to stars, which are just isolated points of light whose intrinsic luminosity is unknown. Are they small dots just out of reach? Or are they large objects at a much greater distance? In later chapters we shall see how we can deduce the distance to celestial objects by using the simple techniques mentioned above.

1.2 DISTANCE UNITS

The word 'distance' comes from a Latin word meaning 'standing apart', the idea being that two separate objects must occupy different points in space. The distance between two points is a measure of the extent of space separating those points. We all have an intuitive understanding of this* and, as we have seen, we all have the ability to estimate distances in the world around us. To talk *quantitatively* about distances, though, we need to agree

*As we shall see in later chapters, this intuition sometimes proves faulty.

upon a standard measuring rod whose length everyone knows. Then, when we measure a distance, we can express the result in units based upon the common standard.

Mankind has used a baffling number of measuring rods throughout history. Often, these were based upon some feature of the human body. For instance, the method of using the length of the human foot as a standard rod has been employed for over four millennia. (As far as we know, Gudea of Lagash (*fl.* BC 2130) was the first to define a legal foot; in modern units the Lagash foot was 0.26 m long.) Another method was to use the length of the forearm — from the elbow to the tip of the middle finger — as the measuring rod. This defined a unit of length called the cubit. A system of units based on the human body has an obvious and immediate drawback: whose body will define the standard? A king's or a peasant's? In England, King Henry I (1068–1135) had no doubts: he decreed that one yard was the distance from the tip of *his* nose to the end of *his* thumb when *his* arm was outstretched. Such a definition will work so long as everyone agrees with it. Clearly, though, it is not ideal — if only because the unit of length is liable to change when a new king decides that *his* appendages should define the standard. We need something better.

The *metric system* of measurement is now in widespread use. The basic unit of length in the metric system is the *metre*. Unlike the early definitions of the cubit and the yard, the standard metre is rigorously defined. In 1984, the 17th General Conference on Weights and Measures defined the metre to be:

'the length of the path travelled by light in vacuum during a time interval of 1/(299 792 458) second'.

They defined the second as:

'the duration of 9192 631 770 periods of the radiation corresponding to the transition between two hyperfine levels of the ground state of the ^{133}Cs atom'.

No room for confusion there. And unlike the imperial system of length units, which required you to remember bizarre conversion factors (such as a rod, pole or perch contains $5\frac{1}{2}$ yards, a rod contains 25 links, a chain is 4 rods and a league is 3 miles), the metric system is easy to use. To create a new unit, just add zeros to the metre. The kilometre is simply 1000 m, the centimetre is 0.01 m and the millimetre is 0.001 m. Easy.

Nevertheless, despite the simplicity and widespread use of the metric system, I will not use it consistently throughout the book. For one thing, in some circumstances it does not 'sound' right. Herschel's telescopes somehow seem less impressive in metric. More importantly, astronomers themselves generally do not use the metric system when discussing cosmic distances. The reason is that distances in astronomy are *huge*, and so the metre is an inappropriate unit. For instance, you must add sixteen zeros to the metre when discussing the distance to the *nearest* star. (In strict metric notation the nearest star is 40 petametre away, which provides us with little or no insight. The use of exponential notation is somewhat better. We can write the distance to the nearest star as 4×10^{16} m, and this at least gives us some feel for the size of the number. But few astronomers bother. They prefer to define more appropriate units.)

Bearing in mind Emerson's famous dictum about foolish consistency, I shall use the most appropriate distance units for each step on the distance ladder. No confusion should arise by using different units in different contexts. We do this all the time. In horseracing, for instance, people talk of a six-furlong race rather than a 1320-yard race, or its metric equivalent, even though few of us use this unit in any other walk of life. (A furlong was originally a 'furrow long': the length of a furrow ploughed across a standard-sized square field of ten acres. Since the size of an acre varied from place to place, and from time to time, from the ninth century onwards the furlong was defined to be equal to yet another measure of length, the stadium, which was one eighth of a mile.) We use a furlong in this context from a mixture of convenience and contingency. Similarly, a unit like the fathom (defined to be six feet in length, but originally the length spanned by outstretched arms) is still often heard in nautical terminology but not in everyday conversation. In addition, new practices and technologies mean that unusual distance units can come into common parlance. For instance, the widespread use of desktop publishing programs on personal computers mean that people are quite happy to express lengths in terms of points and picas, units that would once have been considered obscure. Again, convenience plays a major part in this. The pica and the point are the right size to discuss printing terms: it is better to give the line width of this book as 355 points than as 1.25×10^{-4} km.

So which distance units should we use in astronomy? When discussing the size of the Earth, both the kilometre and the mile are convenient units and I will use both. Distances are so large in the solar system that neither the mile nor the kilometre are particularly convenient units. A new unit, the *astronomical unit*, is preferable. When discussing distances to stars, where a billion miles is less than a stone's throw, the *light year* becomes a useful unit. It is clearly more convenient to give the distance to the nearest star as 4.39 light years than as 2.5×10^{13} miles. (The light year makes use of yet another means of measuring distance: if we know how fast something moves we can express distance in terms of time. Thus in the context of air travel it may make sense to say that New York is six hours distant from London. Similarly, by road, Cardiff is four hours distant from Sheffield. We know the speed of light with far more precision than we know the speed of any aircraft or car, so we may use the distance that light travels in one year as a standard.) Astronomers also use the *parsec* when discussing stellar distances; I will give such distances in both light years and parsecs. Still larger distances require the *kiloparsec* or *megaparsec*.

1.3 MEASUREMENT UNCERTAINTY

Whenever a scientist performs an experiment to measure some quantity there is *always* some error or uncertainty associated with the experiment. This is as true for a 'simple' experiment (such as measuring the distance between two objects) as it is for a 'complex' experiment (such as measuring the rest mass of the muon neutrino). The uncertainty can arise from many sources. The measuring instrument may be inappropriate for the task, or it may be mis-calibrated, or it may be malfunctioning in some way. The observer may be mistaken, or incompetent, or sloppy. The quantity being observed may change randomly, or perhaps systematically, between observations. For all these reasons, scientists do not just publish the value of the quantity they measure; they also publish the attendant uncertainty in the quantity. For example, they would not say 'the distance to object X is 1 m'. Rather, they would say something like 'its distance is 1 ± 0.001 m', where the symbol $\pm$ means 'plus or minus'. The quoted uncertainty is a *standard deviation*, σ, of the mean. So the experimental result 1 ± 0.001 m for the distance means that 68% of the time the measured value differed from the true value by within 0.001 m, 95% of the time within 0.002 m, and 99.7% of the time within 0.003 m. (These particular values arise because uncertainties in measurements are usually well described by the *normal distribution*. Statisticians tell us that if an event is drawn from a normal distribution, there is a 68.27% chance that it lies within 1σ of the mean, a 95.45% chance that it lies within 2σ of the mean, and a 99.73% chance that it lies within 3σ of the mean.)

Scientists are not simply hedging their bets when they attach uncertainties to a measurement. A realistic understanding of the uncertainty in a measured value is as important as the value itself. The goal of the experimental scientist is not to discover the 'true' value of a quantity, but to make the uncertainty in that value as small as possible.

Measurement uncertainties are of two types. *Random* uncertainty — so called because it produces a scatter in the observed value of a quantity — is quite easy to treat with standard statistical techniques. The usual advice is simply to make many measurements of the same quantity and then average the results. For example, suppose you have to measure the length of a wooden desk using a steel rule of length 1 m. Every millimetre of the rule is finely etched. You might know that the 'actual' or 'true' value you are trying to measure is 2 m, but each time you make the measurement you will get a slightly different result. For instance, since the measuring rod has only millimetre graduations you may find yourself 'rounding up' to the nearest millimetre on some occasions and 'rounding down' on others. Your eyes will be in a slightly different position each time you take a reading, leading to an optical parallax that will introduce some variation in the observations. Environmental factors may introduce further uncertainties: the wooden desk may swell in damp weather and contract in dry weather. Nevertheless, after making many trials, and averaging the different measurements, you might conclude that the desk has a length of 2 ± 0.0008 m, which is probably good enough for most purposes.

The skilful experimental scientist can usually recognise random uncertainties and then take steps to reduce them by improving the experiment. In the example above, a laser-based length-measuring device will produce better estimates than a steel ruler; controlling the humidity around the desk will reduce a possible cause of uncertainty; and so on. The skilful practitioner will therefore obtain an answer that is both *accurate* and *precise*.

In everyday language accuracy and precision mean more or less the same thing. In science, though, they have very different meanings*. An accurate set of measurements gives us an estimate that is close to the true value of the quantity we are measuring. A precise measurement is one where the uncertainty in the estimated value is small. In the above example, our estimate of the length of the table is accurate because the 'true' value of the length lay in our estimated range. Our estimate is precise because the uncertainty is just 0.04% of the result.

Experimental scientists always hope to obtain an accurate and precise result from their measurements. Nevertheless, it is quite possible to obtain a result that is precise (i.e. has

---

*Examples from other fields can also highlight the difference between accuracy and precision. In 1654, for instance, the Anglican archbishop of Armagh, James Ussher, used Biblical references to calculate that the universe began at 9 A.M. on 23 October 4004 BC. This was a spectacularly precise conclusion. It was just as spectacularly inaccurate.

a small uncertainty) but inaccurate (i.e. wrong). The commonest way this can happen has to do with the second type of uncertainty: *systematic* uncertainty. Suppose in the above example that you take steps to minimise all possible sources of error. In particular, suppose that you use a laser-based device to measure length. Finally, suppose that the device is perfect — except that it was mis-calibrated at the factory. When it measures a distance of 1 m it registers 0.5 m. (Of course, you would immediately notice such a large mis-calibration. You would not necessarily spot an error of 1 mm, though, and the following argument would be the same.) This time when you perform the experiment you might obtain precise results — precise, because every measurement will be consistent except for a very small random uncertainty — but each measurement will be systematically incorrect by a factor of two. Your best estimate of the length of the table might be 4 m, with an uncertainty of 0.1 mm. This would be a precise yet incorrect estimate. (It is equally possible to make an estimate that is accurate and yet imprecise. Perhaps most measurements fall into this category.) Systematic errors are in general more difficult to discover and weed out than random errors.

The idea of a standard measuring rod causing such large systematic errors may seem contrived, but exactly such a thing happened in astronomy. As we shall see in a later chapter, in the 1940s astronomers realised that their favourite rod for measuring large distances had been mis-calibrated. The distance estimates that these rods produced were quite precise, but they were inaccurate. When the systematic uncertainties were fully understood, the universe doubled in size.

Although for clarity I will often quote just the value for the distance to an object, it is important to remember that there is *always* an uncertainty associated with that value. For the distances discussed in the first few chapters, the uncertainty is usually so small that we can neglect it. For larger distances, though, the uncertainty becomes large. On the largest possible scales, the uncertainty is about 50 per cent: it is as if astronomers do not yet know if their metre sticks are 1 m or 1.5 m long.

1.4 OUTLINE OF THE BOOK

By the time we reach the final chapter of this book we will be in a position to answer the question: how big is the observable universe? Some commentators have accused scientists of arrogance for even asking such a question. This is not quite fair. In science the degree of difficulty of a question has more to do with complexity than with size. The universe may be very large, but on the largest scales it seems to be very simple. Understanding a small yet complex structure like the human brain is a much more challenging problem. In any case, philosophers throughout the ages have argued about the size of the universe. For example, Titus Lucretius Carus (*c*.99BC–*c*.55BC), one of the few Roman philosophers, thought that the universe was infinite in extent. For Lucretius, the notion of an 'edge' to space, seemingly inherent in a finite universe, was absurd:

> 'Let us assume for the moment that the universe is limited. If a man advances
> so that he is at the very edge of the extreme boundary and hurls a swift spear,
> do you prefer that this spear, hurled with great force, go whither it was sent
> and fly far, or do you think that something can stop it and stand in its way?'

The notion of an infinite universe held its own paradoxes, though. In 1826, the German astronomer Heinrich Wilhelm Matthäus Olbers (1758–1840) argued that, if an infinite number of stars were sprinkled through an infinite universe, the combined light from the stars would make the night sky blindingly bright. The paradoxes described by Lucretius and Olbers seem impossible to resolve; and yet, over the past 80 years, astronomers have resolved them. They answered Lucretius by showing how the universe can be finite in extent and yet possess no boundary. They answered Olbers by showing how only a finite amount of starlight has had time to reach us here on Earth. So: how big is the universe?

There are two aspects to this question. The first is practical. It rephrases the question as: how far away are the objects that we can see? This book is an account of how astronomers have answered that question by building a 'distance ladder'. If we know what to expect of objects on one rung of the ladder we obtain clues that let us reach out and understand objects on the next rung. In this way we can go from the diameter of the Earth up to the diameter of the visible universe itself. Each chapter of the book describes one rung of the distance ladder.

The second aspect to the question is philosophical. Lucretius and Olbers were asking: how big is the universe — not just the observable universe — *really*? The final chapter of the book describes some recent ideas about the size of the universe.

CHAPTER SUMMARY

- We use several techniques to estimate distance in everyday life. Examples include: perspective, parallax, and the use of standard rods and standard candles.

- We can choose between many different units to express the outcome of a distance measurement. In this book we use the most appropriate unit for the distance scale under discussion. (The alternative is to express all distances in the correct SI unit, which is the metre.)

- All distance measurements have an uncertainty attached to them.

QUESTIONS AND PROBLEMS

1.1 It is said that in Tibet there used to be a length unit called a 'cup of tea'. It was the distance that one could walk with a freshly boiled cup of tea before it became cool enough to drink. Write an account of some of the other unusual distance units that have been used at various times and in various cultures.

1.2 Suppose you want to find the area of your living room carpet. The only measuring instrument you have is a cloth tape measure, of the type used by tailors. List some of the errors that will effect your final result.

1.3 What types of distance-measuring techniques are used in the animal kingdom? (Consider the echolocation techniques of bats and dolphins, for example.)

1.4 All scientists have to be aware of the systematic and random uncertainties in their work. What are the particular difficulties faced by astronomers?

1.5 Write an account of γ-ray bursters, with particular reference to the various distance scales that have been suggested for them in the 30 years since their discovery.

1.6 Mention is made in the text of Olber's paradox. Write an account of the paradox.

FURTHER READING

GENERAL
Hubel D H (1988) *Eye, Brain, and Vision* (Freeman: New York)
— One of the outstanding *Scientific American Library* series, written by a Nobel Prize-winning neurophysiologist. It describes how the brain handles visual information, including distance information.

Rowan-Robinson M (1985) *The Cosmological Distance Ladder* (Freeman: New York)
— Covers the same ground as this book, but is pitched at a slightly higher level. It is also now rather dated: the last 13 years have seen much progress in the field of distance determination in astronomy. Professor Rowan-Robinson's other books are also recommended for their insight into what it is like to *do* astronomy.

http://antwrp.gsfc.nasa.gov/apod/astropix.html
— The Astronomy Picture of the Day site. It contains images of many objects mentioned in this book.

GAMMA-RAY BURSTS
Costa E *et al.* (1997) Discovery of an X-ray afterglow associated with the γ-ray burst of 28 February 1997. *Nature* **387** 783–785
— The first detection of an X-ray afterglow associated with a particular GRB.

Fishman G J and Meegan C A (1995) Gamma-ray bursts. *Ann. Rev. Astron. Astrophys.* **33** 415–458
— A comprehensive technical review of pre-*BeppoSAX* data on γ-ray bursts, with an emphasis on observations of their temporal, spectral and global distribution properties.

Frail D A *et al.* (1997) The radio afterglow from the γ-ray burst of 8 May 1997. *Nature* **387** 261–263
— The first radio counterpart of a GRB.

Klebesadel R W, Strong I B and Olsen R A (1973) Observation of gamma-ray bursts of cosmic origin. *Astrophys. J.* **182** L85–L88
— The paper that announced the discovery of γ-ray bursters.

Kulkarni S R *et al.* (1998) Identification of a host galaxy at redshift $z = 3.42$ for the γ-ray burst of 14 December 1997. *Nature* **393** 35–39
— Proof that GRB971214 is at a vast distance. For a short while, it must have been the brightest object in the universe.

Metzger M R *et al.* (1997) Spectral constraints on the redshift of the optical counterpart to the γ-ray burst of 8 May 1997. *Nature* **387** 878–880
— The first direct measurement of the distance to a γ-ray burst.

Paczyński B (1998) Are gamma-ray bursts in star-forming regions? *Astrophys. J.* **494** L45–L48
— The author proposes the name 'hypernova' for the burst/afterglow event, and suggests that γ-ray bursters cannot be formed by the popular model of merging neutron stars.

Roth J (1996) Gamma-ray bursts: a growing enigma. *Sky & Telescope* **92** (3), 32–34
— A non-technical pre-*BeppoSAX* account of these mysterious events.

van Paradijs J *et al.* (1997) Transient optical emission from the error-box of the γ-ray burst of 28 February 1997. *Nature* **386** 686–688
— The paper that described the first optical source associated with a GRB.

Wijers R A M J, Rees M J and Mészáros P (1997) Shocked by GRB 970228: the afterglow of a cosmological fireball. *MNRAS* **288** L51–L56
— Observations of GRB 970228 compared with predictions from the relativistic blast wave model.

http://antwrp.gsfc.nasa.gov/diamond_jubilee/debate95.html
— An interesting site devoted to a debate on the distance scale to γ-ray bursters.

http://www.batse.msfc.nasa.gov/
— The *BATSE* (Burst and Transient Source Experiment) home page.

MEASUREMENTS AND UNITS

Barlow R J (1989) *Statistics* (Wiley: Chichester)
— Subtitled 'A guide to the use of statistical methods in the physical sciences', this book does the impossible: it makes statistics seem interesting!

Lyons L (1991) *A Practical Guide to Data Analysis for Physical Science Students* (Cambridge University Press: Cambridge)
— A brief but very useful text for undergraduates.

Rüeger J M (1989) *Electronic Distance Measurement* (Berlin: Springer)
— An undergraduate textbook on how to measure distance.

Skinner F G (1967) *Weights and Measures* (HMSO: London)
— The fascinating story of the ancient origins of units, and their development in Great Britain.

Zupko R E (1968) *A Dictionary of English Weights and Measures* (University of Wisconsin Press: Madison, WI)
— Every imaginable unit, from acers to zemes, is in this book.

GENERAL HISTORY OF ASTRONOMY

Asimov I (1972) *Asimov's Biographical Encyclopedia of Science and Technology* (Pan: London)
— Contains biographical information on many of the astronomers mentioned in this book.

Berry A (1961) *A Short History of Astronomy* (Dover: New York)
— A reprint of a book originally published in 1898.

Leverington D (1995) *A History of Astronomy from 1890 to the Present* (Berlin: Springer)
— The book is divided into 17 chapters, each of which describes the history of a particular astronomical topic.

Pannekoek A (1989) *A History of Astronomy* (Dover: New York)
— A reprint of a book originally published in 1961.

2

First step: the Earth

The earliest civilisations developed fascinating myths about cosmology, but they had no real understanding of the universe in which we live. They had no understanding even of the true nature of the Earth. Early thinkers believed that we live on a plane that stretches out into the distance until it somehow meets the sky. Nevertheless, these people developed the rudiments of geometry and angular measure. In the hands of later thinkers these became tools for calculating the size of the Earth — and the rest of the known universe.

2.1 THE BEGINNINGS OF GEOMETRY

The Nile, the world's longest river, starts in Lake Tana in Ethiopia and Lakes Albert and Victoria in Uganda and, as it snakes northward to the Mediterranean Sea, passes through the desert lands of Sudan and Egypt. The Nile was the only steady supply of fresh water for Ancient Egyptians. Without it, the sands of the Sahara would have swallowed their homes. But the Nile gave them even more than water. Once every year, the melting of snow high in the mountains of Ethiopia caused the Nile to flood. The floods deposited large amounts of thick black silt on either side of its course, and the silt made the land fertile. This natural fertilisation was so rich that it was often possible to grow several crops in one year. Ancient Egypt owed its existence as a habitable country to the Nile.

After the floods it was necessary to resurvey the land for tax purposes. It was almost inevitable that the Egyptians took the first steps in *geometry*. (The very word 'geometry' comes from two Greek words meaning 'earth' and 'measure'.) The Egyptians took a practical approach to geometry. They had specialist surveyors, men called rope stretchers, who worked with a taut rope that had knots marked at equal intervals. Rope stretchers measured distance simply by stretching their ropes between two points and counting the knots. This approach clearly worked: the magnificence of the pyramids, built 4500 years ago, is proof of that. But the Egyptian geometers never took the intellectual step of forming an abstract system of geometry. They were not interested in mathematics for its own sake.

A civilisation even older than the Egyptian was that of the Babylonians, who lived in what is now Iraq. The Babylonians built nothing that could rival the pyramids, but they left behind a rich legacy of mathematics. They developed an intricate sexagesimal system of arithmetic — in other words they used 60, rather than 10, as their base for counting.

They chose 60 because it has many proper divisors (2, 3, 4, 5, 6, 10, 12, 15, 20 and 30 all divide evenly into 60). If you divide a circle into 60 equal parts, then you can express a half circle, or a third, or a quarter, and so on up to a thirtieth, as a whole number of sixtieths. This was useful for the Babylonians, who found it hard to work with fractions.

The number 360, which is 60×6, is even easier to work with. For this reason, the Babylonians decreed that there were exactly 360 days in the year. (They probably knew that the year is a little over 365 days long, but the convenience of using 360 outweighed other considerations.) In one year the Sun seems to make one full circle across the sky. Just as the Babylonians divided this particular circle into 360 equal parts, they chose to divide *all* circles into 360 equal parts. Today we call such equal parts *degrees*, the symbol for which is °. They divided each degree into a further 60 equal parts. We call these smaller subdivisions *minutes* (or, if confusion might arise with the unit of time, *minutes of arc*). The symbol for the minute of arc is ′. Each minute was subdivided into a further 60 equal parts. We call these fine subdivisions *seconds* (or *seconds of arc*). The symbol for the second of arc is ″. We still use the Babylonian system when we measure angles*.

As we saw in the previous chapter, we can give the apparent size of an object in angular measure. For instance, on average the full Moon subtends an angle of 31′. In other words, if you draw a line from your eye to each edge of the full Moon the lines make an angle of 31 minutes of arc. Using angular measure to specify an apparent size is much better than using linear measure, because it makes no assumption about distance. An object can be large and distant — like the Moon — or small and nearby — like your thumb — and still subtend an angle of 31′. If you use linear measure to give the size of the Moon, and say something like 'it's about 30 cm across', you are automatically fixing a particular distance to the Moon. To look as large as the full Moon, a circle of diameter 30 cm has to be only 33 m away. Clearly, the Moon is farther away than 33 m. The use of angular measure obviates this difficulty. (A useful set of figures to remember is that, when your arm is outstretched: the width of your index finger is about 2°; the width of your fist is about 10°; and the width of your spread hand, from little finger to thumb, is about 20°.)

We can also measure angles in *radians*. One complete revolution is defined as containing 2π radians, so

$$2\pi \text{ rad} = 360° \qquad \text{or} \qquad 1 \text{ rad} = 57.3°.$$

Example 2.1 Radian measure

Convert 1″ into radian measure.

Solution.

$$1° = \frac{\pi}{180} \text{ rad} \quad \text{so} \quad 1'' = \left(\frac{1}{60 \times 60}\right)\left(\frac{\pi}{180}\right) \text{ rad} = 4.85 \times 10^{-6} \text{ rad}.$$

---

*It is strange to think that, in an age in which we have rationalised most measures, we still use an arbitrary system of angular measurement that dates back 5000 years.

That there are 360 degrees in a circle is a completely arbitrary decision; radian measure is the 'natural' way to measure angles. Even so, astronomers nearly always express angles in terms of degrees, minutes and seconds.

2.2   A FLAT EARTH

The philosopher Thales of Miletus (624 BC–546 BC) is almost a mythical figure. The later Greeks considered him to be the founder of mathematics, science and philosophy, and attributed to him all sorts of fundamental discoveries in every branch of knowledge. Particularly important for our purposes are his contributions to geometry. In fact, Thales may not have discovered the geometric propositions with which he is credited, but it does seem likely that he organised the geometric knowledge of his day. And while the Greeks, like the Egyptians, generally scorned the practical application of mathematics, two tales about Thales show the power of Greek geometry. They illustrate how geometry can be used to calculate distances on a flat Earth (which is what the ancient Greeks thought they inhabited).

The first story tells how Thales calculated the height of the Great Pyramid at Giza. It seems impossible to measure the height of a pyramid directly. Even if you climb to the top of the pyramid, marking your ascent with a measuring rod or using the rope stretching technique of the Egyptians, you will obtain only the length of the side of the pyramid, and not the pyramid's height. To calculate the height, Thales used the following proposition: two *similar* triangles (triangles with identical angles) have their three sides in proportion. Figure 2.1 illustrates the idea.

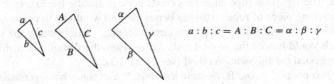

$$a : b : c = A : B : C = \alpha : \beta : \gamma$$

FIGURE 2.1: Three *similar* triangles. They have identical angles, so their sides must be in proportion.

Legend has it that Thales placed his staff at the edge of the shadow cast by the pyramid (see figure 2.2). He knew that he was dealing with similar triangles. He therefore argued that the ratio of the height of the pyramid to the height of his staff was the same as the ratio of the length of the pyramid's shadow to the length of the shadow cast by his staff. He already knew the distance along the base of the Great Pyramid; the rope stretchers had determined it to be 230 m (in our units). He knew that his staff was 2 m long. All he had to do was measure the length of the shadow of the pyramid (the distance from the tip of the shadow to the centre of the base of the pyramid) and the length of the shadow cast by his staff. It turned out that the tip of the shadow was 105 m from the edge of the base of the pyramid, and the shadow of the staff was 3 m long. The similarity of the triangles

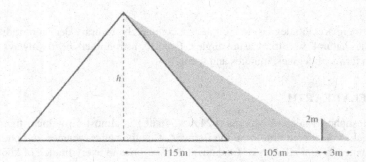

FIGURE 2.2: How Thales measured the height of the Great Pyramid at Giza. The shadow cast by the pyramid and the shadow cast by the staff form similar triangles. Thales could measure all the lengths involving the staff, and could measure the length of the pyramid's shadow. Similarity gave him the pyramid's length. (Not to scale.)

gave him the height, h, of the pyramid:

$$\frac{h}{115 + 105} = \frac{2}{3}$$

thus

$$h = \frac{220 \times 2}{3} = 147 \,\text{m}.$$

The second story about Thales is his use of *triangulation*. Suppose you want to measure a large distance — the distance between two villages, maybe, or the distance to a ship out at sea. The Egyptian rope stretchers could perhaps handle the first task if they had a particularly long piece of rope. An even better method would be to count the number of revolutions of a foot wheel with a known circumference*. But neither rope stretching nor foot wheels could handle the second task. To measure the distance to a ship out at sea, Thales suggested the following method (see figure 2.3).

Have two people, A and B, observe the ship, S. Get them both to measure the angle between the ship and the line connecting the two of them. Call these angles a and b. Then measure the distance between A and B in the usual way. We now have enough information to define the triangle ABS. For example, a might be $84°$, b might be $75°$ and the distance between A and B might be one kilometre. Now draw a similar triangle, A'B'S', with angles a and b at A' and B'. For ease of drawing, we must scale down the lengths in the triangle. Suppose we have to reduce all lengths by a factor of 5000, so that the distance A'B' is 20 cm. Since the two triangles are similar, the distance AS divided by the distance AB is the same as the distance A'S' divided by the distance A'B'. But we can measure the distance A'S' directly. If you draw this triangle, you will find that A'S' is 53.8 cm. Therefore AS, the distance from observer A to the ship, is 5000 multiplied by 53.8 cm: 2.69 km.

---

*This method was used in the second half of the seventeenth century to draw maps of England. Using foot wheels, surveyors discovered that the distance between London and the Scottish border at Berwick was 339 miles; previously, they thought that Scotland was just 260 miles from London!

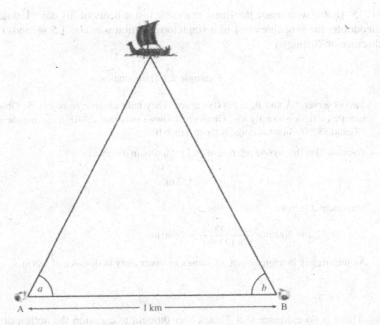

FIGURE 2.3: The method of triangulation enables two observers, A and B, to calculate the distance to a ship
out at sea by measuring the angles *a* and *b*. The method depends upon parallax (see figure 1.3).

Thales had to draw similar triangles to deduce distances. Using results from trigonometry, we can provide explicit formulae for the distances AS and BS:

$$AS = AB \times \frac{\sin b}{\sin(a+b)}$$

$$\quad (2.1)$$

$$BS = AB \times \frac{\sin a}{\sin(a+b)}.$$

The method of triangulation works fine in theory. In practice, its effectiveness depends on how accurately we can measure the angles *a* and *b*. The ancient Greeks had
no instruments to help them measure angles, and this limited the range over which they
could use triangulation. For example, if the distance between A and B is no more than
a few hundred metres, and the distance to the ship is more than about a kilometre, then
the angles *a* and *b* are close to 90°. An error of just half a degree then causes an error of
several tens of metres in the calculated distance.

(For this reason, modern surveyors — who to this day employ the method of triangulation — use a *theodolite* to measure angles. The invention of the theodolite predates that
of the telescope, but it was only in the eighteenth century that theodolites became precision instruments. This was mainly due to the work of the Yorkshireman Jesse Ramsden

(1735–1800), who made the finest scientific instruments of his day. Using a Ramsden theodolite, the probable error of a single observation was about 5 seconds of arc over a distance of 70 miles.)

Example 2.2 Triangulation

Two observers, A and B, are 300 m apart. They both observe an object, S. Observer A is sure that ∠BAS is exactly 85 . Observer B knows only that ∠ABS lies somewhere between 88 and 88 30′. In what range is the distance BS?

Solution. Use the second relation of (2.1). Minimum distance:

$$BS_{min} = 300\,m \times \frac{\sin 85}{\sin 173} = 2452\,m.$$

Maximum distance:

$$BS_{max} = 300\,m \times \frac{\sin 85}{\sin 173\ 30'} = 2640\,m.$$

An uncertainty in angle of just 30′ causes an uncertainty in distance of 188 m.

There is no evidence that Thales ever thought to question the notion of a flat Earth. In this regard, he made no advance on the Egyptians and Babylonians. Others, though, realised that they could use his methods of similarity and triangulation to investigate the size and shape of the Earth.

2.3   A SPHERICAL EARTH

As far as we know, the first person to suggest that the Earth's surface was anything other than flat was the Greek philosopher Anaximander of Miletus (611 BC–546 BC), who was a pupil of Thales. Anaximander proposed that we live on the surface of a cylinder that is curved north and south. This would explain why travellers heading north find that some stars appear from behind the northern horizon while other stars disappear beyond the southern horizon. When moving south the same thing happens in reverse: new stars appear from behind the southern horizon while other stars disappear beyond the northern horizon.

The first person to suggest that the Earth was a sphere was the Greek philosopher Pythagoras of Samos (c.582 BC–c.497 BC). He gave two pieces of evidence that showed the Earth's curvature was spherical, rather than cylindrical as Anaximander supposed.

First, people living by the coast noticed that ships heading out to sea did not simply become smaller and smaller until they vanished into a point. Instead, ships always disappeared while still perceptibly larger than a point, and their hulls always disappeared first. Furthermore, it did not matter in which direction the ships sailed: they always disappeared from view in the same way. This is exactly what would happen if the Earth were spherical, but not if it were flat or cylindrical.

Second, it seemed clear to Greek astronomers that a lunar eclipse was best explained as the result of Earth's shadow falling on the Moon. In every lunar eclipse ever seen, the Earth's shadow was circular in cross section even though the relative positions of the Sun, Moon and Earth were different during each eclipse. The only solid that casts a shadow with a circular cross section in all directions is a sphere. (The irascible and eccentric astronomer Fritz Zwicky often employed a colourful illustration of this idea. He called many of his colleagues 'spherical bastards' because, whichever way he looked at them, they were still bastards.)

The arguments for a spherical Earth were overwhelming. Since about 350 BC few educated people in the West have ever seriously questioned the idea, even though direct proof that the Earth is spherical took almost 2000 years.

If the Earth is a sphere the obvious question is: what is its radius? The first to suggest an answer based upon observation was the Greek philosopher Eratosthenes of Cyrene (c.276 BC–c.196 BC), around 240 BC.

Eratosthenes was a man of wide-ranging interests. His nickname was Beta, the second letter of the Greek alphabet, because in many of his fields of study he was the second best scholar in the world. He was therefore an ideal person to direct the library at Alexandria, a position bestowed upon him by Ptolemy III. While working at Alexandria, someone told him that on the summer solstice, 22 June, the noonday Sun was at the zenith in the southern city of Syene (what is now Aswan). In other words, a stick placed upright in the ground at Syene, at that particular time, cast no shadow. Eratosthenes carried out the same experiment at the library. In Alexandria, at noon on the solstice, an upright stick cast a short shadow. The length of the shadow indicated that the Sun was a little over 7° south of the zenith.

These observations were yet more proof that the Earth was not flat. Vertical sticks at Alexandria and Syene pointed in different directions, something that would not happen on a flat Earth. But Eratosthenes went further. He realised that his observations would let him deduce the circumference of the Earth. Figure 2.4 shows how.

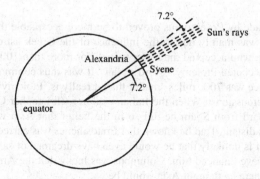

FIGURE 2.4: How Eratosthenes measured the circumference of the Earth. When the Sun is directly overhead in Syene, it is 7.2° south of the zenith in Alexandria. This happens because the Earth is curved. The angle subtended by Syene and Alexandria at the centre of the Earth is 7.2° or 1/50 of a circle. Eratosthenes knew the distance between the two towns, and so he could calculate the circumference of the Earth.

His experiment showed that the angle between Syene and Alexandria, at the centre of the Earth, was about 7.2°: 1/50 of a full circle. The distance between the two towns was thus about 1/50 of the circumference of the Earth. Eratosthenes knew that Alexandria was 5000 stadia to the north of Syene*. So the circumference of the Earth was 5000 × 50 = 250 000 stadia — about 25 000 miles. The radius of the Earth was thus about 4000 miles.

The value obtained by Eratosthenes is surprisingly close to the real value. For some Greeks, though, it was too large to be true. The Greek philosopher Poseidonius of Apamea (c.135 BC–c.50 BC) repeated the experiment, but instead of using light rays from the Sun he instead used the position of the star Canopus. In theory this should have been an improvement. In practice, since Poseidonius neglected to allow for the shift in the star's position due to atmospheric refraction, his value for the Earth's radius was too low. Taking one stade to be one tenth of a mile, he obtained a value for the Earth's radius of 2800 miles, and thus a circumference of about 18 000 miles.

Example 2.3 The radius of the Earth

An observer at A notes that a vertical stick 2.74 m high casts a shadow of length 1 m. At B, 555 km to the north, a stick of height 2.58 m casts a shadow of length 1.2 m. Deduce a value for the radius of the Earth.

Solution. We have $R_\delta = AB/(\alpha - \beta)$, where AB is the distance between A and B, and $\alpha - \beta$ is the difference in latitude (in radians).

The difference in latitude is

$$\tan^{-1}\tfrac{1.2}{2.58} - \tan^{-1}\tfrac{1}{2.74} = 25° - 20° = 0.0873 \text{ rad.}$$

Thus

$$R_\delta = 555 \text{ km}/0.0873 = 6357 \text{ km.}$$

The estimate made by Poseidonius proved to be more acceptable than that made by Eratosthenes. This was mainly due to the influence of the Greek astronomer Claudius Ptolemy (*fl.*AD 140), who accepted the lower value. For more than 1000 years, scholars all over the world accepted Ptolemy's word as law. It was thus commonly held that the Earth's circumference was 7000 miles smaller than it really is. Ptolemy's mistake was to have important repercussions. When the Italian explorer Christopher Columbus (1451–1506) sailed westward from Spain he did so in the belief that Asia was only three or four thousand miles distant. Had he known that Eratosthenes was correct, and that it was 12 000 miles west, it is unlikely that he would even have dreamed of sailing. It is certain that no one would have financed him. Columbus was lucky that the Americas happen to occupy the space where he thought Asia would be.

---

*Historians are not certain of the size of the stade that Eratosthenes used. Several different units called the stade were in use in Greek and Roman times. However, it seems probable that, for Eratosthenes, one stade was about one tenth of a mile.

It was not until 1522 that mankind had direct *proof* that the Earth was round. The proof came from an epic voyage led by the Portuguese navigator Ferdinand Magellan (1480–1521), who sailed from Spain in August 1519 with five ships under his command. Magellan sailed to Brazil, and then headed south*. At the tip of South America, Magellan discovered a strait that gave access to a new body of water that he called the Pacific Ocean. His ships weathered severe storms while passing through the strait to reach the Pacific. They then sailed for 99 days (during which time the crews were close to starvation, and ate rats, ox hides and sawdust to survive) before reaching the island of Guam in March 1521. After replenishing their stocks on Guam they sailed westward again until they reached the Philippine Islands. Magellan unfortunately became embroiled in local politics, and in April 1521 he died during a skirmish between rival native tribes. Only two ships left the Philippines, and one of those sank after reaching Indonesia. The remaining ship, *Victory*, continued sailing westward under the captaincy of Juan Sebastian del Cano (1460–1526), and finally returned to Spain in September 1522. The eighteen men on board were the first to circumnavigate the world and prove it was round. In doing so, they showed that Eratosthenes had been correct. The circumference of the Earth is indeed about 25 000 miles.

2.4   SATELLITE LASER RANGING

We no longer need epic voyages to furnish proof of a spherical Earth. Photographs from artificial satellites prove the matter so much more easily. Furthermore, the recent technique of *satellite laser ranging* (SLR) lets scientists measure the size of the Earth with unprecedented accuracy.

Laser light is quite unlike ordinary light. Ordinary light, from the Sun or from a source like a torch, is a jumble of different colours and the light waves are all out of step with one another. Laser light, on the other hand, consists of just one colour and all the waves are 'in phase'. (The difference is like that between a crowd of people leaving a cinema and a regiment of soldiers on parade. Each member of the crowd wears different coloured clothes, and because they each amble off in different directions the crowd soon disperses. Each member of the regiment wears identical clothes, and because they all march in step the regiment keeps its shape.) A pencil-thin laser beam travels over large distances and remains pencil thin. Astronomers make use of this property in SLR systems.

Consider the set-up in figure 2.5. A transmitter sends a short pulse of laser light to a satellite. The satellite carries retro-reflectors — small pieces of glass or quartz in the shape of a cut-off corner of a cube. The retro-reflectors act like the cats' eyes on a motorway. Because the laser beam remains tightly focused, the retro-reflectors are able to bounce the light pulse back to a receiver mounted alongside the transmitter. An event timer driven by an atomic clock measures the 'time-of-flight': the time taken for the light pulse to reach the satellite and return. Atomic clocks can measure the round-trip time to an accuracy of

---

*In these deep southern latitudes, Magellan and his men became the first Europeans to describe two foggy patches of light in the night sky, close to the Milky Way. These patches are now known as the Large and Small Magellanic Clouds, and they play an important part of the story told later in this book.

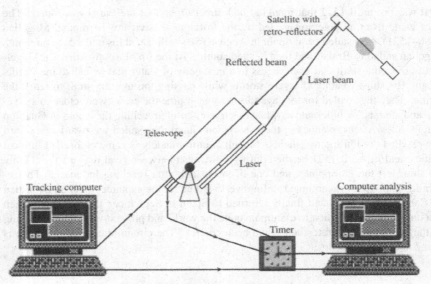

FIGURE 2.5: A simplified diagram of a satellite laser ranging system.

50 ps (5×10^{-17} s) or better. Finally, since we know the speed of light*, simple arithmetic gives us the distance from the laser to the satellite. (These distances are typically several thousands of miles. For instance, *Lageos* — the laser geodynamic satellite launched in 1976 by NASA — orbits at 3700 miles. The Russian satellites *Etalon I* and *Etalon II* orbit at 11 800 miles.)

A single distance measurement is not particularly useful. Combine the measurements from SLR stations around the world, however, and you can uncover a wealth of information about the Earth. For instance, the distance measurements are so precise that geologists can watch the movement of tectonic plates and the deformation of the Earth's crust in seismically active regions like California. Physicists are also interested in SLR data. The orbits of *Lageos*, and its sister satellite *Lageos II* which was launched in 1992, are known so precisely that they can be used to test a prediction of general relativity called gravitomagnetism.

A technique related to SLR is that of the global positioning system (GPS). Any observer on the Earth's surface can uniquely define his location in three dimensions by determining the distance between himself and three transmitters whose positions are known. The global positioning system identifies an unknown position on the Earth's surface by ranging from the known positions of satellites that transmit coded signals. A GPS-based system may soon be part of every car — becoming lost will be a thing of the past.

---

*All forms of electromagnetic radiation — radio, microwave, infrared, visible light, ultraviolet, X-rays and γ-rays — travel through a vacuum at the same speed: 299 792 458 m s^{-1}. In fact, the speed of light is *defined* to have this value. Light waves travel more slowly through air, but by an amount that we can account for.

Satellite laser ranging is sensitive enough to measure how much the Earth departs from sphericity. Because the Earth rotates around its axis, centrifugal forces cause the equator to 'bulge': the equator is very slightly farther from the centre of the Earth than are the north and south poles. (To use the technical term for such a slightly flattened sphere, the Earth is an 'oblate spheroid'.) There is thus no such thing as 'the' radius of the Earth. The *average* length of the radius is 3958.89 miles, corresponding to a circumference of 24 902.4 miles.

CHAPTER SUMMARY

- Triangulation lets us calculate distances by measuring angles. In a triangle ABS, if we know the distance AB and if we can measure the angle a at A and b at B, then:

$$AS = AB \times \frac{\sin b}{\sin(a + b)}$$

$$BS = AB \times \frac{\sin a}{\sin(a + b)}.$$

- Eratosthenes used the length of shadows cast at the same time at different places to deduce the circumference of the Earth.

- The Earth is an oblate spheroid. The mean length of the radius is 3958.89 miles.

- We can measure the precise dimensions of the Earth using the techniques of satellite laser ranging and the global positioning system.

QUESTIONS AND PROBLEMS

2.1 How did Eratosthenes know the distance between Alexandria and Syrene?

2.2 If you have access to e-mail, try to contact fellow students in another country and repeat Eratosthenes' experiment. What value do you obtain for the radius of the Earth?

2.3 The Moon looks much bigger when it is close to the horizon than when it is high in the sky. Propose a simple experiment to show that this must be some sort of visual illusion rather than a real effect. What do you think is the cause of this illusion?

2.4 Compare the accuracy of modern laser-based distance measurement techniques with the theodolites built by Ramsden.

2.5 The first experimental evidence that the Earth is an irregular spheroid came from pendulum observations. A pendulum adjusted so that it swings seconds at Paris, say, must be shortened before it swings seconds at the equator. Explain this effect; if a pendulum has a period of 1 s at the equator, what is its period at the north pole?

2.6 How far away from your eye must you hold a coin of diameter 1 cm so that it just covers the disc of the full Moon? 1.1 m

2.7 The metre was originally defined as one ten-millionth of the distance from the north pole to the equator. How big was the original kilometre in terms of miles? (Assume a spherical Earth with a circumference of 24 900 miles.) 1 km = 0.6225 miles

FURTHER READING

GENERAL
Chapman A (1995) *Dividing the Circle* (Wiley–Praxis: Chichester)
— A thorough account of the development of angular measurement in astronomy in the period 1500–1850 AD.

Webster R S (1975) Jesse Ramsden. In *Dictionary of Scientific Biography* ed. C. C. Gillispie (Scribner's: New York)
— A brief biography of the finest instrument maker of the 18th century; Ramsden's precision instruments made modern surveying possible.

SLR AND RELATED TECHNOLOGIES
Dixon T H (1991) An introduction to the global positioning system and some geological applications. *Rev. Geophys.* **29** 249–276
— A useful introduction to GPS, particularly for those readers interested in geology.

Ewing C E and Mitchell M M (1970) *Introduction to Geodesy* (American Elsevier: New York)
— Although dated, this remains an accessible introduction to the problem of measuring the Earth's dimensions.

Hager B H, King R W and Murray M H (1991) Measurement of crustal deformation using GPS. *Ann. Rev. Geophys.* **95** 2679–2699
— How the techniques of GPS can be used to measure the movement of the Earth's crust.

Kaplan E D (1996) *Understanding GPS* (Artech House: Boston)
— The theory and application of GPS, at a level suitable for students.

Kovalevsky J (1995) *Modern Astrometry* (Springer: Berlin)
— A very useful book for finding details on all aspects of precision astronomy. Perhaps of more use for later steps on the distance ladder, but the sections on satellite laser ranging are also good. For the advanced reader.

Leick A (1990) *GPS Satellite Surveying* (Wiley–Interscience: New York)
— A comprehensive review of the global positioning system.

3

Second step: the solar system

We can measure the size of the Earth, and identify positions on its surface, to essentially any degree of accuracy we require. But this is just the first step on the distance ladder. Even the earliest civilisations realised that there was more to the universe than the Earth. There are the various heavenly objects, most notably the Sun and Moon. How far away are they?

3.1 THE PLANETS AND THE CELESTIAL SPHERE

To the ancient Greeks the night sky seemed to be an inverted bowl resting on a flat plate. Sprinkled over the inner surface of the bowl, arranged in patterns that did not change from one year to the next, were the stars. During the course of a night the stars wheeled around a fixed point in the sky, as if the bowl were rotating around an axis passing through that point and the observer's position. The bowl's rotation caused some stars to sink below the western horizon and others to rise above the eastern horizon, suggesting that the bowl was part of a star-covered sphere. The Greeks called this the *celestial sphere.* The points around which the sphere rotated were the *celestial poles.* We now know that the stars are at different distances from us and that the celestial sphere is simply a convenient illusion. For most of history, though, people regarded the celestial sphere as a real object. It was usually thought of as a solid sphere of crystal to which the stars were somehow attached. (The early Greek astronomers pointed out an interesting fact. You are always at the centre of the sphere, no matter where on Earth you happen to be, since otherwise the constellations would appear to change size as they moved across the sky. This implies that the radius of the celestial sphere is very much larger than the radius of the Earth. In other words, the distance to the stars is much greater than 4000 miles.)

The celestial sphere turns once every 23 hours 56 minutes and 4 seconds, so if a particular star rises above the horizon at midnight one evening, the following evening it will rise 3 minutes and 56 seconds before midnight. The Sun takes slightly longer to make one full revolution. Suppose that one evening the Sun sets just as a particular star rises. The next evening, as that star rises, the Sun will be a few minutes away from setting. The following evening the Sun will be even further away from setting as that star rises. Thus the Sun moves across the celestial sphere relative to the stars. It makes one complete circuit of the celestial sphere in one year, always moving west to east. The Moon also wanders across the celestial sphere, relative to the stars, as do various other celestial bodies: Mercury, Venus, Mars, Jupiter and Saturn. (These names are Latin translations of the corresponding Greek names: Hermes, Aphrodite, Ares, Zeus and Krones.) To distinguish these moving objects from the *fixed stars*, the Greeks called them *planets*, from a word meaning 'wanderer'.

The methods we use to estimate distances in daily life indicate that the planets all lie at different distances. For instance, the Moon sometimes moves in front of the stars and planets, but we *never* see the stars and planets move in front of the Moon. Therefore the Moon must be the closest of the celestial objects. The planets all move at different speeds, which suggests that they are not all at the same distance. Saturn crawls across the celestial sphere: it returns to the same place among the stars after 29.5 years. The detailed motion of Saturn relative to the stars is complicated, but in terms of its *average* daily motion across the sky we can say that it moves 2' per day. Jupiter moves slightly faster. It takes 11.9 years to make one full lap of the celestial sphere, which means it has an average daily motion of 5'. Mars has an average daily motion of 31'27". The average speed of the Sun is 59'8" per day, while Venus moves at an average speed of 1°36' per day. Mercury races across the sky at an average rate of 4°5'33" per day. The Moon, which is the closest celestial body, also moves the fastest. It has an average daily motion of 13°10'35". The Greeks used this information to order the planets in terms of their distance from Earth. The Moon was closest, then came Mercury, Venus, the Sun, Mars, Jupiter and Saturn.

(This ordering of planetary distances explains the order of the days of the week. Ancient astrologers assigned one of the seven planets to 'rule' each hour of the day. Saturn, the farthest planet, ruled the first hour of the first day. Jupiter, the next farthest planet, ruled the second hour of the first day. Mars ruled the third hour, and so on. This cycle repeated itself every seven hours, which meant that the first hour of successive days was ruled by Saturn, the Sun, the Moon, Mars, Mercury, Jupiter and Venus. The day was named after the ruler of the first hour. We can recognise the planetary influence on the names of the first three days — Saturday, Sunday, Monday — quite easily. The names of the other days are derived from the Teutonic equivalents of the Latin gods: Wodan replaces Mercury, Thues replaces Mars, Thor replaces Jupiter and Freia replaces Venus.)

In modern terminology the planets are the large bodies — Mercury, Venus, Earth, Mars, Jupiter and Saturn — that orbit the Sun. From now on I will use the word planet in its modern sense. The Moon is a natural satellite of the Earth, rather than a planet in its own right. The Sun, of course, is a star. Since the planets and their satellites dance attendance upon the Sun, the whole system is called the *solar system*, the word 'Sol' being Latin for 'Sun'.

The only other astronomical objects known to the Greeks — *comets* and *meteors* — followed no sensible pattern. Meteors were thought to be somehow terrestrial, rather than celestial, in origin. The Greek word 'meteor' simply meant 'something high up', and the Greeks used it to include clouds, lightning, rainbows and similar phenomena. We retain this meaning when we refer to *meteorology*, the study of all that happens in the atmosphere up to a height of about ten miles. Comets (originally 'aster kometes', or 'long-haired star', referring to a comet's characteristic tail) were a mystery to all ancient peoples. The appearance of a comet was sure to cause widespread superstitious panic.

If meteors and comets were unpredictable, at least the motion of the Sun, Moon and planets seemed to have some sort of regularity. Perhaps it was possible to understand the regularity behind the cosmos. This hope led the Greek astronomers to try to build a dynamic model of the universe — a model that would give the distance and position of the planets at any time. This grand endeavour dominated astronomy for centuries.

3.2 ARISTARCHUS, HIPPARCHUS AND PTOLEMY

Many scientists contributed to the development of Greek astronomy, but three names are prominent: Aristarchus, Hipparchus and Ptolemy.

3.2.1 Aristarchus

Aristarchus of Samos (*c.*320 BC–*c.*250 BC) was the most imaginative and original of all the Greek astronomers. From the modern point of view he was also the most successful. More than 17 centuries before the idea became popular, he taught that the motions of the planets are easier to explain if the Earth moves around the Sun rather than *vice versa*. He also presented arguments to determine the size and distance of various objects in the solar system. Given the crude level of technology available, the arguments were very effective.

First, he gave an argument, known as the *lunar dichotomy method*, that let him express the Earth–Sun distance in terms of the Earth–Moon distance. (The mean Earth–Sun distance became known as the *astronomical unit*. As we shall see, the problem of obtaining an accurate value for the astronomical unit challenged astronomers for more than 2000 years.) Figure 3.1 illustrates the idea.

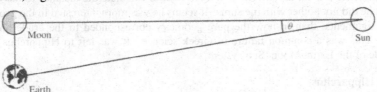

FIGURE 3.1: By measuring the angle between the Moon and the Sun, when the Moon is exactly half full, we immediately know θ, the angle subtended by the Earth–Moon distance at the Sun. Simple trigonometry lets us calculate the Earth–Sun distance in terms of the Earth–Moon distance and θ. Aristarchus estimated θ to be 3°. Note the similarity to figure 2.3. The difference here is that we already know that one of the angles in the triangle is a right angle. (Not drawn to scale.)

Aristarchus began his argument with the (correct) assumption that the Moon shines by reflecting light from the Sun. When the Moon is exactly half full, the Earth, Moon and Sun form a right angle. If we can measure the angle between the centre of the Moon and the centre of the Sun, we immediately know θ: the angle subtended by the Earth–Moon distance (EM) at the Sun. From figure 3.1 we see that the Earth–Sun distance (ES) is given by

$$ES = EM/\sin\theta.$$

Aristarchus estimated θ to be 3°, from which he deduced that the Sun was 19 times more distant than the Moon.

The lunar dichotomy method gave Aristarchus the relative distances in the Earth–Moon–Sun system. To set the scale of the system he needed to know the distance to the Moon. To obtain this, he produced another geometrical argument: the eclipse diagram. Consider figure 3.2. It shows the Moon in the shadow cone of the Earth. By timing the duration of lunar eclipses, we can calculate the width of the Earth's shadow cone at the distance of the Moon. We can measure the Moon's apparent diameter, and we can take it to be equal to the Sun's apparent diameter (a happy coincidence that makes for spectacular total solar eclipses). Finally, if we assume that the Sun is 19 times more distant than the Moon, we need only use some simple geometry to deduce all the absolute distances in the Earth–Moon–Sun system.

FIGURE 3.2: The eclipse diagram of Aristarchus. By measuring the time the Moon spends in the shadow cone of the Earth during a lunar eclipse, and by measuring various angles in this diagram, it is possible to estimate the distance to the Moon. (Not drawn to scale.)

Aristarchus had all the information he needed to calculate the distance to the Moon, and yet he did not bother with the sums! It seems he was more interested in the geometry of his arguments, than in how the pure geometry corresponded to the crude physical world. This was a common failure of Greek science. It was left to Hipparchus to set the scale of the Earth–Moon–Sun system.

3.2.2   Hipparchus

Hipparchus of Nicaea (*c.*190 BC–*c.*120 BC) was the greatest astronomer of the ancient world. Among his many contributions to astronomy, he discovered the precession of the equinoxes. He compiled the first star catalogue: a list of the position and brightness of 1080 stars, and in doing so founded the science of *astrometry*, the measurement of the angles between celestial objects, which plays an important role in later chapters. He was also the first to apply the method of parallax to determine the distance to a heavenly body.

The parallax of the Moon. As we saw on page 5, the parallax effect occurs because observers at different locations see the same object from different perspectives. The size of the effect depends on the baseline between the two observers and the distance of the object being observed. If you can measure the parallactic shift, and if you know the size of the baseline between the observers, simple trigonometry gives you the distance to the object. For a remote object to show a parallax, the baseline between the two observers must be large. For the Moon, which is very distant indeed, a suitable baseline is the radius of the Earth. The *horizontal* parallax of an object is the parallactic shift measured by observers at the Earth's pole and equator. It is a measure of how big the Earth appears as viewed from that object. See figure 3.3. (Whenever astronomers talk of the parallax of

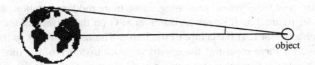

FIGURE 3.3: The horizontal parallax of an object in the solar system is the parallax of that object as viewed from the Earth's pole and its equator. It is thus a measure of how large the Earth appears from that object. The smaller the parallax, the more distant the object (and the smaller the Earth will seem to an observer on that object). Note the similarity between this figure and figure 1.3.

a body in the solar system, they always mean the horizontal parallax of that body.)

It was impractical for the Greeks to place observers at the poles and the equator. For one thing, they were poor explorers. Although Hipparchus himself was the first to work out the properties of the polar regions of a spherical Earth, no Greek ventured anywhere near the poles. This was not necessary, though. Hipparchus realised that as long as the observers were at very different latitudes — not necessarily pole and equator — they might have a large enough baseline to detect the Moon's parallax. If the observers made simultaneous observations, then one could quite easily convert the parallax between these two latitudes into the horizontal parallax. The distance to the Moon follows immediately. But how could Hipparchus be sure that these widely separated observers watched the Moon at the same time?

On 14 March 189 BC (perhaps around about the time that Hipparchus was born), observers at Hellespont recorded a total eclipse of the Sun. Observers at Alexandria recorded a partial eclipse: the Moon covered only $\frac{4}{5}$ of the Sun's disc. Hipparchus argued that this difference was due to the parallax of the Moon between Hellespont and Alexandria. See figure 3.4.

FIGURE 3.4: The observer at A sees a total eclipse of the Sun. The observer at B sees only a partial eclipse. The difference between the two observations is a parallax effect.

At some times of the month the Moon is closer than at other times, so its apparent diameter varies. When the Moon is farthest from us its apparent diameter is 29′22″, and when it is closest its apparent diameter is 33′31″. Hipparchus measured its average apparent diameter to be 33′15″. By implication, the Sun's apparent diameter was also 33′15″. One fifth of the Sun's apparent diameter was therefore about 6′40″. This was the parallax of the Moon between Hellespont at latitude 41° and Alexandria at latitude 31°. Since Hipparchus knew the elevation of the Moon at these locations, he had enough information to calculate the Moon's distance.

We can understand Hipparchus' reasoning, using more modern notation, as follows.

First, consider figure 3.5. It shows an observer at O on the surface of the Earth, and a hypothetical observer at C at the centre of the Earth. The zenith at O is in the direction OZ, and the observer there measures the *apparent zenith distance* of the moon, $\angle ZOM$, to be z_O. The observer at C measures the *true zenith distance* of the Moon, $\angle ZCM$, to be ω.

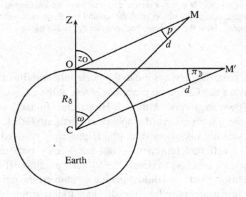

FIGURE 3.5: The horizontal parallax of the Moon.

The *geocentric distance* to the Moon is d, and $\angle OMC$ is the parallax angle p — the angle subtended at the Moon by the baseline OC. (OC is the radius of the Earth, R_δ.) From $\triangle OCM$ we have

$$\frac{\sin p}{R_\delta} = \frac{\sin(180° - z_O)}{d}$$

so

$$\sin p = \frac{R_\delta}{d} \sin z_O. \tag{3.1}$$

If the Moon is rising or setting, so that it is on the observer's horizon, its apparent zenith distance is $90°$. In this case, as shown in the figure, $\angle COM' = 90°$ and $CM' = d$. We denote the parallax angle $OM'C$ in this special case by the symbol $\pi_{\mathcal{D}}$. It is the horizontal parallax of the Moon. (Remember that $\pi_{\mathcal{D}}$ here is a *variable*; it has nothing to do with the constant $\pi = 3.141\cdots$.) So we can write

$$\sin \pi_{\mathcal{D}} = \frac{R_\delta}{d}$$

which lets us write (3.1) in the form $\sin p = \sin \pi_{\mathcal{D}} \sin z_O$. Now p and $\pi_{\mathcal{D}}$ are both small angles; if they were not small, we would not have to go to all this trouble in order to find the Moon's parallax! We can therefore make the small-angle approximations $\sin \pi_{\mathcal{D}} \approx \pi_{\mathcal{D}}$

and $\sin p \approx p$ and write

$$p = \pi_{\mathbb{D}} \sin z_{\mathrm{O}}. \tag{3.2}$$

Now, to use the Hipparchus method of measuring the distance to the Moon we must arrange for two observers at A and B to make simultaneous observations of the Moon. (Since the Moon is not a point source of light, the observers need to agree which part of the Moon to observe: perhaps a particular crater, or the limb of the Moon.) A and B have the same longitude but must be widely separated in latitude. Suppose A has latitude ϕ_{A} N and B has latitude ϕ_{B} S, and suppose that the apparent zenith distance of the Moon is z_{A} at A and z_{B} at B. Denote the true zenith distances by ω_{A} and ω_{B}. Finally, denote the angles of parallax for A and B by p_{A} and p_{B} respectively. (See figure 3.6.)

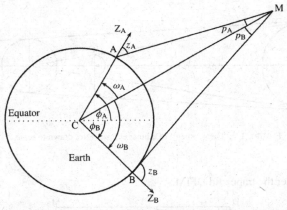

FIGURE 3.6: The diagram shows a method for measuring the horizontal parallax of the Moon. In essence, this is the method employed by Hipparchus to deduce the distance to the Moon.

From $\triangle$ACM we have $p_{\mathrm{A}} = z_{\mathrm{A}} - \omega_{\mathrm{A}}$; from $\triangle$BCM we have $p_{\mathrm{B}} = z_{\mathrm{B}} - \omega_{\mathrm{B}}$. Therefore

$$p_{\mathrm{A}} + p_{\mathrm{B}} = z_{\mathrm{A}} + z_{\mathrm{B}} - (\omega_{\mathrm{A}} + \omega_{\mathrm{B}}).$$

But $\omega_{\mathrm{A}} + \omega_{\mathrm{B}} = \phi_{\mathrm{A}} + |\phi_{\mathrm{B}}|$, where $|\phi_{\mathrm{B}}|$ denotes the modulus of ϕ_{B}, and so

$$p_{\mathrm{A}} + p_{\mathrm{B}} = z_{\mathrm{A}} + z_{\mathrm{B}} - (\phi_{\mathrm{A}} + |\phi_{\mathrm{B}}|).$$

From (3.2) we know that

$$p_{\mathrm{A}} + p_{\mathrm{B}} = \pi_{\mathbb{D}}(\sin z_{\mathrm{A}} + \sin z_{\mathrm{B}})$$

which gives us

$$\pi_{\mathbb{D}} = \frac{z_{\mathrm{A}} + z_{\mathrm{B}} - (\phi_{\mathrm{A}} + |\phi_{\mathrm{B}}|)}{\sin z_{\mathrm{A}} + \sin z_{\mathrm{B}}}. \tag{3.3}$$

All the variables on the right-hand side of (3.3) are measurable, and so we can deduce the horizontal parallax of the Moon. The geocentric distance CM follows immediately, if we know the radius of the Earth. (In practice the two observers at A and B are unlikely to be on the same meridian of longitude, and allowance must be made for this.)

Hipparchus found that the Moon's least distance was 71 ER and its greatest distance was 83 ER.

Refinement of the eclipse diagram. Hipparchus knew that his parallax calculation could not produce a precise estimate of the Moon's distance; the observations he had to work with were too crude. He therefore decided to refine the eclipse method of Aristarchus.

Figure 3.7 shows the Moon in the Earth's shadow cone. SB is the Sun's radius, TH is the Earth's radius and DE is the Moon's radius. We want to find TM (or TD): the distance to the Moon.

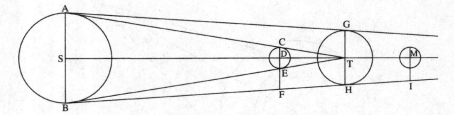

FIGURE 3.7: The geometry of a lunar eclipse. (Not drawn to scale.)

First, consider the trapezoid DFIM.

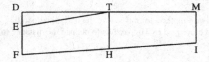

It is clear that MI + DF = 2TH, so

$$DF = 2TH - MI.$$

We deduce from studies of lunar eclipses that the radius of the Earth's shadow cone at the Moon's distance is 2.5 times the radius of the Moon, i.e. MI = 2.5DE, so

$$DF = 2TH - 2.5DE.$$

Since EF = DF − DE we obtain

$$EF = 2TH - 3.5DE.$$

Now consider the trapezoid SBHT.

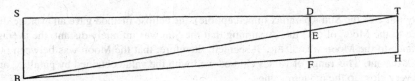

We have

$$\frac{TH}{EF} = \frac{TB}{EB}, \quad \text{and} \quad \frac{TB}{EB} = \frac{TS}{DS} \quad \text{so} \quad \frac{TH}{EF} = \frac{TS}{DS}.$$

But DS = TS − TD, so

$$\frac{TH}{EF} = \frac{TS}{TS - TD}$$

which produces, after some manipulation,

$$TD = \frac{TS(TH - EF)}{TH}.$$

Substituting the expression for EF we obtain

$$TD = \frac{TS(TH - 2TH + 3.5DE)}{TH} = \frac{TS(3.5DE - TH)}{TH}.$$

Now DE is the Moon's radius, which Hipparchus determined to be $\frac{1}{1300}$th part of its orbit; so

$$DE = \frac{2\pi TD}{1300}$$

which gives us

$$TD = TS\left(\frac{\frac{7\pi TD}{1300} - TH}{TH} \right)$$

which, after some manipulation, produces

$$TD = \frac{TS}{\frac{7\pi}{1300}\frac{TS}{TH} - 1}.$$

If we express everything in terms of the Earth's radius, we can write TH = 1 and we get

$$TD = \frac{TS}{\frac{7\pi}{1300}TS - 1}. \tag{3.4}$$

What value of TS, the distance to the Sun, should we use in (3.4)? Hipparchus assumed several different values for the distance to the Sun and, for each value, calculated the Moon's distance. Assuming that the Sun was 490 ER distant (close, but still far enough away to make the Sun's parallax imperceptible), the eclipse method gave an average distance to the Moon of 67.3 ER. Assuming that the Sun was infinitely distant, the average distance to the Moon was 59 ER. It seemed, therefore, that the Moon was between 60–67 ER distant. This range of values chimed well with the value obtained by parallax, and it is very close to the modern value.

Although Hipparchus obtained an excellent estimate of the distance to the Moon, he was unable to improve upon previous estimates of the distance to the Sun. Since the Sun's parallax was too tiny to measure, he had only the lunar dichotomy method to work with. The theory behind the method is fine, but it does not work in practice. Not only is the angle too tiny to measure accurately with the naked eye, but it is impossible to decide the exact moment when to make the measurement. Remember that the method depends upon observing the position of the Sun when the Moon is half full. However, the surface of the Moon is not smooth. Shadows from lunar mountains cause the terminator (the boundary line between the day and night hemispheres of the Moon) to be jagged, so we can never be entirely sure when the Moon is half full.

3.2.3　Ptolemy

Almost everything we know about Hipparchus comes down to us by way of Ptolemy. Ptolemy unified much of the astronomy of his day into a consistent world picture. Most of his system was based on the work of earlier astronomers, primarily Hipparchus, and this has led many commentators to question Ptolemy's own contribution. This seems rather harsh: Ptolemy built a system of astronomy that dominated astronomical thinking for 1400 years. In this sense, he was one of the most influential scientists who ever lived.

The Ptolemaic system was complicated. It had the Earth at rest at the centre of the universe. Nested around the Earth were several crystal spheres, called *deferents*. The Sun, the Moon and the planets were each associated with a particular deferent, and as each deferent revolved around a point in space close to the Earth's centre it carried its planet with it. A planet was not attached to its deferent. Rather, each planet was part of a smaller sphere called an *epicycle*, and it was the centre of the epicycle that was attached to the deferent. So each planet moved in a circle as the epicycle turned, and also moved in a larger circle as the deferent turned. By adjusting the speeds of rotation of the two spheres, and where necessary by adding more epicycles, Ptolemy could duplicate the observed motion of the heavenly bodies.

Rather like the arcane conspiracy theories that spring up whenever a famous person dies, the Ptolemaic system of epicycles upon epicycles upon deferents was too complex to be true. On the other hand, it was useful in predicting such things as the times of eclipses. Although the system outlined by Aristarchus was conceptually simpler than that of Ptolemy, Aristarchus did not bother to use his model to predict anything useful. So it was the ugly Ptolemaic system, rather than the elegant Aristarchan system, that people accepted as a true model of the universe.

Ptolemy used his model to calculate the distance of the various heavenly bodies. (See table 3.1.) The planetary distances came from a peculiar mix of geometry and philosophy, and had little scientific merit. His estimate of the distance to the fixed stars — about 60 million miles — was equally without foundation, but the figure is interesting since it

TABLE 3.1: The mean distance to the Sun, Moon and planets in the Ptolemaic system. (In converting Ptolemy's distance estimates into 'modern' units I have taken the value of the Earth's radius that Ptolemy himself used: 28 667 stades, which probably corresponds to about 2867 miles.)

Object	Mean distance (ER)	Mean distance (miles)
Moon	48	137 616
Mercury	115	329 700
Venus	623	1 784 700
Sun	1210	3 469 000
Mars	5040	14 496 700
Jupiter	11 503	32 979 100
Saturn	17 026	48 813 500
Fixed stars	20 000	57 340 000

is effectively his estimate of the size of the universe. The important value in Ptolemy's scale of distances was the distance to the Sun.

To calculate the Sun's distance, Ptolemy first estimated the distance to the Moon and then used this value in the eclipse diagram. This was the opposite of how Hipparchus used the eclipse diagram. Hipparchus had used a sound method: he *assumed* a large value for the distance to the Sun in order to calculate the distance to the Moon, and then checked that his result was insensitive to the exact value chosen for the Sun's distance. The way Ptolemy used the eclipse diagram was poor science: small variations in the value taken for the Moon's distance produce large changes in the calculated distance to the Sun. Even so, Ptolemy's estimate of the Sun's distance — about 3 500 000 miles — was accepted throughout medieval times.

3.3   COPERNICUS, TYCHO AND KEPLER

The sixteenth century produced three astronomers who were the equal of any of the Greeks: Copernicus, Tycho and Kepler. By the time of Kepler's death, the Ptolemaic view of the solar system was obsolete.

3.3.1   Copernicus

The Polish astronomer Nicolas Copernicus (1473–1543) found he could produce tables of planetary positions more easily if he assumed that the Earth revolved around the Sun, rather than *vice versa*. He produced a working model of the heavens based on just four simple assumptions.

- The Earth rotates on its axis once a day.

- The Earth revolves around the Sun and makes one revolution per year.

- The other planets also revolve around the Sun.

- The Moon revolves around the Earth and makes one revolution per month.

Nicolas Copernicus

For Copernicus, the observed motions of the celestial bodies were not due solely to real motions, but were in large part due to the motion of the Earth carrying the observer along with it.

The Copernican system suffered from one of the drawbacks of the old system. Copernicus had all the planets move in perfect coplanar circles, so he had to use epicycles to make theory match observation. Copernicus, indeed, used more epicycles than Ptolemy. Nevertheless, in all other respects the Copernican system was so much simpler than the Ptolemaic system, and it solved so many puzzling aspects of planetary motion, that people came to regard it as more than just a useful tool for calculating planetary positions: it described the Universe as it really was. The Earth *really did* move around the Sun.

The order of the planets, starting from the Sun, was: Mercury, Venus, Earth, Mars, Jupiter, Saturn. Planets with orbits inside the Earth's orbit were called *inferior* planets, and those with orbits outside the Earth's orbit were called *superior* planets. The knowledge of planetary orbits was no better in Copernicus' day than it had been in the time of Ptolemy, but the new Copernican viewpoint made it possible to deduce the relative distances of the planets from the Sun. Consider, for instance, figure 3.8, which shows the circular orbits of an inferior planet, P, and the Earth, E, around the Sun, S.

The angle SEP is called the *elongation* of the planet, which varies between 0° (when the planet is in *conjunction*) and a maximum angle, less than 90°, called its *greatest elongation*. A moment's thought is enough to convince you that the planet is at greatest elongation when the line EP is tangential to its orbit. In other words, at greatest elongation $\angle$SPE = 90°. We can therefore write

$$SP = SE \times \sin(SEP_{max}). \tag{3.5}$$

If we set SE = 1, and thus express all distances in terms of astronomical units (i.e. the Earth–Sun distance), we simply need to measure the planet's greatest elongation in order to determine its distance from the Sun.

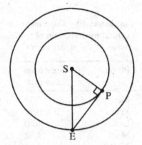

FIGURE 3.8: The orbits of the Earth and an inferior planet in the Copernican system.

The analysis for a superior planet is more involved, but we can obtain a relation similar to (3.5) that gives us the distance of the planet from Sun in astronomical units. So the Copernican model allows us to build an accurate scale model of the whole solar system by making careful measurements of planetary positions relative to the Sun.

Example 3.1 The greatest elongation of Venus

You measure the elongation of Venus on a series of nights around its greatest elongation. From your measurements you decide that the greatest elongation of Venus is $46°18'$. How far is Venus from the Sun?

Solution. From (3.5), the distance of Venus from the Sun, when expressed in astronomical units, is just $\sin 46°18'$. In other words, it is at a distance of 0.723 AU.

Table 3.2 gives the relative distances of the planets from the Sun, as deduced by Copernicus. Note that to turn these relative distances into absolute distances we need to obtain a value for the astronomical unit.

Copernicus based his model on a mean solar distance of 1142 ER, but this value was not obtained from any new technique or method. The Copernican model, however, made an interesting prediction regarding distances. Suppose an astronomer measures the position of a fixed star. Six months later, according to Copernicus, the Earth will have moved through half its orbit. It will be 2284 ER — about 6 500 000 miles — away from where it was. This is a huge baseline for observing. The astronomer measuring the position of the same star six months later should thus see a parallactic shift. In other words, the fixed stars should show an annual parallax: they should move in tiny circles, mirroring the motion of the Earth about the Sun. No annual stellar parallax had ever been found, and some astronomers considered this to be a fatal flaw in the Copernican system. Copernicus and his followers argued that the stars were too far away to show any parallax.

TABLE 3.2: Relative planetary distances as deduced by Copernicus on the basis of coplanar circular orbits. Each planetary distance is derived from a careful measurement of the angle between planet and Sun at various points of the planet's orbit.

Planet	Distance from Sun (AU)
Mercury	0.38
Venus	0.72
Earth	1.00
Mars	1.52
Jupiter	5.22
Saturn	9.17

3.3.2 Tycho

The Danish astronomer Tycho Brahe (1546–1601) was, along with Hipparchus, the greatest of the naked-eye astronomers.

Tycho (he is almost always referred to by his first name, which is a latinized version of the Danish Tyge) was the son of a nobleman. He seems to have been a particularly spoiled example of the species. When he was 19 he fought a midnight duel over a minor point of mathematics. His opponent cut off Tycho's nose, and for the rest of his life Tycho wore a false nose made of metal. This incident did nothing to improve his disposition. Throughout his life he treated his colleagues and servants harshly, and he quarrelled with everybody. However, his talent for observing the sky compensated for his lack of charm.

Tycho became famous in 1572, when he discovered a *nova* — a new star. This was important in its own right, but Tycho went further and made careful accounts of the nova's position. He showed that its parallax was imperceptible, and thus that it was very distant. After this discovery the King of Denmark became Tycho's patron, and subsidised the construction of an observatory on the island of Hven. This was the first true astronomical observatory. Tycho fitted it with the best instruments available, and he used them to make the best ever naked-eye observations of the sky. His tables of the Sun's motion were better than anything that had gone before, and he determined the length of the year to within one second. He also prepared accurate tables of planetary motion. And whereas Hipparchus fixed the positions of stars to an accuracy of about 1°, Tycho fixed their positions to an accuracy of 1′ — the angle made by a man's height at a distance of 3.5 miles. His continued observations of the stars demonstrated without doubt that their annual parallax had to be less than 1′. In the Copernican system this meant that the fixed stars had to be *at least* 7 850 000 ER away. (Tycho thought such large distances were absurd, and argued against the Copernican system. He was the last great astronomer to do so.)

Tycho met with an unfortunate end. He drank too much at a banquet given by the King. Rather than commit the social gaffe of asking to be excused, he stayed at the table until the King rose, and his bladder ruptured. As he died, in great pain, he said 'Oh, that it may not appear that I have lived in vain'. He did not live in vain. Upon his death, his wealth of observational data passed into the hands of his assistant, the German astronomer Johann Kepler (1571–1630).

Tycho Brahe

3.3.3 Kepler

Tycho set his successor a difficult task: to understand the motion of the planets as revealed by his observations. Kepler worked at the task with great patience. After much toil he realised that the Copernican system of multiplying epicycles would not work. The predictions of such a model simply did not correspond with Tycho's precise observations. Eventually, Kepler broke with the belief of two millennia and announced that the planets do not move in perfect circles; they move in ellipses. This was the first of Kepler's three *laws of planetary motion*. (An explanation of these laws of planetary motion came a century after Kepler, with the work of the English scientist Isaac Newton (1642–1727). Newton showed that the planets move in elliptical orbits in response to the gravitational force of the Sun. Kepler's laws are an inevitable consequence of gravity.)

Kepler's laws are as follows.

1. The orbit of each planet is an ellipse, with the Sun at one focus of the ellipse. (Since planetary orbits are elliptical, planets are at different distances from the Sun at different points in their orbit. The mean distance of a planet from the Sun is the semi-major axis of its orbital ellipse.)

2. The radius vector — an imaginary straight line joining the centre of a planet to the centre of the Sun — sweeps out equal areas in equal times. (This means that a planet moves quickest when it is closest to the Sun.)

3. The square of the orbital period of a planet — in other words the time it takes to make one trip around the Sun — is proportional to the cube of its mean distance from the Sun. The constant of proportionality is the same for all planets. (Newton showed that the constant of proportionality is not quite the same for each planet, but for our present purposes we do not need to consider the correction factor.)

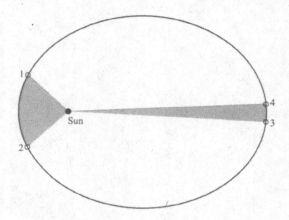

FIGURE 3.9: This diagram shows the elliptical orbit of a planet around the Sun. According to Kepler's first law, the Sun is at one focus of the ellipse. The planet takes the same time to move from point 1 to point 2 as it takes to move from point 3 to point 4. According to Kepler's second law, a line joining a planet to the Sun sweeps out equal areas in equal times. The shaded areas in the diagram are therefore equal. (Note that none of the planets has an orbit that is as elliptical as the one shown in the diagram.)

Figure 3.9 illustrates the first and second laws.

Kepler's third law is the most interesting from the point of view of determining distances in the solar system. (Note that Kepler's laws refer only to planets orbiting the Sun. They say *nothing* about the fixed stars. Kepler, in fact, held rather primitive views about the distance to the stars.) Suppose we express a planet's orbital or *sidereal* period, T, in terms of the Earth's sidereal period. In other words, we measure sidereal periods in terms of years. And suppose we express a planet's average distance from the Sun, D, in terms of the astronomical unit — the average Earth–Sun distance. We can then write Kepler's third law as

$$T^2 = D^3.$$

A planet's orbital period is easily measured: it is almost always easier to time something than it is to find a distance. Once we know the periods, the third law gives us the relative average distances of the planets from the Sun. We can build a scale model of the solar system much more quickly and easily than using the method developed by Copernicus.

Table 3.3 gives the average distance from the Sun of each of the planets known to Kepler. Figure 3.10 shows a scaled-down representation of the planetary orbits.

Thanks to Kepler, astronomers then knew the *relative* distances of the planets from the Sun. If they could measure the astronomical unit all the other distances would follow; they would have the set the scale of the solar system.

Kepler had as little idea of the value of the astronomical unit as did Tycho or Copernicus. The situation had not improved since Aristarchus proposed the eclipse method. It was the invention of the telescope that revolutionised mankind's understanding of scale of the solar system, just as it revolutionised the rest of astronomy.

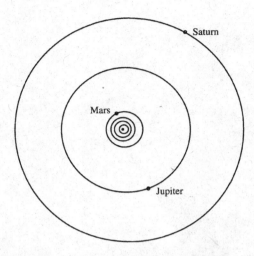

FIGURE 3.10: The orbits of the planets, drawn to scale. The innermost planets, not named on the diagram, are Mercury, Venus and Earth, in increasing distance from the Sun. Some of the orbits, particularly that of Mercury, are noticeably elliptical. Some of the orbits are almost circular (at this scale, the orbit of Venus is indistinguishable from a circle).

TABLE 3.3: Kepler's third law shows a relationship between a planet's sidereal period T (in years) and its average distance D from the Sun (in AU): $T^2 = D^3$. Measuring the sidereal period of the planets thus gives us a scale model of the solar system.

Planet	Sidereal period (years)	Distance from Sun (AU)
Mercury	0.241	0.387
Venus	0.615	0.723
Earth	1	1
Mars	1.881	1.524
Jupiter	11.858	5.200
Saturn	29.424	9.531

3.4 THE TELESCOPE

No one knows for certain when the first telescope was made, nor who invented it. The English surveyor Leonard Digges (*c.*1520–*c.*1559), the man who invented the theodolite, may also have made a telescope around 1550. The credit, though, usually goes to Hans Lippershey (*c.*1570–*c.*1619), a Dutch spectacles maker from the town of Middelburg in the Zeeland region of the Netherlands*. The story is that one day in 1608 Lippershey's

---

*Other people also claimed to have invented the telescope. For instance, Zacharias Janssen (1580–*c.*1638), a neighbour of Lippershey, claimed to have made a telescope as early as 1604. It is entirely possible that he did so. Indeed, the discovery must surely have been made several times by lens-makers.

Johann Kepler

apprentice, fooling around with some lenses, held one lens close up to his eye and another at arm's length. The apprentice found to his surprise that a distant weathervane seemed much larger and closer than it really was. He told Lippershey, who immediately realised the importance of the discovery. Lippershey also realised that it was impractical to stand around holding a pair of lenses. He therefore made a metal tube into which he put the lenses at the right position to create the magnifying effect. Lippershey called the device a *looker*. Other people called it a *perspective glass*, an *optic tube* or an *optic glass*. In 1612, Ionnes Demisiani (*d.*1619), a Greek poet and mathematician, suggested that the device be called a *telescope* (from Greek words meaning 'to see at a distance').

There is less controversy surrounding the first use of the telescope in astronomy. Although the English mathematician Thomas Harriot (1560–1621) and the German astronomer Simon Marius (1570–1624) may have observed the skies with a telescope slightly before the Italian physicist Galileo, it was Galileo Galilei (1564–1642) who was the first to truly understand the significance of telescopic observations.

Reports of the new device reached Galileo early in 1609, and he immediately set about building his own instrument. He soon had a telescope that made objects appear three times nearer (and therefore three times bigger in breadth and height). Galileo of course understood the military importance of his device: the farther away one can see the enemy, the more time one has to prepare for battle. He therefore travelled to Venice, where he demonstrated his telescope to the Doge and the Signoria. They peered through it and saw ships that were invisible to their naked eyes. Suitably impressed, they doubled Galileo's salary.

The newly wealthy Galileo began to grind bigger lenses for use in better telescopes. Within weeks he had telescopes that magnified by a factor of 33. Then, when he turned his instruments to the night sky, he made a series of startling discoveries.

Galileo Galilei

For instance, when Galileo looked at the Moon he saw craters. The craters disfigured the smooth face of the Moon and caused the terminator to be jagged. This discovery shattered the long-held belief that celestial bodies are somehow perfect: in some ways, the Moon is just like the Earth. Galileo then studied the planets. They appeared not as dots through his telescope, but as tiny discs. The conclusion seemed to be that the planets are like the Moon and the Earth — but very distant. Most importantly, on 7 January 1610, Galileo aimed his telescope at the planet Jupiter. He saw three faint points of light close to the planet. The next night he noticed that the points of light were in different positions relative to Jupiter. On 13 January he saw a fourth point of light. He decided that all four objects were revolving around Jupiter. They were *satellites* of Jupiter. This was a major discovery: the solar system contained more objects than anyone had supposed. The discovery was important for another reason. It disproved the doctrine that the Earth was the only centre of motion in the universe. Although none of these observations proved that Copernicus was right, they made the Copernican model of the solar system — one in which the Earth was just another planet orbiting the Sun — seem very plausible.

Technology did not stand still. Telescopes became bigger and better, and in 1638 the English astronomer William Gascoigne (1612–1644) invented the *micrometer,* a device for measuring small angular separations. An astronomer, using a telescope along with a micrometer, could determine the positions of celestial objects with unprecedented accuracy. The science of astrometry had come of age. Now that astronomers could pinpoint the position of planets on the celestial sphere they could begin to look for the parallaxes of the planets. Two widely separated observers, by recording different positions for the same planet, could determine the horizontal parallax — and thus the distance — of that planet. And thanks to Kepler's third law this would set the scale of the solar system.

3.5   PLANETARY PARALLAXES

Figure 3.11 shows Mars at *opposition* (in other words, when it is directly opposite the Sun in the sky). The Earth–Mars distance at opposition is only 0.37 times that of the average Earth–Sun distance. At opposition the parallax of Mars is thus $1/0.37 = 2.5$ times greater than the Sun's parallax. Although no progress had been made since antiquity on the problem of measuring the parallax of the Sun, by 1670 astronomers thought they might have the technology to measure the parallax of Mars.

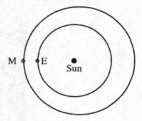

FIGURE 3.11: When Mars is at opposition it is close to Earth, and thus has a large parallax.

The French Academy of Sciences, urged on by the Italian–French astronomer Giovanni Domenico Cassini (1625–1712), decided to finance an expedition to Cayenne to observe the 1672 opposition of Mars. Cassini hoped that the location of Cayenne, just 5° north of the equator, was far enough away from Paris (latitude 48°52′ N) for parallax effects to be evident. On 8 February 1672, the young French astronomer Jean Richer (1630–1696) sailed for Cayenne armed with several large telescopes and measuring arcs, and a list from the Academy of the observations he was required to make. Cassini stayed in Paris to make the same observations.

Meanwhile, in England, the young astronomer John Flamsteed (1646–1719) was also busy. He learned that Mars would pass close to three dim stars in the early part of October 1672. By using these three stars as reference markers, Flamsteed hoped to measure the position of Mars with great accuracy. On the night of 6 October 1672, he measured the angular distance of Mars from two of the three stars during an interval of six hours and ten minutes. In that time, Mars moved 20′ in longitude and 1′13.5″ in latitude. From this, Flamsteed used a difficult geometrical procedure to calculate the parallax. When he finally published his conclusions in a journal, in July 1673, he stated that the parallax of Mars at opposition was no greater than 25″. From this it followed that the parallax of the Sun was at most 10″. In turn, this meant that the Sun was at least 21 000 ER distant. Cassini, using similar measurements made from Paris, concluded that the average solar distance was 22 000 ER. Could this be simple coincidence? The scientific community waited in suspense for the return of Richer, and news of his observations.

Richer returned in August 1673. After analysing the data from Paris and Cayenne, Cassini concluded that the parallax of Mars was 25″ and thus that the parallax of the Sun was 9.5″. The value of the astronomical unit was 21 600 ER, or 85.6 million miles. According to Kepler's third law, the average distance of Saturn from the Sun was a staggering 820 million miles. Suddenly the universe was 20 times larger than had been thought.

Cassini's estimate was better than all previous estimates. Nevertheless, although many contemporary astronomers accepted Cassini's work, not all of them did. Cassini's harshest critic was the English astronomer Edmond Halley (1656–1742).

Halley argued that the technology available to Cassini and Richer was crude. He did not believe that they could measure the parallax of Mars with sufficient precision to make a reliable estimate of the astronomical unit. Halley believed that an accurate value for the astronomical unit could best be obtained by measuring the parallax not of Mars, but of Venus. He further pointed out that this would be easiest during a *transit* of Venus (in other words, when it crosses the face of the Sun). During a transit, Venus is at its closest to Earth; it is even closer to us than Mars is at opposition. When Venus transits its parallax is almost three times greater than the Sun's parallax. As early as 1679, Halley pointed out that Venus would cross the face of the Sun on 6 June 1761 and 3 June 1769*. He urged astronomers, as yet unborn, to observe these transits. He wanted observers at different places on Earth to time, as accurately as possible, the moment when Venus entered (or left) the Sun's disc. The difference in timing between, say, Rome and Moscow, would give the parallax of Venus between those two cities, which in turn would give the distance to the planet. He reasoned that timing these events would be easier than measuring tiny angles on the celestial sphere, and therefore give more accurate results than Cassini or Flamsteed could manage.

We can understand the details of Halley's argument by referring to figure 3.12, which shows the transits A′ and B′ as seen by widely separated observers at A and B. Suppose the baseline between the two observatories, A and B, is b. Let the Earth–Sun distance be d_{ES}, the Venus–Sun distance be d_{VS} and the Earth–Venus distance be d_{EV}.

We can obtain an expression for the Earth–Sun distance, in terms of measurable quantities, as follows.

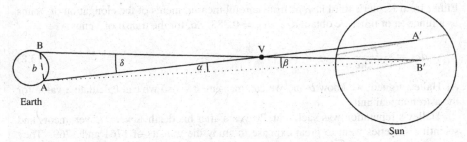

FIGURE 3.12: The observer at A sees Venus transit across the Sun's face along path A′. The observer at B sees Venus transit across the Sun's face along path B′. (Not drawn to scale.)

---

*Because the orbital paths of Venus and Earth are inclined to one another, transits of Venus are rare. They occur in pairs eight years apart, and the interval between the latter of one pair and the earlier of the next pair is alternately 105.5 years and 121.5 years. There were transits in 1874 and 1882, and there will be transits again in 2004 and 2012.

First, as figure 3.12 makes clear, we can write

$$d_{ES} = \frac{b}{\beta}.$$ (3.6)

Using

$$d_{EV} = \frac{b}{\delta}$$

we obtain the following expression for β:

$$\beta = d_{EV} \times \frac{\delta}{d_{ES}}.$$

Now we need to obtain an expression for α. Again using figure 3.12 we see that

$$\alpha = \frac{A'B'}{d_{ES}} = d_{VS} \times \frac{\delta}{d_{ES}}.$$

So we can write

$$\beta = \alpha\left(\frac{d_{EV}}{d_{VS}}\right).$$

Substituting this expression for β back into (3.6) we see that

$$d_{ES} = \frac{b}{\alpha(d_{EV}/d_{VS})}.$$ (3.7)

Either from Kepler's third law, or from careful measurements of the elongation of Venus as a function of time, we obtain $d_{EV}/d_{VS} = 0.383$. So, for the transit of Venus:

$$d_{ES} = \frac{b}{0.383\alpha}.$$ (3.8)

As Halley argued, we know b and we can measure α — so we can calculate a value for the astronomical unit.

Halley's reputation was such that, 20 years after his death, several governments and scientific societies went to great expense to study the transits of 1761 and 1769. They set up observatories all over the world: from Siberia to California and from Hudson Bay to Madras. The greatest of all explorers, Captain James Cook (1728–1779), sailed to Tahiti to study the 1769 transit. It is a measure of the importance attached to Cook's work that the French government — who at that time were at war with England — told all its men-of-war to allow his ships safe passage. Astronomers travelled thousands of miles to observe two events that would last just a few hours each.

(The unluckiest observer of all time was a French astronomer with the grand name of Guillaume Joseph Hyacinthe Jean Babtiste Le Gentil de la Galaisiré (1725–1792). Le Gentil gained a commission from the French Academy of Sciences to observe the

1761 transit from Pondicherry in India. Just before he arrived there in 1760, the English captured the town and he could only watch the transit from his ship. From such a vantage point he was unable to make any scientifically significant observations. Rather than waste the trip, he decided to stay in the East and wait for the 1769 transit. During the eight years between transits Le Gentil did some useful astronomy, and as an offshoot of his work he calculated that Manila would be an excellent place from which to observe the 3 June 1769 transit. Unfortunately the French Academy ordered him back to Pondicherry. While Manila was clear on the day of the transit, clouds obscured the Sun at Pondicherry and observations were impossible. Le Gentil's run of bad luck had not ended. He arrived back in Paris in 1771 after an absence of almost 12 years to find his relatives dividing his estate — they believed him to be dead. Even after expensive legal action, which left him responsible for court costs, he did not receive all the money owed him. Perhaps not surprisingly, he gave up astronomy.)

Even if Le Gentil had observed the transits, his measurements would not have been as useful as Halley had hoped. The situation was not as clear cut as the above example assumes. Because the motion of Venus is so slow, it was impossible to time with any great accuracy the moments of outer contact (the moment when the black disc of Venus first touches the Sun's border, and the moment when it finally disappears from the Sun's face). This had, in fact, been expected. Better results were expected for the timings of inner contact (the moment when the entire disc of the planet first becomes visible on the Sun's face, and the last moment when the whole of the disc may be seen on the Sun's face). Unfortunately, due to optical effects, the planet's disc looked like a teardrop connected to the Sun's border by a thin black thread — the so-called 'black drop'. When the thread broke, Venus was some way inside the Sun's border. It was impossible to say exactly when the moment of inner contact had started. The timings of astronomers standing beside one another differed by tens of seconds.

Although the hoped-for accuracy was missing, data from the transits made it clear that the parallax of the Sun was somewhere in the range 8–9″. This was an improvement on Cassini's value. In 1822, and again in 1835, the German astronomer Johann Franz Encke (1791–1865) re-analysed the whole set of observations and concluded that the parallax of the Sun was 8.57″. This corresponded to a distance of 95 370 000 miles. As we shall soon see, this was just 2.6% higher than the true value.

The first person to determine the value of the astronomical unit with modern accuracy was the Scottish astronomer David Gill (1843–1914).

Gill revived the idea of measuring the parallax of Mars at opposition, and in 1877 he observed the planet from the island of Ascension. This gave him a better value for the astronomical unit than anyone had obtained before. His work also showed that better observations of Mars would not necessarily lead to a better value for the astronomical unit. He found that Mars showed a perceptible disc in his telescope, with a boundary that was fuzzy due to the planet's atmosphere. This limited the precision with which he could determine the planet's position and hence its parallax.

Gill therefore decided to measure the parallaxes of *asteroids*. The asteroids are small rocky bodies, most of which inhabit the space between the orbits of Mars and Jupiter. The Italian astronomer Giuseppe Piazzi (1746–1826) discovered the first asteroid, Ceres, in 1801. Ceres is the largest of the asteroids, yet it is only about 700 miles in diameter.

Example 3.2 The transit of Venus

The 3 June 1769 transit of Venus was observed in Sweden and Tahiti (among other places). Swedish astronomers timed the transit as lasting 5 h 53 m 14 s; for astronomers in Tahiti the transit took 5 h 30 m 4 s. The Sun's full diameter is 32′, which Venus would transit in 8 h. The baseline between Sweden and Tahiti is 13 400 km. How big is the astronomical unit?

Solution. The situation is as shown in the following diagram:

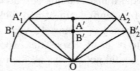

We need to calculate α, which is A′O − B′O. From the duration of transit A′ (observed at Tahiti) we can calculate the angular distance $A'_1 A'_2$:

$$A'_1 A'_2 = 32' \times \tfrac{5.501}{8} = 22.004'.$$

We can do the same for the angular distance $B'_1 B'_2$:

$$B'_1 B'_2 = 32' \times \tfrac{5.887}{8} = 23.548'.$$

From the diagram above,

$$\alpha = A'O - B'O = \sqrt{(A'_1 O)^2 - (\tfrac{1}{2} A'_1 A'_2)^2} - \sqrt{(B'_1 O)^2 - (\tfrac{1}{2} B'_1 B'_2)^2}$$

and $A'_1 O = B'_1 O = 16'$. Therefore

$$\alpha = \sqrt{16^2 - 11.002^2} - \sqrt{16^2 - 11.774^2} = 0.783'$$
$$= 2.28 \times 10^{-4}\,\text{rad.}$$

Substituting this value for α, and the given value of b, into (3.8) we get

$$d_{ES} = \frac{13\,400}{2.28 \times 10^{-4} \times 0.383} = 1.5 \times 10^8\,\text{km.}$$

Even in large telescopes such small objects show up only as points of light, so Gill knew he could locate the positions of asteroids with great precision. Between 1888–1889, Gill obtained the parallax of the asteroids Victoria, Sappho and Iris. From this work he deduced a value for the Sun's parallax of 8.8″ — corresponding to an average Earth–Sun distance of just under 93 million miles*.

---

*The parallax of the Sun is 8.794″. Compare this with the resolving power of the normal eye, which is 100″. Even the theoretical resolving power of the human eye, which only few individuals ever approach, is just 20″. This means that we cannot measure the Sun's parallax with the naked eye. (It also means that someone with perfect eyesight, viewing the Earth from a hypothetical vantage point on the Sun, would see the Earth as a point of light rather than as a disk.)

The asteroids that Gill studied were more distant than either Mars or Venus, so their parallaxes were small and difficult to measure. In 1898, however, the German astronomer Karl Gustav Witt (1866–1946) discovered the asteroid Eros. Eros has an orbit lying between Mars and Earth, so at closest approach it has a large and easily measured parallax. Furthermore, since it is small it remains as a point of light in a telescope. Astronomers could fix its position with pinpoint precision. In 1931 it was due to pass within 16 million miles of the Earth, and the English astronomer Harold Spencer Jones (1890–1960) led a long programme of study of the asteroid. From his observations of Eros, in 1941 Spencer Jones published a value of the Sun's parallax of $8.790'' \pm 0.001''$. (It is interesting to note that his value was too small by four times its claimed probable error.) This was about as good as the parallax method allows. Further improvement needed a new method of distance determination.

3.6   THE MODERN VIEW OF THE SOLAR SYSTEM

After Cassini's work, it was clear that the solar system was much larger than the ancient Greeks had supposed. The system itself, though, was something Aristarchus would have recognised: six planets — the outermost being Saturn — revolved around the Sun. In 1781, a musician did something no other person had done in history: he discovered a new planet. The solar system was not just bigger than the Greeks had thought. It also contained more planets.

William Herschel (1738–1822) was born in Hanover. At age 15 he became a musician, an oboist, in the Hanoverian army. He served briefly in the Seven Years' War but, due to his delicate health, his parents sent him to England to seek employment as a musician. Herschel settled in the spa town of Bath, where he became a successful musician and music teacher. His passion, though, was astronomy. Aided by his devoted sister Caroline Lucretia Herschel (1750–1848), he spent hours grinding lenses and making telescopes. His first telescopes were failures, but he persevered. In March 1774, when he was 36, he at last had a telescope with which he could make useful observations.

Herschel studied the heavens with an energy unmatched by anyone before or since. He would often hurry out of a theatre during the intervals between acts to spend a few moments gazing up at the sky. But Herschel was no dilettante. He was systematic in his observations and thorough in his analyses of them. It was these qualities that made him perhaps the greatest observational astronomer of all time. It is no surprise that he was the first to discover a new planet.

On 13 March 1781, while observing faint stars in the neighbourhood of η Geminorum, Herschel noticed a star that seemed brighter than the rest. He suspected it might be a comet, a suspicion that strengthened in the following weeks when he found that the object moved relative to the fixed stars at a rate of about $1''$ per day. Unlike ordinary comets, though, this had no tail but it did have a well-defined disk. Herschel told the Astronomer Royal, Nevil Maskelyne (1732–1811), of this unusual new object, and astronomers throughout Europe studied it in order to compute its orbit. After several months, the Swedish astronomer Anders Johan Lexell (1740–1784) succeeded in calculating the orbit. He announced that it could not be a comet: it was a new planet. It revolved around

William Herschel

the Sun at an average distance of about 19 AU — twice as far as Saturn. The solar system had doubled in size.

Herschel's discovery was unique. King George III rewarded Herschel with a position as a royal astronomer and a salary of 200 pounds. Herschel could give up music and concentrate on astronomy. (In gratitude, Herschel tried to exercise his discoverer's right of naming by calling the new planet *Georgium Sidus*, or *George's Star*. Astronomers on the continent never accepted this name, and the new planet was eventually called Uranus.)

After several decades of careful study it became clear that there were small irregularities in the motion of Uranus. The predicted orbit and the actual orbit did not quite agree. It was unthinkable that Newton's law of gravitation might be wrong, so the English astronomer John Couch Adams (1819–1892) and the French astronomer Urbain Jean Joseph Leverrier (1811–1877) independently proposed the existence of a planet beyond Uranus. They argued that the gravitational influence of such a planet could explain the orbital anomalies of Uranus. They calculated an orbit for their hypothetical planet and predicted its position. The German astronomer Johann Gottfried Galle (1812–1910) undertook a search for the planet and, on 23 September 1846, the very first evening of his search, he found the planet close to the predicted spot. This new planet, Neptune, was at an average distance of 30 AU from the Sun.

The ninth and last planet to be discovered was Pluto. In 1929, the American astronomer Clyde William Tombaugh (1906–1997) obtained a job as an assistant at the Lowell Observatory in Arizona, where he was given the task of searching for planets beyond Neptune. The search involved comparing photographs of the same regions of sky taken on different nights. The positions of the fixed stars on the plates would not change from night to night, but a planet would have slightly different positions on each plate as it wandered across the sky. Thus, if two plates taken on different nights were flashed alternately on a screen, a planet would appear to flicker back and forth. It was hard work. Tombaugh had to compare hundreds of plates in his search for a new planet, and each plate held anything from 50 000 to 400 000 stars. After more than a year of patient toil, Tombaugh found a flicker on two of the plates. The Lowell astronomers tracked the object for a month and computed its orbit. On 13 March 1930, they announced the discovery of Pluto. This, the most distant planet, was at an average distance from the Sun of nearly 40 AU.

We now know of dozens of major bodies in the solar system: the Sun, the nine planets and the many large moons that belong to the planets. In addition, we know of thousands of asteroids and comets. The largest object in the solar system is of course the Sun, which has a diameter of 865 000 miles. The planets range in size from the mighty Jupiter, with a diameter of 88 700 miles, to the tiny Pluto with a diameter of only 1420 miles. (Planetary diameters thus do not constitute a standard rod. If we discovered a new planet we could not use its apparent angular diameter to calculate its distance.)

Although no more planets have been discovered, astronomers have made huge strides in determining accurately the distance to the planets. By the mid-1960s astronomers knew the size of the astronomical unit to an accuracy of one part in a million — which meant that the uncertainty in the average Earth–Sun distance was under 100 miles! This huge improvement in accuracy was due to the *radar method*.

The radar method is similar in concept to the method of satellite laser ranging discussed on page 21. The main difference is that radio waves replace the laser, and the signal bounces off a planet rather than an artificial satellite. Groups of Russian, American and British scientists bounced radio waves off Venus (in 1961), Mercury (in 1962) and Mars (in 1965) and detected the echoes. Simply by measuring the time taken for these waves to make the journey from transmitter to planet to receiver, and dividing by the speed of light, gave them the round-trip distance to those planets. The method produced better results than parallax could ever achieve.

Example 3.3 The radar method

An observatory sends a radio signal to Mercury. Exactly 21.1108 minutes later it detects the echo. How far away is Mercury?

Solution. Light takes 1266.648 s to make the round-trip distance from observatory to Mercury and back again. Light travels at 2.99792458×10^5 km s^{-1}. The distance to Mercury is thus

$$2.99792458 \times 10^5 \times \tfrac{1}{2}(1266.648) = 1.8987 \times 10^8 \text{ km.}$$

The radar method of determining distance is now so advanced that it can map certain features on the surface of Venus — a surface that is hidden from view due to a permanent cloud cover. A radar signal bounced off the summit of a Venusian mountain has less far to travel than a signal bounced off the floor of a Venusian valley, and so it takes less time to make the round trip. The time difference is only a few millionths of a second, but this poses no problems for modern radio observatories.

The Solar System Dynamics Group at NASA's Jet Propulsion Laboratory has software that can determine the distance to Venus, or any other nearby body, at any minute of the day. The Horizons On-Line Ephemeris System is the same software that JPL uses when sending spacecraft to bodies in the solar system. As I write this, for instance, it takes light 12.6255 minutes to make the one-way trip to Venus. This means that the planet is now 141.115 million miles away.

The drawback with the radar method is that we cannot use it for planets more distant than Jupiter. The farther the radio pulses have to travel the weaker they become, and beyond Jupiter the echoes are too faint to detect. At present, we must measure the distance to the outer planets directly by using spacecraft. Only Pluto has yet to be visited by probes.

TABLE 3.4: Modern values for planetary distances.

Planet	Mean distance from Sun (AU)	Mean distance from Sun (miles)	Mean distance from Sun (km)
Mercury	0.387	35 980,000	57 910 000
Venus	0.723	67 210 000	108 200 000
Earth	1.000	92 957 000	149 600 000
Mars	1.524	141 670 000	227 990 000
Jupiter	5.200	483 380 000	777 920 000
Saturn	9.531	885 970 000	1 425 800 000
Uranus	19.142	1 779 400 000	2 863 600 000
Neptune	29.928	2 782 000 000	4 477 200 000
Pluto	39.475	3 669 500 000	5 905 400 000

Table 3.4 gives the distance of each of the planets from the Sun, as determined by the radar method and spacecraft data. Note that these are *average* distances. Since the planets move in elliptical orbits, at times of closest approach to the Sun (i.e. at *perihelion*) they can be much closer than at times of farthest distance (i.e. at *aphelion*). For instance, at perihelion the Earth is about 3.4% closer to the Sun than at aphelion — 91.4 million miles as opposed to 94.5 million miles. (Incidentally, the distance of the Earth from the Sun has nothing to do with the seasons. The time of closest approach occurs during the northern hemisphere's winter.)

3.7   THE ASTRONOMICAL UNIT

Newton derived the general form of Kepler's third law. A planet of mass m orbiting the Sun of mass $m_\odot$ at a mean distance a and with a sidereal period T, satisfies the equation

$$k^2(m + m_\odot)T^2 = 4\pi^2 a^3. \tag{3.9}$$

The German mathematician Johann Karl Friedrich Gauss (1777–1855) used a system of units in which the unit of mass is the solar mass, the unit of time is the mean solar day, and the unit of distance is the Earth's mean distance from the Sun. In these units, (3.9) becomes

$$k^2(1 + m)T^2 = 4\pi^2 a^3. \tag{3.10}$$

The constant k in (3.10) is known as the *Gaussian constant of gravitation*. If we apply (3.10) to Earth, so that

$$T = 365.256\,3835 \text{ days} \qquad m = 1/354\,710 \qquad a = 1$$

then

$$k = 0.017\,202\,098\,95.$$

The quantity a, the semi-major axis of the Earth's orbit, was defined to be the astronomical unit. The problem with this definition, as Gauss himself recognised, is that when a quantity such as the sidereal year is determined with more accuracy the value of k must be recalculated. To avoid having to do this, astronomers have retained the original value of k and have instead redefined the astronomical unit:

> 1 AU is the distance at which a massless particle in an unperturbed circular orbit about the Sun would have a mean daily motion of k radians per day, where $k = 0.017\,202\,098\,95$.

This definition makes no reference to the Earth, and so we are now free to treat the Earth like any other planet. It turns out that, as derived from Kepler's third law, the semi-major axis of the Earth's orbit is $1.000\,000\,03$ AU. In fact, due to the gravitational influence of the other planets, a better approximation to the Earth's orbit is given by an ellipse with a semi-major axis of $1.000\,0002$ AU. In either case, the definition of the AU as being the mean Earth–Sun distance is an approximation — but a very good one.

The size of the astronomical unit is $1.495\,978\,70 \times 10^{11}$ m.

3.8   THE OUTSKIRTS OF THE SOLAR SYSTEM

Figure 3.13 is a representation of the orbits of the outer planets, up to Pluto. Are there any planets more distant than Pluto*? Since 1930, many people have searched for the so-called Planet X, a hypothetical tenth planet that lies beyond the orbit of Pluto. So far the search has revealed nothing, and Pluto may well be the outermost planet. Even so, astronomers believe that the solar system extends far beyond Pluto.

The region between 30–100 AU from the Sun is called the *Kuiper belt*, after the Dutch–American astronomer Gerard Kuiper (1905–1973). The Kuiper belt contains many icy bodies, and is probably the source of short-period comets like Halley's comet. But even 100 AU does not mark the limit of the solar system.

In 1950, the Dutch astronomer Jan Hendrik Oort (1900–1992) suggested that 10^{12} or more primordial icy comet nuclei, left over from the birth of the planets, form a vast halo around the outskirts of the solar system. Because of their small size and low reflectivity no comet has been seen at distances beyond about Saturn, so the existence of the Oort Cloud is still conjectural. Nevertheless, astronomers have good reason to believe that the Oort Cloud exists, and extends with decreasing density from 30 000 AU out to a distance of perhaps 100 000 AU: 10^{13} miles! This is the edge of the solar system.

---

*As figure 3.13 shows, Pluto's orbit is so eccentric that at times Neptune is the outermost planet. This was recently the case, in fact. Pluto regained its status as the outermost planet in 1999.

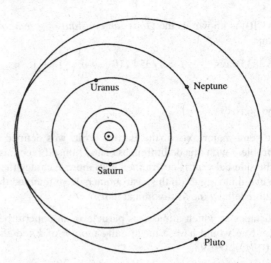

FIGURE 3.13: A diagram of the outer solar system, drawn to scale. The innermost orbit shown in the diagram is that of Mars. At this scale the orbits of the Earth, Venus and Mercury are too small to represent. On the other hand, the orbits of the Oort Cloud comets are too large to fit on the page. Notice the eccentricity of Pluto's orbit. The orbit of Pluto occasionally takes it inside the orbit of Neptune, so at times Neptune is the outermost planet.

CHAPTER SUMMARY

- The eclipse diagram provided the Greeks with a good estimate of the lunar distance. The lunar dichotomy method in principle provides a means of calculating the solar distance (otherwise known as the astronomical unit), but does not work in practice.

- The horizontal parallax of an object is the parallax of that object measured by observers at the Earth's pole and equator.

- The Moon is the only major body in the solar system with a horizontal parallax large enough to be measured without the aid of a telescope. Hipparchus used observations of a solar eclipse to calculate the Moon's parallax.

- The Copernican model of the solar system lets us estimate an inferior planetary distance by careful measurements of the planet's greatest elongation.

- Kepler's third law is a relationship between the sidereal period, T, of a planet and its mean distance, d, from the Sun: $T^2 \propto d^3$.

- Early attempts to set the distance scale of the solar system included measurements of the parallax of Mars at opposition, the parallax of Venus during transit, and the parallax of Eros at perigee.

- Distances in the solar system are now best determined by using the radar method.

- The size of the astronomical unit is $1.495\,978\,70 \times 10^{11}$ m.

QUESTIONS AND PROBLEMS

3.1 Aristarchus is often called 'the Copernicus of Antiquity'. Is this fair? Or should Copernicus be called 'the Aristarchus of the Renaissance'?

3.2 Take the series 0, 3, 6, 12, 24, ⋯ . If we add 4 to each term we get 4, 7, 10, 16, 28, ⋯ . Now divide each term by 10 to get 0.4, 0.7, 1.0, 1.6, 2.8, ⋯ . Compare these numbers with the relative planetary distances given in table 3.3. What do you notice? (Remember that there is an asteroid belt!) This series is called *Bode's law,* after the German astronomer Johann Elert Bode (1747–1826). Do you think it is a law or a mathematical curiosity? Investigate Bode's law for the major moons of Jupiter and Saturn; does it hold for these systems too?

3.3 The Danish astronomer Olaus Römer (1644–1710) noticed that Jupiter's satellites were occulted by the planet about 16.67 minutes later when Jupiter was near superior conjunction than when Jupiter was at opposition. Following Römer, explain this observation and use it to calculate the speed of light.

3.4 Obtain an expression like (3.5) that provides the distance to a superior planet in terms of astronomical units.

3.5 Using the values given in table 3.4, calculate the mean horizontal parallax of each of the planets.

3.6 If planets move in coplanar circular orbits, what is the maximum possible elongation of Mercury?   22.8°

3.7 For a Venusian astronomer, which planets are inferior and which are superior? For the same observer, what is the greatest elongation of Mercury?   Only Mercury is inferior; 32.4°

3.8 When Eros is at perigee (its closest approach to Earth) an observer at A measures the angle between Eros and a star S to be 12″. An observer at B measures the angle between Eros and the star S to be 8″. If A and B are separated by 2170 km, what is the minimum Earth–Eros distance? (Assume that A, B, Eros and the star all lie in the same plane.) 2.238×10^7 km

3.9 When the two planets are closest, how big does the Earth appear to someone who is standing on the surface of Mars?   A disc of diameter 47.4″

3.10 For an astronomer based on the surface of Mars, what is the greatest angular separation of the Earth from the Sun?   41°

3.11 At $t = 0$ you send a radar signal toward Venus; 4 minutes 32 seconds later you detect the radar echo. How far away is Venus? 4.08×10^{10} m

3.12 A fictitious planet orbits at a mean distance of 100 AU from the Sun. How long does it take to revolve once around the Sun?   1000 years

FURTHER READING

BIOGRAPHY AND HISTORY

Beer A and Beer P (eds) (1975) *Kepler: Four Hundred Years* (Pergamon: Oxford)
— The large (1034-page) proceedings of a conference held in honour of Kepler. The book forms volume 18 of the journal *Vistas in Astronomy*. Volume 17 of the journal is a similar publication devoted to Copernicus.

Chapin S L (1973) Le Gentil. In *Dictionary of Scientific Biography* ed. C. C. Gillispie (Scribner's: New York)
— A brief biography and bibliography of the unlucky Le Gentil.

Drake S (1978) *Galileo at Work: His Scientific Biography* (Dover: New York)
— Details of the scientific achievements of Galileo, including his many contributions to astronomy.

Gingerich O (1985) Did Copernicus owe a debt to Aristarchus? *J. Hist. Astron.* **16** 37–42
— An interesting discussion of the extent to which Copernicus may have been influenced by Aristarchus.

King H C (1955) *The History of the Telescope* (Griffin: London)
— A comprehensive account of the evolution of the telescope.

Koestler A (1959) *The Sleepwalkers* (Hutchinson: London)
— A classic book about Copernicus, Kepler and Galileo.

Kuhn T S (1957) *The Copernican Revolution* (Harvard University Press: Cambridge, MA)
— An important book on the philosophy of science.

Shapley H and Howarth H E (1929) *A Source Book in Astronomy* (McGraw-Hill: New York)
— An invaluable collection of astronomical greats. Contributions relevant to this chapter include classic writings by Copernicus, Brahe, Kepler, Galileo and Halley.

Toomer G J (1980) Hipparchus. In *Dictionary of Scientific Biography* ed. C. C. Gillispie (Scribner's: New York)
— Perhaps the best biography of Hipparchus. The author also has an interesting biography of Ptolemy in the same publication.

THE SOLAR SYSTEM

Christiansen E H and Hamblin W K (1995) *Exploring the Planets* (Prentice–Hall: Englewood Cliffs, NJ)
— A useful textbook for those interested in the geology of the planets.

Lang K R (1995) *Sun, Earth and Sky* (Springer: Berlin)
— A beautifully illustrated book about the Sun, and its importance to conditions here on Earth.

Littmann M (1988) *Planets Beyond* (Wiley: New York)
— A readable, non-mathematical account of the discovery of the outer solar system.

Morrison D and Owen T (1996) *The Planetary System* (Addison–Wesley: New York)
— An up to date undergraduate textbook.

Ostro S J (1993) Planetary radar astronomy. *Rev. Mod. Phys.* **65** 1235–1280
— This review outlines the techniques of radar astronomy and describes the principal observational results.

Shirley J H and Fairbridge R W (1997) *Encyclopaedia of Planetary Sciences* (Chapman and Hall: London)
— Many distinguished authors have contributed articles to this volume, which covers all aspects of solar system studies.

Stix M (1990) *The Sun* (Springer: Berlin)
— An introduction to all aspects of solar phenomena.

http://ssd.jpl.nasa.gov/
— The home page of the Solar System Dynamics Group of the Jet Propulsion Laboratory. The site provides in-depth information on solar system objects. It also describes the Horizon On-Line Ephemeris System.

http://nssdc.gsfc.nasa.gov/planetary/planetfact.html
— Fact sheets for every planet and for many minor bodies in the solar system.

http://www.ex.ac.uk/tnp/   (there are mirror sites all over the world)
— *The Nine Planets* by William Arnett. An extensive multimedia tour of the solar system.

THE ASTRONOMICAL UNIT

Atkinson R d'E (1982) The Eros parallax, 1930–31. *J. Hist. Astron.* **13** 77–83
— The author examines the ambitious programme of Spencer Jones to determine the solar parallax from the parallax of Eros.

Shapiro I I (1963) Radar determination of the astronomical unit. *Bull. Astron.* **25** 177
— This paper describes the early attempts to determine planetary distances with the radar method.

Van Helden A (1985) *Measuring the Universe: Cosmic Dimensions from Aristarchus to Halley* (University of Chicago Press: Chicago)
— The author traces, in great detail, mankind's struggle to measure the value of the astronomical unit.

4

Third step: nearby stars

We know the size of the Earth and we know the size of the solar system. To take the next step on the distance ladder we must resolve the distance scale of the stars. At first sight this seems to be an impossible task since we have so little information about stars. They seem to be just isolated points of light. All we know for sure is that some are brighter than others, some have distinctive colours, and all appear to have a fixed position on the celestial sphere. How can this meagre amount of information tell us anything about the stars or their distance? In fact, the key to this step of the distance ladder is simply to make as accurate a determination of stellar positions as possible. The rest follows directly from Thales and Copernicus.

4.1 POSITIONAL ASTRONOMY

The celestial sphere, you will recall, is centred upon the Earth's centre and its radius is indefinitely large. We wish to be able to fix the position of a star on the inner surface of the sphere. First, we need to define an appropriate coordinate system in which to express our results. There are a number of ways we could proceed. Of the several systems in use the most important is the *equatorial system* of coordinates, which mirrors the latitude and longitude system on the Earth.

Imagine for a moment that we have stopped the Earth's rotation. Project the terrestrial longitude–latitude coordinate grid onto the surface of the sphere. The Earth's equatorial plane intersects the celestial sphere in a great circle called the *celestial equator*. The Earth's north and south poles of rotation intersect the sphere at the north and south *celestial poles*. Finally, parallels of latitude project onto small circles concentric to the poles and meridians of longitude appear as so-called *hour circles* on the celestial sphere. Now let the Earth resume its rotation. The grid of the celestial equatorial coordinate system begins to rotate westward, at the rate of one revolution per day. But since the stars also rotate about the Earth at the same rate, we can give the position of each star in terms of a fixed pair of equatorial coordinates.

In the equatorial system, we specify the position of a star by giving its *declination* (δ) and its *right ascension* (α). See figure 4.1.

Declination is the celestial equivalent of terrestrial latitude: it is the smallest angular distance from the celestial equator to a star along the hour circle passing through the star.

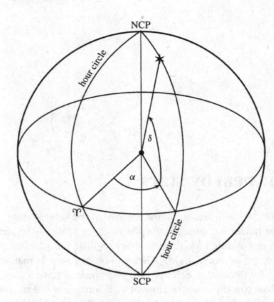

FIGURE 4.1: Right ascension (α) and declination (δ) specify the position of a star on the celestial sphere.

We measure declination in degrees (°), minutes (′) and seconds (″) — the same as with latitude. It ranges from 0° (the celestial equator) to +90° (the north celestial pole) to −90° (the south celestial pole).

Right ascension is the celestial equivalent of terrestrial longitude. Just as we need to define a prime meridian for longitude (for historical reasons the zero of longitude passes through Greenwich in London), we need to define a prime meridian for right ascension. The point of origin of right ascension is chosen to be the *first point of Aries* (♈), otherwise known as the *vernal equinox*, which is defined by the intersection of the celestial equator and the *ecliptic**. The right ascension of a star is then the angular distance eastward along the celestial sphere from ♈ to the hour circle containing the star. It is also equal to the sidereal time when the star crosses the meridian. Since we can measure right ascension simply with a clock, it has the units of time (^{h} ^{m} ^{s}). It ranges from $0^h 0^m 0^s$ to $23^h 59^m 59^s$. Clearly, a rotation of 15° corresponds to one hour of time.

So just as we can specify a location on Earth by giving its latitude and longitude (for example, Cairo is at about (30°03′ N, 31°15′ E)) we can specify a location on the celestial sphere by giving its right ascension and declination (for example, Sirius is at about ($6^h 43^m$, −16°39′)). The equatorial coordinates of a star change slightly with time because of the precession of the equinoxes. As Hipparchus discovered, the vernal equinox moves backwards along the ecliptic at a rate of about 50″ per year; ♈ completes one full

---

*The ecliptic is the great circle where the orbital plane of the Earth intersects the celestial sphere. It thus describes the apparent annual path of the Sun in the sky. The plane of the ecliptic is at an angle of about 23°17′ to the plane of the celestial sphere. This angle is called the *obliquity of the ecliptic*.

lap of the sky in about 26 000 years. Since the right ascension and declination of ♈ changes each year, the right ascensions and declinations of the stars also change. When we give the position of a star, therefore, we must specify a particular epoch for which we have the position of ♈. Formulae then exist that allow us to calculate the position of a star at some later date.

The equatorial system is not the only coordinate system in use. For instance, the *ecliptic coordinate system* is very useful for describing the motion of bodies in the solar system. In this case the ecliptic, rather than the celestial sphere, is the primary circle. The poles are the *north ecliptic pole* and the *south ecliptic pole*. In the ecliptic system, *celestial longitude* (λ) is the angular distance — in degrees — measured eastward along the ecliptic from ♈. *Celestial latitude* (β) is the angle in degrees from the ecliptic (measured positively towards the north ecliptic pole). The conversion from (α, δ) to (λ, β) and *vice versa* is straightforward.

With these definitions we are now in a position to develop the science of astrometry. First, we make raw observations of the declinations and right ascensions of the stars. Then we apply corrections to take account of instrument errors. We also correct for atmospheric refraction: the Earth's atmosphere bends light rays from a star by an amount that depends on the zenith distance of the star. We must also even correct for our height above sea level. Finally, we must specify the epoch at which we make our observations so that we can allow for the precession of the equinoxes. When all this is completed we will have a catalogue of precise stellar positions — and we can then search for small changes in those positions that might be due to parallax. And parallax, as we have seen, is the key to distance determination.

4.2   THE PARALLACTIC ELLIPSE

The stars are very distant; just *how* distant became clear once astronomers accepted the Copernican model of the solar system. According to Copernicus, the Earth changes its position by 2 AU every six months. Therefore, due to parallax effects, the position of a star in January should be different from the position of the same star in July. Similarly, the position of a star in April should be different from its position in October. A star should mirror the motion of the Earth around the Sun, and move in an ellipse on the celestial sphere. Figure 4.2 shows this effect.

In more formal terms, the stars should show an *annual* parallax. The annual parallax of a star is the angle subtended by the radius of the Earth's orbit as seen from the star. This is the same as the definition of horizontal parallax (see page 29), with the radius of the Earth replaced by the radius of the Earth's orbit.

Figure 4.3 shows the parallax geometry for a star, at distance d, that lies at the pole of the ecliptic. From the figure we see that the star appears to move in a circle. (Of course, the star actually moves in an ellipse, but since the eccentricity of the Earth's orbit is so small, and since the stars are so far away, the apparent motion is indistinguishable from a circle.) The angular radius of the star's parallax circle, π_*, is given by

$$\sin \pi_* = \frac{R}{d} \tag{4.1}$$

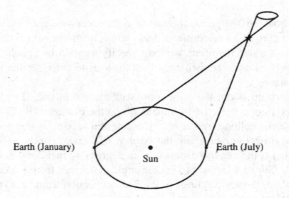

FIGURE 4.2: The long baseline (186 million miles) produced by the Earth's motion around the Sun means that a star should show an annual parallax. (The eccentricity of the Earth's orbit is exaggerated.)

where R is the radius of the Earth's orbit. Since we know that the stars are very much more distant than 1 AU, we know that π_* is a small angle. So if we measure π_* in seconds of arc, and if we note that 1 rad $\equiv 206\,265''$, we can write (4.1) as

$$\pi_*'' = 206\,265 \frac{R}{d}. \tag{4.2}$$

The quantity π_* is called the *trigonometric parallax* or *annual parallax* (or just parallax) of a star. Equation (4.2) is clearly very important, because it enables us to calculate the distance of a star if we can measure its parallax.

Figure 4.3 shows a star at the ecliptic pole. If the star lies at celestial latitude β, away from the pole, its parallactic motion will be different. As figure 4.4 illustrates, in this case the star will move in an ellipse of semi-major axis π_* ($= 206\,265'' R/d$) and of semi-minor axis $\pi_* \sin \beta$. Its trigonometric parallax is still π_*; trigonometric parallax is an unambiguous distance indicator.

Example 4.1 Parallactic motion of a star

A star has right ascension $1^\mathrm{h}51^\mathrm{m}38^\mathrm{s}$ and declination $11°28'37''$. Describe the star's annual parallactic motion.

Solution. Most astronomy textbooks contain formulae for converting from equatorial to ecliptic coordinates. If we use these formulae, we see that the star has celestial latitude $0°$. In other words it lies in the ecliptic plane.

Since $\beta = 0$, we see that the star moves through an angle $\pm \pi_*$, to and fro along a straight line, where

$$\pi = \frac{R}{d}.$$

(R is 1 AU and d is the distance to the star.)

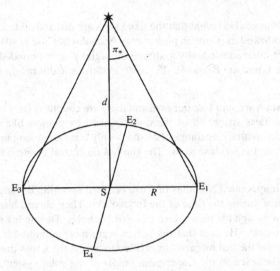

FIGURE 4.3: A star at the ecliptic pole appears to move in a parallactic circle of angular radius $\pi_* = R/d$.

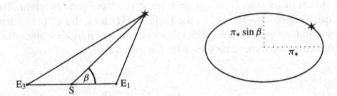

FIGURE 4.4: A star that lies at celestial latitude β moves in an ellipse. The semi-major axis of the ellipse is π_*, which is the trigonometric parallax of the star. Compare this with figure 4.3.

4.3 EARLY ATTEMPTS TO MEASURE PARALLAX

Copernicus thought that the Earth's orbit was only 1100 times wider than the Earth itself. Even using this gross underestimate, with such a large baseline to work with we might expect annual parallax to be conspicuous. A major problem for the Copernican system was that no one had ever seen a star move from its fixed position on the celestial sphere. Copernicus could offer only one explanation: the stars must be so far away that parallax effects are too tiny to measure.

Copernicus shared the general belief of his time that the stars are all the same distance from us. Even Kepler, a century after Copernicus, believed that the stars all occupy a shell just two miles thick. The English mathematician Thomas Digges (1543–1595), the son of Leonard Digges, was one of the first to popularise the Copernican model, but he made one important change. He taught that the fixed stars are distributed throughout infinite space. In this he followed the German philosopher Nicholas of Cusa (1401–1464), who held that the stars are other suns equally scattered throughout an unbounded universe. The Italian philosopher Giordano Bruno (1548–1600) enthusiastically taught the ideas of Nicholas. To Bruno:

> 'the stars are suns like our own and there are countless suns freely suspended in limitless space, all of them surrounded by planets like our own Earth, peopled with living things. The sun is only one star among many, singled out because it is so close to us. The sun has no central position in the boundless infinite.'

(The church appointed Nicholas of Cusa cardinal in 1448. Bruno, though, had the misfortune to live during the time of the Inquisition. They burned him alive at the stake for heresy, even though his views were those of Nicholas. During his seven-year trial Bruno refused to recant. He said that his judges were more afraid of him than he was of his judges. Even at the last moment, he refused to accept the Cross they held out to him.)

Galileo believed in the Copernican model of the solar system, but also in the ideas of Bruno and Digges. When Galileo pointed his telescope to the sky he discovered the existence of many stars that were too faint for the naked eye to see. For instance, the Pleiades is a cluster of six stars — seven, or more, if you have good eyesight. Through his crude telescope, Galileo saw 36 stars in the Pleiades. He took this as proof that the stars are all at different distances from us. The telescope brought into view stars that were too distant, and therefore too dim, to be seen with the naked eye.

FIGURE 4.5: The differential method of parallax. Initially, an astronomer observes the nearby star A to be at A′ — to the right of the distant star B. Six months later, the astronomer sees star A to be at A″ — to the left of star B. This motion of star A relative to star B is easier to observe than the same motion relative to an arbitrary point in space.

The idea that the stars are at different distances led Galileo to suggest the *differential method* of parallax. Consider figure 4.5. It shows a bright star, A, close in the sky to a dim star, B. If the stars are all the same brightness then star A must be closer than star B. (This assumption will occur again and again throughout the book. It could be wrong. The dim star might be dim because it is indeed distant, or it might be nearby and appear dim *because it really is dim*. In the absence of evidence one way or the other, Galileo could only assume that 'bright means near; dim means far'.) The idea is to observe the position of A relative to B. Star B is assumed to be so remote that its parallactic shift is imperceptible. In that case, any change in position would be due to the annual parallax of star A. For instance, in January we might see A to the right of B. Six months later, we might see A to the left of B. The point is that we do not have to fix the position of the stars with great accuracy for this method to work. It is much easier to see small positional changes in such a system than the same changes relative to an unmarked point like the celestial pole.

Galileo never found a stellar parallax. His failure meant either that Copernicus was wrong, and the Earth did not move around the Sun, or that the stars were very distant indeed. Galileo chose the latter explanation. Further confirmation for this choice came when he noticed that none of his telescopes magnified the stars into visible discs, but instead just made them brighter. This effect is exactly what would be expected if the stars were extremely distant. Nevertheless, to *prove* that Copernicus was right astronomers would have to detect an annual parallax.

The English physicist Robert Hooke (1635–1703) made the first 'modern' attempt to detect an annual parallax — 'modern' in the sense that he understood the practical difficulties of taking measurements over an annual cycle. For instance, instruments made of metal might expand during a hot summer and contract during a cold winter. Instruments made of wood might swell during a wet winter and shrink during a dry summer. These effects would be small, of course, but the angles that had to be measured were also small. Hooke understood the need for precision astrometry.

Hooke decided to look for the parallax of the star γ Draconis. There were three reasons for this. First, it is a relatively bright star. He hoped that 'bright means near', and the nearer the star the larger the parallax. Second, it is well placed for observing from London, where he lived and worked. Third, it passed directly overhead so he could observe it without the complicating effects of atmospheric refraction.

In 1669, Hooke built a telescope in his London lodgings with the sole purpose of observing γ Draconis. The telescope was of his own design. It consisted of two tubes: an outer tube fixed to the roof and an inner tube that held the lens. He set the outer tube in the vertical direction, so that he could watch γ Draconis as it passed overhead. A micrometer, located by two 12-yard long vertical plumb lines attached to either side of the outer tube, allowed him to measure the star's precise zenith distance. The whole set-up was protected from the weather by a lid that could be opened by pulling on a string.

It was an ingenious device, but Hooke found nothing but trouble with it. The tube would bend during warm days, ruining the careful alignment that he needed. Later he discovered a slight warp in the roof of his house. During four months in 1669 he made only four observations: not only was the weather bad, but poor health struck him down. When he recovered enough to work, the lens accidentally broke.

From his sparse observations Hooke derived a parallax of between 27–30″ (corresponding to a distance to γ Draconis of about 7000 AU). Few, if any, of Hooke's colleagues took his claim seriously. Four data points proved nothing. Besides, his colleagues pointed out that he had not used the differential method of parallax, so that the angular measurements required were very fine — too fine to be made with his apparatus.

If the differential method of parallax was not used, and stellar positions were instead taken against some arbitrary fixed point, then astronomers needed exceptionally precise values for the positions of the stars. Flamsteed, the first Astronomer Royal, devoted himself to measuring the positions of the stars with a precision never before achieved. Flamsteed was a slow but meticulous observer. By 1712, he had fixed the positions of 3000 stars to within 10″ — a precision six times better than Tycho achieved. This was the first great star catalogue of the telescopic age. While carrying out this work, Flamsteed

Isaac Newton

made a series of observations of Polaris, the Pole Star. He studied the star for seven years, and he thought he had detected its annual parallax. He published this finding in 1699. At first, Flamsteed's observations carried more weight than Hooke's. Unfortunately, as Cassini quickly pointed out, the Astronomer Royal had made a basic geometric error, and had not found an annual parallax after all. The distance to the stars was *still* a matter of conjecture.

If two great scientists like Hooke and Flamsteed had tried and failed to detect an annual parallax, perhaps another method was needed to calculate the distance to the stars.

4.4   THE DISTANCE TO SIRIUS

The Dutchman Christiaan Huygens (1626–1695) was a versatile scientist. In 1656, for instance, he made the first the pendulum clock, an invention that transformed the art of making exact astronomical measurements. Huygens was also one of the finest telescope makers of his era, and with his telescopes he made several important discoveries (such as his discovery of Titan, the largest satellite of Saturn). He also made, in 1698, the first reasonable estimate of the distance to a star.

Huygens used an ingenious argument to estimate the distance to Sirius, the brightest star in the sky. He first made a tiny pinhole in a dark room, and let light from the Sun shine through the hole. The hole was much brighter than any star and, hard as he tried, it was impossible to make a pinhole small enough so that it looked like a star. He therefore used a simple optical arrangement to magnify the Sun's image from the hole until its diameter was 27 664 times that of the pinhole. The magnified image then appeared as bright as the star Sirius. Huygens concluded that, if Sirius had the same intrinsic brightness as the Sun, then it must be 27 664 times more distant than the Sun. In other words, he estimated Sirius to be at a distance of 27 664 AU.

This was a difficult experiment to perform. Huygens had to compare the brightness of the magnified image from a pinhole in a darkened room with how bright he *remembered* Sirius to be! So while his method was certainly ingenious, astronomers were cautious about his conclusion. His estimate was not influential.

A better estimate of the distance to Sirius came from Isaac Newton. Newton was without doubt the greatest scientific genius who has ever lived. His famous book *Principia* marks the birth of our modern scientific understanding of the universe. He made seminal contributions to mathematics, experimental physics and theoretical physics, and also revolutionised astronomy. For instance, with his law of universal gravitation he finally explained the motions of the planets, which Kepler had described a century earlier. In developing the reflecting telescope he pioneered an instrument that would eventually be much more powerful than refracting telescopes. His discovery that white light can be split into different colours forms the basis of spectral analysis, from which much of our knowledge of the universe derives. His achievements are far too many to list. One of the least of his accomplishments was an argument, given in *Principia*, for estimating the distance to Sirius.

Newton began by making the usual assumption that all stars, including the Sun, emit the same amount of light. In other words, he supposed that all stars have the same *intrinsic* brightness. He also assumed that there is no loss of light as it travels through space. If these assumptions are correct, then the *apparent* brightness of stars gives their relative distance from the Earth.

To make this clear, imagine for a moment that the Sun and Sirius are at the same distance from us. If they have the same intrinsic brightness then they will appear equally bright. Now suppose we move Sirius back twice as far away as the Sun. The amount of light captured by our eyes — in other words the apparent brightness — falls off as the *square* of the distance. So Sirius now seems $2 \times 2 = 4$ times less bright than the Sun. If we move it back three times as far as the Sun, it seems $3 \times 3 = 9$ times less bright than the Sun. We have reduced the problem of comparing the distances of the Sun and Sirius to the problem of comparing their apparent brightnesses.

So far, we have said nothing new. This is exactly the problem that Huygens tried to solve. Newton, though, came up with a better way of comparing the brightness of the two stars. He noted that Sirius seems to be about as bright as the planet Saturn. Since Saturn emits no light of its own — it is bright only because it reflects light from the Sun — its brightness $L_\hbar$ is given by the product of three terms:

$$L_\hbar = L_\odot \times f \times A. \tag{4.3}$$

The first term, $L_\odot$, is the amount of light emitted by the Sun. The second term, f, is the fraction of that light intercepted by Saturn. The third term, A, is the fraction of intercepted light that Saturn reflects back into space.

Newton set the first of these terms equal to 1, and thus expressed his results in terms of the brightness of the Sun. In order to calculate the second of these terms, namely the fraction of sunlight intercepted by Saturn, Newton needed two numbers: the distance of Saturn from the Sun, and the diameter of Saturn. The 1672 observations of Mars at opposition gave him a reasonable value for the distance of Saturn from the Sun. From the observed angular size of the planet he could deduce a reasonable value for its diameter.

It was then just a matter of arithmetic to calculate the fraction of the Sun's radiation that Saturn intercepts.

Now consider the third term. The reflecting power of a planet — the ratio of the amount of reflected light to the amount of light falling onto the planet — is known as its *albedo*. The albedo of any object depends upon the material that makes up its surface. A perfect reflector has $A = 1$, since it reflects 100% of the light that shines upon it. An object with a sooty surface has $A \approx 0$. What is the albedo of Saturn? Newton, of course, had no idea of the surface structure of Saturn. He guessed that it might consist of some type of rocky material, and used a value of 0.25 for the albedo. (We now know that its albedo is 0.76.)

As example 4.2 shows, Newton used (4.3) to deduce that Sirius was at a distance of about 800 000 AU. (This was a remarkably good estimate, considering that he had no idea of what a 'reasonable' estimate might be.) From (4.2) we see that the parallax of a star at a distance of 800 000 AU is 0.26″. That is about the width of your thumb at a distance of 12 miles. This explained why an annual parallax had not been found. The instruments of the early eighteenth century could not possibly detect such a small parallax.

For the first time mankind had a reasonable idea of the distance scale of the stars. The stars were more than 100 000 times farther away than the Sun. With such large distances involved, what possible hope was there of detecting an annual parallax?

Example 4.2 The distance to Sirius

Saturn appears to be about as bright as Sirius. (In fact, Sirius is slightly brighter.) According to Newton, the apparent angular radius of Saturn as seen from the Sun is 9″. If Saturn's albedo is 0.25, estimate the distance to Sirius.

Solution. Let Saturn's radius be $R_\saturn$ and $d_\saturn$ be its distance from the Sun. The area of Saturn's disk shown to the Sun is $\pi R_\saturn^2$. At the distance of Saturn, the total surface area over which the Sun's radiation is spread is $4\pi d_\saturn^2$. The fraction of total sunlight intercepted by Saturn is therefore $\frac{1}{4}(R_\saturn/d_\saturn)^2$. But $R_\saturn/d_\saturn$ is the sine of Saturn's angular radius as seen from the Sun, i.e. $R_\saturn/d_\saturn = \sin 9''$. The fraction of total sunlight intercepted by Saturn is therefore $\frac{1}{4}\sin^2 9''$. Since only one side of the Sun shines on Saturn at any one time, the planet must receive $\frac{1}{2}\sin^2 9''$ of the sunlight coming from the hemisphere facing it. Finally, since $A = \frac{1}{4}$, Saturn reflects $\frac{1}{8}\sin^2 9'' = 2.38 \times 10^{-10}$ of the sunlight coming from one hemisphere.

Apparent brightness falls off as the square of distance, so if the Sun were $1/\sqrt{2.38 \times 10^{-10}} = 64\,823$ times more distant than Saturn it would appear as bright as Saturn. Assuming that the Sun and Sirius have the same intrinsic brightness, Sirius must be about 65 000 times farther away than Saturn.

Sirius is actually brighter than Saturn. (Saturn varies markedly in brightness; most of the variation depends upon the orientation of its ring system as seen from Earth. But at all times Sirius is brighter.) A reasonable estimate of the distance to Sirius, therefore, is that it is 100 000 times more distant than Saturn. We know that Saturn is about 8 AU away. Therefore Sirius is at a distance of about 800 000 AU.

4.5   THE SEARCH FOR PARALLAX

Despite Newton's demonstration that the annual parallax of the stars must be tiny, astronomers continued to search for parallax. Indeed, it became something of a holy grail for them. After Newton's death, several generations of astronomers joined the search. They all failed. But these were heroic failures. Although astronomers did not find a parallax, they often made interesting discoveries that advanced our understanding of the stars (and eventually led to the detection of parallax). In this section I describe three of these discoveries, each of which play a later part in the story.

4.5.1   Halley and the proper motion of stars

Edmond Halley understood the implication of Newton's estimate. If stellar parallaxes were indeed of the order of 1″ then, in order to detect parallax, astronomers needed to fix the position of stars with better than second-of-arc precision. Halley therefore decided to check the precision of the star catalogue prepared by Flamsteed (his predecessor as Astronomer Royal). He began, in 1718, by comparing the catalogue positions with those reported by the ancient Greeks. In most cases he found excellent agréement. However, four bright stars — Sirius, Arcturus, Betelgeuse and Aldeberan — were up to half a degree away from their ancient positions. The Greeks did not have telescopes to help them compile their catalogues; even so, it was hard to believe that they could be so wrong with such prominent stars. When he checked further, he found that Sirius had moved since the time of Tycho, just a century and a half before. It was impossible to believe that both the Greeks *and* Tycho had been wrong in the case of Sirius. Therefore Sirius (and probably the other three stars as well) *had* shifted position. Halley concluded that the 'fixed stars' were not fixed after all. They had a *proper motion*. They only appeared motionless because of their great distance from us.

If we suppose, for simplicity, that all stars move at more or less the same speed, then the shift in position due to proper motion will be greatest for the nearest stars. It seemed to Halley, therefore, that Sirius, Arcturus, Betelgeuse and Aldeberan were probably among the closest stars.

Halley's discovery of proper motion was significant for two reasons. First, it implied that there was no celestial sphere to which stars were attached. Stars were almost certainly scattered through space, just as Nicholas of Cusa had taught three centuries before. Second, bright stars with a large proper motion were probably close by, and were therefore the best candidates to show an annual parallax.

4.5.2   Bradley and the aberration of light

Another who took up the challenge of finding an annual parallax was Samuel Molyneux (1689–1728), a gifted amateur astronomer. Molyneux copied Hooke in commissioning a special telescope to observe γ Draconis as it passed overhead. In 1725 he fixed the telescope, which was 8 yards long and 3.7 inches in diameter, to a chimney stack within his house. Before starting his observations he enlisted the help of James Bradley (1693–1762), Savilian professor of astronomy at Oxford.

Bradley and Molyneux spent the whole of November 1725 checking the accuracy of the new telescope. This was necessary because the telescope mounting was extremely delicate (three men standing nearby generated enough heat and air movement to disturb the plumb line defining the vertical). That done, on 14 December 1725 Bradley began a series of observations of γ Draconis. (He worked alone on this. Molyneux had to return to work at the Admiralty.) After a few nights, Bradley noticed that the star passed overhead a little lower than its original position. He observed the star night after night, for several months. Each night it transited a little lower than the previous night. Eventually it reached a minimum value of declination. Thereafter it transited a little higher each night, and in one year it made an oscillation with an amplitude of about 20″. This looked like parallax. Unfortunately, the star reached its largest value of declination in September and its smallest value in March. For the shift in position to be due to annual parallax, the star should have reached its largest declination in June and its smallest declination in December. The effect Bradley found was therefore three months out of phase with the parallactic motion he was looking for. It could not be parallax. So what was he seeing?

Bradley fought unsuccessfully with the problem for several months until, one afternoon in the summer of 1728, he went sailing in a pleasure boat on the River Thames. He noticed that the pennant on top of the mast changed direction each time the boat turned. What determined the direction of the pennant was the *relative* motion of the boat and wind, not the direction of the wind alone. Bradley thought of an analogy to explain his observation. Suppose you are outside in the rain with an umbrella for protection. There is no wind, so the raindrops fall vertically. If you stand still then you must hold the umbrella directly over your head in order to keep dry. Now suppose that you start to walk. If you keep the umbrella directly overhead you will move into some raindrops that have just cleared the umbrella. In order to keep dry you must tilt the umbrella slightly in the direction of your motion. The angle of tilt of the umbrella depends on the *relative* speed between you and the raindrops. The faster you walk, or the slower the raindrops fall, the more you must tilt the umbrella. A similar argument explained his observations of γ Draconis. Light from the star fell on the Earth with some velocity. At the same time, the Earth moved around the Sun with some other velocity. Like an umbrella in a rainstorm, the telescope had to point slightly in the direction of motion of the Earth in order to catch the starlight directly. Bradley called this effect the *aberration of light*.

The discovery of aberration was important for several reasons. First, it confirmed Römer's conclusion that the speed of light was not infinite. Second, measuring the tilt of the telescope produced the ratio of the speed of light to the speed of the Earth in its orbit. Since the Earth's orbital speed was known, Bradley could estimate the speed of light. (Bradley's measurements suggested that light would cross the diameter of Earth's orbit in 16 minutes 26 seconds. The value he used for the astronomical unit — the value obtained by Cassini — was too small. If we use the modern value for the astronomical unit, then Bradley's estimate for c turns out to be $283\,000\,\mathrm{km\,s^{-1}}$.) Third, Bradley had demonstrated, just as clearly as if he had discovered an annual parallax, that Copernicus was right. The Earth indeed moves around the Sun. Fourth, astronomers now knew that they had to account for aberration if they hoped to fix the positions of stars with precision*. (After he succeeded Halley as Astronomer Royal, Bradley discovered another effect that

---

*The rotation of the Earth about its axis also gives rise to aberration. Similarly, a telescope on board an orbiting satellite suffers from an extra aberration effect. This must be taken into account when fixing the positions of stars.

caused a small, regular shift in the position of stars: *nutation*. Nutation is the wobble of the Earth's axis caused by the gravitational attraction of the Moon. Astrometry must take into account both aberration and nutation.) Fifth, the very fact that he did not detect an annual parallax for γ Draconis implied that its parallax was less than 1″. This in turn suggested that γ Draconis was at a distance of at least 200 000 AU — a result that chimed nicely with Newton's estimate of 800 000 AU for the distance to Sirius.

4.5.3   Herschel and binary stars

Herschel searched for an annual parallax by returning to Galileo's suggestion. He decided to use the differential method of parallax. Herschel hunted for *double stars*: pairs of stars that are so close together that they appear to the naked eye as a single object. Using a micrometer attached to his telescope, Herschel could measure the angle between such stars. He hoped that even a slight change in their relative position might be noticeable.

With his usual thoroughness, Herschel found lots of double stars. In 1782, he published a catalogue of 269 systems that contained two stars separated by less than 2″. (No fewer than 227 of these double stars were newly discovered by him.) Two years later he published a second catalogue of 434 double stars. Herschel now had a problem. Were these double stars simply 'line of sight' coincidences, with one star much more distant than the other, or were they really physically close together in space? If the latter was the case, then he could not use them to detect parallax. The Earth's motion around the Sun would affect them both equally, which ruined the whole point of Galileo's idea.

As early as 1767, the English priest-come-scientist John Michell (1724–1793) had shown that double stars were, in all likelihood, physically close to each other in space. Consider the bright star Castor, which consists of two stars, A and B, separated by less than 5″. Michell calculated that the probability of their being so close by chance alignment was 1 in 300 000. Since Herschel had found hundreds of similar systems, it seemed even more likely that double stars were physically close together in space.

From a series of observations of Castor, Herschel proved that the two stars A and B were indeed physically close to each other, since they revolved around one another with a period of several hundred years. This double star was an example of a *binary star*. (Later, more detailed observations of binary star systems showed that gravity governed their motion. This was the first direct indication that Newton's law of universal gravitation really was universal — it governed the movement of distant stars as well as the planets in our solar system.)

By 1804, Herschel had found five more binary stars. Although he could not prove that all double stars were binary systems, it was likely that at least some of them (and perhaps most of them) were binaries. And binaries, unfortunately, were worthless for the differential method of parallax. Once again, an astronomer set out to detect an annual parallax but instead found something else.

4.6   PARALLAX — AT LAST!

During the 1830s, events combined to make the observation of an annual parallax seem feasible. Astronomers had bigger and better instruments, and they had definite criteria for

Example 4.3 The method of dynamic parallax

Two stars, of masses M_1 and M_2, form a binary system. They orbit one another with a period of T years, and the semi-major axis of their orbit is a''. Obtain an expression for the trigonometric parallax, and thus distance, of the binary system.

Solution. The general form of Kepler's third law is

$$(M_1 + M_2)T^2 = A^3 \qquad\qquad (*)$$

where masses are measured in solar masses, the orbital semi-major axis (A) is measured in astronomical units, and the orbital period (T) is measured in years. Now, simple geometry tells us that

$$A = \frac{a''}{\pi''} \qquad\qquad (**)$$

where π'' is the parallax. Substituting $(**)$ in $(*)$, and rearranging, we obtain:

$$\pi'' = \frac{a''}{[(M_1 + M_2)T^2]^{1/3}}.$$

We can thus measure the parallax (and hence distance) of a binary system if we can measure a'', T and the total mass of the system. This is the method of *dynamic* parallax.

choosing stars that were likely to be nearby.

The German optician Joseph von Fraunhofer (1787–1826) had a reputation for crafting astronomical instruments of the highest quality. Friedrich Georg Wilhelm von Struve (1793–1864) and Friedrich Wilhelm Bessel (1784–1846), both of whom were German astronomers, had access to instruments made by Fraunhofer. Struve used the world's largest and best refracting telescope at Dorpat in Estonia. Bessel, working at Königsberg in Prussia, used a large *heliometer* to measure small angles with great precision. Furthermore, both Struve and Bessel focused their effort — and their telescopes — on stars that were likely to be close.

Struve clearly set out three criteria for proximity.

- A star is likely to be near if it is bright.

- A star is likely to be near if it has a large proper motion.

- A binary star is likely to be near if the two component stars appear widely separated for the time they take to complete one orbit.

Moreover, in order to use the differential method of parallax, it made sense to study stars that had close companions. Struve went on to list the stars he thought were closest. (Struve chose his criteria well. His list contains several stars that we now know to be nearby.)

Struve decided to study Vega. It was bright, it had a large proper motion, and it was close enough to another star for him to make precision observations. (He was confident that Vega was not a binary system because the other star was faint, and it did not share Vega's large proper motion.) Between November 1835 and December 1836 he made 17 observations, an analysis of which indicated that Vega had a parallax of 0.125″ with an uncertainty of 0.05″. This is almost exactly the modern value. Struve published his results in 1837. Unfortunately from the point of view of his place in the record books he promised to continue his observations, and in 1840 he published the results of 96 observations made up to August 1838. The parallax he obtained this time was more than twice the original result, which cast doubt on both values. The astronomical community was not convinced.

Bessel, meanwhile, chose to study 61 Cygni, the 'flying star'. This star, being quite faint, scored poorly on Struve's first criterion for closeness. On the other hand, back in 1812, Bessel had shown that 61 Cygni had the largest proper motion of all the stars yet measured: more than 5″ per year. It thus came top on the second of Struve's criteria for closeness. Furthermore, it had two nearby companions, one of which was 8′ away and the other 12′ away, so that he could make accurate measurements. (The reference stars did not share the large proper motion of 61 Cygni, and so were presumably farther away from us.) In addition, 61 Cygni was near the pole, so he could observe it throughout the year at a large distance from the horizon. To Bessel, then, it seemed the perfect star to study in order to detect a parallax.

Bessel began his observations of 61 Cygni in September 1834, but other work, including observations of the returning Halley's comet, interrupted his programme. He returned to the task in 1837, perhaps encouraged by Struve's preliminary data on Vega. For more than a year he charted the position of 61 Cygni in relation to the two reference stars, making 16 or more painstaking observations of the star every night. He then had to analyse his data carefully, making allowance for all effects that were not parallax. At the end of 1838, Bessel announced that over a period of one year 61 Cygni made a small ellipse in the sky. The greatest displacement from the average position was just 0.31″ with an error of 0.02″. This tiny motion of 61 Cygni was a direct consequence of Earth's motion around the Sun. Bessel had finally discovered an annual parallax.

Although Struve's measurements of Vega were made prior to Bessel's measurements of 61 Cygni, it was Bessel's result that convinced fellow astronomers that a parallax had been properly and accurately measured. His frequent and consistent observations coincided exactly with what was predicted by theory. And so it was Bessel who was deservedly credited for discovering stellar parallax.

It often happens in science that a discovery is 'in the air': there are many examples of scientists quite independently making the same discovery at about the same time. So it was with parallax. After centuries of waiting, within weeks astronomers had *three* measurements of a stellar parallax. Predating the observations of both Struve and Bessel were those made by Thomas Henderson (1798–1844), a Scottish astronomer.

As director of the Cape of Good Hope Observatory in South Africa, Henderson was in a position to observe the southern star α Centauri. This star satisfied all three of Struve's criteria for closeness. It was third brightest star in the sky. It had the large proper motion of 3.7″ per year, much larger than any other bright star. Finally, it was a binary star with a wide angular separation between the components. Henderson completed his observations in 1833, and analysed his data upon his return to Scotland later that year. He completed his calculations before Bessel, and derived a parallax for α Centauri of 1.16″ with an error of 0.11″. Before publishing his result, however, he asked a colleague to check his work. In the end, he published several weeks after Bessel.

By the end of 1840, then, three stellar parallaxes had been found. In order of publication they were the following.

- Bessel's result of 0.31″ for 61 Cygni (modern value: 0.287″).

- Henderson's result of 1.26″ for α Centauri (modern value: 0.742″).

- Struve's result of 0.2619″ for Vega (modern value: 0.125″).

And once we know the parallax of a star, we can immediately calculate its distance from Earth.

Vega, 61 Cygni and α Centauri were presumably three of the *closest* stars, and yet their distances were measured in hundreds of thousands of astronomical units or tens of trillions of miles. Clearly, the mile and the astronomical unit would become as unwieldy as the inch or the centimetre when discussing stellar distances. The need for new distance units was apparent.

Example 4.4 The distance of nearby stars

Calculate the distance — in both astronomical units and miles — of α Centauri (parallax 0.742″), 61 Cygni (parallax 0.287″) and Vega (parallax 0.125″).

Solution. From (4.2), the distance d of a star, expressed in astronomical units, is

$$d = \frac{206\,265''}{\pi''}.$$

Therefore α Centauri is $206\,265/0.742 = 278\,000$ AU distant. A similar calculation tells us that 61 Cygni is $719\,000$ AU distant and Vega is $1\,650\,000$ AU distant.

One astronomical unit is about 9.3×10^7 miles. So α Centauri is 2.6×10^{13} miles away, 61 Cygni is 6.7×10^{13} miles away and Vega is 1.5×10^{14} miles away.

Astronomers use two distance units when discussing stellar distances: the *parsec* and the *light year*.

One parsec (pc) is the distance at which a star would have an annual *parallax* of one *sec*ond of arc. If a star has parallax π'', then its distance d in parsecs is given by

$$d = \frac{1}{\pi''}. \tag{4.4}$$

As we shall soon see, no star (except for the Sun) has a parallax as large as 1″. In other words, all stars are more distant than 1 pc.

Example 4.5 The parsec

Calculate the distance in parsecs to α Centauri, 61 Cygni and Vega.

Solution. We have $d = 1/\pi''$. So, for α Centauri, $d = 1/0.742 = 1.35$ pc; for 61 Cygni, $d = 1/0.287 = 3.48$ pc; and for Vega, $d = 1/0.125 = 8$ pc.

The light year is perhaps a better unit. I have already mentioned that light travels at exactly $299\,792.458\,\text{km s}^{-1}$. Since the speed is constant, light travel time may be used as a distance measure. For instance, 1 light second is the distance that light travels in one second: $299\,792.458$ km or $186\,282$ miles. The distance to the Moon is thus about 1.3 light seconds. The Sun is more distant. Light from the surface of the Sun takes 8 minutes 19 seconds to reach us, so it is 8.3 light minutes away*. The stars are *much* more distant, so we need to use the light year (ly). In one year, light travels 9.45×10^{12} km (which is about 5.88×10^{12} miles). So α Centauri is 4.39 ly away, 61 Cygni is 11.35 ly away, and Vega is 24.30 ly away. Clearly, 1 pc is equivalent to 3.26 ly. Put another way, if a star were 3.26 ly away it would have a parallax of exactly $1''$.

4.7   THE DISTANCE AND BRIGHTNESS OF NEARBY STARS

It is clear to anyone who looks up at the night sky that the stars differ in brightness. A few stars are as brilliant as the planets. Most stars are so faint they are barely visible. These are descriptive terms, though. How can we quantify the brightness of stars?

The key to quantifying the observed brightnesses of stars is to realise that when we talk of the brightness of a star we are really talking about the *energy flux F* — the energy crossing a unit area of our detector per unit time — received from that star. Astronomers now have many sensitive instruments for measuring flux, but before such instruments were available astronomers had to make use of a different kind of device: the human eye. Hipparchus divided the stars that were visible to the naked eye into six classes, called *magnitudes*. A bright star was of first magnitude and a faint star was of sixth magnitude. The *apparent magnitude scale* is thus 'backwards', rather like the scoring method in golf. The *smaller* the magnitude, the *brighter* the star. (The magnitude of a star depends on the instrument we use to measure it. The human eye is most sensitive to yellow–green light near a wavelength of 550 nm; early photographic plates were most sensitive to blue light at 450 nm. So the visual (V) magnitude differs from the blue (B) magnitude, and indeed from the ultraviolet (U) and infrared (I) magnitudes. Unless stated otherwise, magnitudes in this book are V magnitudes. Note that we can quantify the colour of a star by giving the difference between its B and V magnitudes; B − V is called the *colour index*. Colour indices can be measured precisely, even for faint stars, and are extremely important.)

In the 1850s, the German physiologist Ernst Heinrich Weber (1795–1878) and the

---

*When we look at the Sun we see it as it was 8.3 minutes ago. If the Sun suddenly stopped shining we would not know about it for 8.3 minutes. The more distant the object, the greater this *lookback time*. Inevitably, as we look out into the universe we look back in time.

German physicist Gustav Theodor Fechner (1801–1887) investigated the response of the human senses to changes in stimuli. It turns out that our eyes, like our ears, do not respond to flux changes in a simple linear way. They register flux changes on a logarithmic scale. In other words, we perceive equal flux *ratios* to be equal flux *intervals*. (Actually, this is not quite true. Our senses follow power-law rather than logarithmic curves. But the Weber–Fechner logarithmic scale is now so widely used that it is impractical to introduce a more appropriate scale.) So a star that Hipparchus classified as being first magnitude is not five times brighter than a sixth magnitude star: the flux ratio is much bigger than five. In fact, Herschel found that a first-magnitude star is about 100 times brighter than a sixth-magnitude star. The English astronomer Norman Robert Pogson (1829–1891) suggested that this hundredfold increase in flux be *defined* to be a difference of exactly five magnitudes. If two stars have observed energy fluxes F_1 and F_2, and apparent magnitudes m_1 and m_2, then Pogson's definition corresponds to

$$\frac{F_1}{F_2} = 100^{(m_2-m_1)/5}.$$

(4.5)

We can see that (4.5) is correct because every time the difference $m_2 - m_1$ increases by 5, the ratio F_1/F_2 increases by 100.

Equation (4.5) is inconvenient because we usually work in powers of 10 rather than powers of 100. Writing 100 as 10^2, (4.5) becomes

$$\frac{F_1}{F_2} = 10^{(m_2-m_1)/2.5}$$

and taking the logarithm (base 10) of both sides gives us

$$m_1 - m_2 = -2.5 \log_{10}\left(\frac{F_1}{F_2}\right).$$

(4.6)

Equation (4.6) is called *Pogson's equation*. Its importance in distance measurement will soon become clear.

Since a difference of five magnitudes corresponds to a factor of 100 in brightness, a difference of one magnitude corresponds to a factor of 2.512 in brightness. (This is so because $2.512^5 = 100$. Note that the factor 2.512 has nothing to do with the term 2.5 appearing in (4.6) above.) This places the apparent magnitude scale of Hipparchus on firm ground. To set the zero point of the scale, astronomers arbitrarily assigned the bright star Vega with a magnitude of 0.

Once we have quantified the scale in this way we can assign precise values to the apparent magnitude of stars. We can also extend the scale beyond the crude limits drawn by Hipparchus. On this scale, objects brighter than Vega have a negative apparent magnitude. For instance, the star that shines brightest in our skies, Sirius, has an apparent magnitude of −1.44. The brightest planet, Venus, has a peak magnitude of −4.6. The full Moon reaches −12, while the Sun has an apparent magnitude of −26.78. We can extend the scale in the other direction, too. Telescopes, particularly when used with modern imaging techniques, let us record objects far beyond the +6 limit of human vision.

Example 4.6 Apparent magnitude

A star varies in apparent magnitude from +7.1 to +7.8. Find the relative increase in brightness from minimum to maximum.

Solution. The amplitude of magnitude variation is 0.7. We therefore have

$$\log_{10}\left(\frac{F_{max}}{F_{min}}\right) = \frac{0.7}{2.5} = 0.28$$

and so $F_{max}/F_{min} = 10^{0.28} = 1.93$. The star is almost twice as bright at maximum as it is at minimum.

Large ground-based telescopes reach to an apparent magnitude of about +28. Recently, the *Hubble Space Telescope* reached nearly to the 30th magnitude.

More important than apparent magnitude, at least from the point of view of understanding the nature of stars, is *luminosity*. A star's luminosity is the total amount of energy it radiates into space each second. Unfortunately, the apparent magnitude of a star tells us nothing about its luminosity. A highly luminous star will seem faint if it is very far away. Conversely, a star of low luminosity will seem bright if it is close to us. In other words, there must be a link between intrinsic luminosity, apparent magnitude and distance. That link is the *inverse square law*.

According to the inverse square law, if the distance to a light source is doubled the observed flux drops by a factor of $2^2 = 4$. If the distance to the source is increased by a factor of three, the observed flux drops by a factor of $3^2 = 9$. And so on. Figure 4.6 shows how the flux (brightness) and the magnitude of a light source change as the distance between the source and the observer increases.

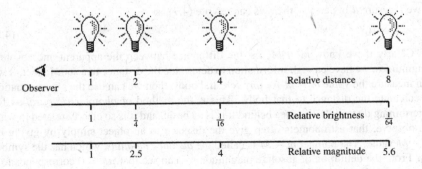

FIGURE 4.6: The same light bulb, placed at different distances, has different observed brightnesses and different apparent magnitudes. Note that the logarithmic magnitude scale has *increasing* values for *decreasing* brightness.

If the stars all had the same luminosity then apparent magnitude would be a pure distance effect. We would be able to use the inverse square law to calculate the distance to a star simply from its observed brightness. Unfortunately, as Herschel's work on binary systems showed, stars differ in luminosity: two stars at the same distance from us can have different apparent magnitudes. The determination of annual parallaxes confirmed this finding. Astronomers had enough information to calculate the luminosity of 61 Cygni, Vega and α Centauri — and they were all different. (A third factor can influence the observed brightness of a star: the existence of interstellar material between us and the star. Clouds of interstellar dust and gas play an important role in dimming the light from certain regions of space. For the moment, though, we will assume that light from the stars travels through a perfect vacuum. In this case, only two factors determine the brightness of a star: its luminosity and its distance.)

How luminous is 61 Cygni compared to α Centauri, say, or the Sun? To describe the intrinsic luminosity of a star, astronomers imagine it to be at some fixed distance from Earth and ask: how bright does it appear? The fixed distance astronomers have chosen is 10 pc. The apparent magnitude a star would have if it was at a distance of 10 pc is called its *absolute* magnitude.

Using the inverse square law we can link apparent magnitude, m, the absolute magnitude, M, and the distance, d, of a star. To do this, consider two identical stars. The first, at distance d, has apparent magnitude m. The second, at the standard distance d_0, has apparent magnitude M. (By definition, M is also the absolute magnitude of both stars.) The energy flux drops as the inverse square of distance, so the ratio of the energy flux of the first star to the second is $(d/d_0)^2$. Putting this into (4.6) produces

$$m - M = 2.5 \log_{10}\left(\frac{d}{d_0}\right)^2.$$

Using the relation $\log x^2 = 2 \log x$, and noting that $d_0 = 10$ pc, produces

$$m - M = 5 \log_{10}\left(\frac{d}{10\,\text{pc}}\right). \tag{4.7}$$

If we measure d in parsecs, then we can rewrite (4.7) as

$$d = 10^{(m-M+5)/5}. \tag{4.8}$$

Clearly if we know $m - M$, i.e. the difference between the apparent and absolute magnitudes, we have enough information to determine the distance to a star. In fact, if we can measure the value $m - M$ for *any* celestial body, then we can use the above equation to calculate the distance to that body. This is the method of *photometric parallax* for determining distance*. The idea behind this is so basic, and it is so clearly related to what we observe, that astronomers often give the distance to an object simply by giving its $m - M$ value. The number $m - M$ is called the *distance modulus*, which has the symbol μ_0. From the definition of absolute magnitude we can see that $\mu_0 = 0$ corresponds to a distance of 10 pc, $\mu_0 = 5$ corresponds to a distance of 100 pc, and so on.

---

*The use of the word 'parallax' here is misleading, since no angular measurements are involved. Astronomers gave it this name simply because, to them, parallax was the basis of all distance-measuring techniques. Parallax became synonymous with distance, regardless of how the distance was measured.

Example 4.7 Absolute magnitude

Which is more luminous — the Sun or Sirius? (Sirius is at a distance of 2.64 pc.)

Solution. From the text, the Sun's apparent magnitude is −26.78 and the apparent magnitude of Sirius is −1.44.

The Sun is at a distance of 1 AU, or $1/206\,265$ pc. Therefore, from (4.7), the absolute magnitude of the Sun is

$$M_\odot = m_\odot - 5 \log_{10}(d_\odot/10\,\text{pc})$$
$$= -26.78 - 5 \log_{10}(1/2.06265 \times 10^6) = 4.79.$$

Similarly, the absolute magnitude of Sirius is $-1.44 - 5 \log_{10} 0.264 = 1.45$. Sirius is thus $10^{(4.79-1.45)/2.5} \approx 20$ times more luminous than the Sun.

Example 4.8 Distance modulus

What is the distance modulus of Sirius?

Solution. From the previous example we know that for Sirius $m = -1.44$ and $M = 1.45$. So $\mu_0 = -1.44 - 1.45 = -2.89$.

As an example of how photometric parallax works, consider 61 Cygni. It has an apparent magnitude of 5.2 and an absolute magnitude of 7.49. From (4.8) we immediately see that $d = 3.48$ pc. This example shows the power of photometric parallax — except, of course, that we have cheated. We knew the absolute magnitude of 61 Cygni only because we *already* knew its distance from parallax measurements. If we know the distance to an object we can calculate its absolute magnitude and thus its distance modulus — but that is no help to us because it is the distance we want to calculate in the first place! Nevertheless, as we shall see in later chapters, there are situations where we can identify a standard candle. We then know the absolute magnitude of a body (or we can at least make a good guess of the absolute magnitude). In those cases we can use (4.8) to calculate the distance. This is one of the most important distance-measuring techniques of all.

4.8   THE REACH OF PARALLAX

How far out in the universe does the technique of trigonometric parallax take us? This is a clear enough question to pose, but the answer is not as straightforward as might be thought. If you look in astronomy textbooks you will see that the maximum effective range of the parallax technique is described as being anything between 25 pc (about 80 ly) to 100 pc (about 325 ly). It is worthwhile understanding why the textbooks differ on such a seemingly fundamental point.

At first sight the answer to the question seems simple. The distance d to a star is just $d = 1/\pi''$, where π'' is its parallax in seconds of arc and d is in parsecs. For example, suppose that we measure the parallax of a star to be $0.01''$. We could say that its distance is just $1/0.01 = 100 \, pc$, but the situation is more complicated than this. A parallax measurement always has an attendant uncertainty, and we must take this uncertainty into account. It may be, for instance, that the uncertainty in this parallax measurement is $\pm 0.003''$. In this case we can say only that the true value of the star's parallax probably lies somewhere in the range $0.007''$–$0.013''$ — so the distance is probably in the range 77–$143 \, pc$. Is such a large uncertainty in the distance acceptable? It may be for some applications, but not for others. If we demand less uncertainty in our distances then we must either reduce the uncertainty in the measurement or we must work with larger measured parallaxes. For instance, if a star has a measured parallax of $0.1''$ and the uncertainty remains $\pm 0.003''$ then the actual parallax probably lies between $0.097''$ and $0.103''$ — so its distance is probably in the range 9.7–$10.3 \, pc$. This is good enough for most purposes, but a distance of about $10 \, pc$ does not take us very far out into the universe.

In other words, what is often important is not just the parallax, π, but the ratio of the uncertainty in the parallax, e_π, to the parallax itself. If we want the ratio e_π/π to be small, we are limited to relatively small distances. If we are willing to accept a larger fractional uncertainty in the parallax, then we can reach to greater distances. There is no commonly agreed definition of what constitutes a 'reasonable' value for the ratio e_π/π, so there is naturally some disagreement in the textbooks about how far we can reach with parallax. Some authors might demand 5% accuracy in distance estimates; others 10%. For some purposes, 20% accuracy might be sufficient. In all cases, the farther out we go the greater the fractional uncertainty in the distance.

4.8.1   The Lutz–Kelker bias

There is another, more subtle, problem that limits the range of the parallax method.

When we convert from a parallax into a distance, the inevitable measurement uncertainty in the parallax can act in two ways. It can make a star that is *truly* far away seem closer, or it can make a star that is *truly* nearby seem farther away. The two effects are not symmetrical because, as we go farther out into space, the number of stars increases. Therefore, given a parallax measurement for some star, it is more likely that it is really a distant star for which measurement errors make it appear nearer, than it is a truly nearby star for which measurement errors make it appear farther away. There is thus a bias to parallax measurements — called the *Lutz–Kelker bias* — that causes us to underestimate the distance to stars. The bias is named after the American astronomer Thomas Edward Lutz (1940–1995) and the American statistician Douglas Herson Kelker (1940–).

Lutz and Kelker calculated the size of the bias for a uniform distribution of stars having all possible luminosities. In this rather artificial situation the rule of thumb is that you should not trust a distance greater than about 0.175 times the theoretical limit of your device. For instance, if your device can measure a parallax with a precision of $0.01''$, the theoretical distance limit is $100 \, pc$. But according to Lutz and Kelker, distances beyond $17.5 \, pc$ (or $57 \, ly$) are unreliable. Similarly, with a precision of $0.001''$, distances beyond $175 \, pc$ ($570 \, ly$) are unreliable.

Fortunately, things are seldom as bad as in the original Lutz–Kelker analysis! The size of the bias for a real sample of stars depends critically upon how the stars are selected. The bias can be very different from case to case, and in some cases it can be negligible. Nevertheless, it is important to be aware that in principle parallaxes will always have a bias of the type described above, even if the magnitude of the bias may be small. Therefore care must be taken when converting from a measured parallax into a distance, especially if the parallaxes have low precision.

If we demand 10% accuracy in the distance to an individual star then the *very best* parallax measurements — which I will now describe — reach out to about 90 pc (300 ly). A star that is closer than that may have its distance measured with greater accuracy. Distances beyond that start to become unreliable.

4.9 SPACE-BASED TELESCOPES

The results of Bessel, Struve and Henderson appeared within a few months of each other, so perhaps astronomers thought they had cracked the problem of determining stellar distances. If so, they were wrong. Work on parallaxes progressed slowly. People at different observatories often studied the same star and obtained completely different values for the parallax. Not until the 1870s did Gill show how to obtain reliable results, and even 70 years after Bessel's pioneering work astronomers knew the distance to only 100 stars. As late as 1952, the *Yale Parallax Catalog* gave parallaxes for only 5822 stars.

The difficulty is that, as we probe out to greater distances, stellar parallaxes become smaller and thus the precision required of the measurements increases. It might be thought that the answer is simply to build bigger telescopes, but such an approach does not help in this case. Starlight travels in a straight line through space for trillions of miles, but in the final few miles of its journey to our eyes it must pass through the Earth's turbulent atmosphere. Atmospheric turbulence blurs the image of a star, and makes it impossible for us to say exactly where a star is on the celestial sphere. (Although a stellar image is fuzzy, we can usually determine the centre of the image quite precisely — which is just as well, because otherwise the parallax method would be *severely* restricted.) The atmosphere also bends light, systematically displacing stars from their true position. This latter effect depends on the detailed physical conditions of the different atmospheric layers. Astronomers can allow for the effect, but the allowance they make is never perfect. The Earth's atmosphere thus limits the precision with which we can make parallax measurements.

Most parallaxes in the 1952 *Yale Parallax Catalog* have a precision of 0.01″. Higher precision came from the 61-inch telescope at the US Naval Observatory at Flagstaff, AZ, which was built specifically to provide high-quality parallaxes. The telescope continues to give excellent parallaxes, approaching a precision of 0.001″. (An angle of 0.001″ is one milliarcsecond, or 1 mas*.) Unfortunately, it can study only a few hundred stars, and the necessary observations of each star can take many years. In order to acquire a precision of 0.001″ for stars over the whole sky we need a telescope that does not suffer from the blurring effects of the atmosphere: we need a space-based telescope.

---

*To appreciate how small this angle is, take a second to glance at someone who is standing a metre away from you. During that time a strand of hair on that person's head will have grown by about 0.001″.

4.9.1   The *Hubble Space Telescope*

In 1946, the American physicist Lyman Spitzer, Jr (1914–1997) wrote a paper entitled 'Astronomical advantages of an extraterrestrial observatory', in which he described the benefits of operating a large optical telescope above the Earth's atmosphere. This was more than a decade before the launch of *Sputnik* ushered in the space age, so Spitzer was far-seeing. Detailed design studies for a space telescope began in the mid-1960s. By 1971, a NASA committee chaired by Spitzer recommended the construction of the Large Space Telescope — an orbiting telescope with a 3-m primary mirror. This was thought to be too costly. In 1977, the US government agreed to fund a Space Telescope with a 2.4-m mirror (they dropped the word 'Large' because of the decrease in size of the primary mirror). In 1983, NASA renamed it the *Hubble Space Telescope*, after an American astronomer who, as we shall see, played a pivotal role in developing our understanding of the size of the universe. Finally, on 24 April 1990, NASA launched the *Hubble Space Telescope*.

Behind the 2.4-m primary mirror, the space telescope carried five instruments: the Wide Field/Planetary Camera, the Faint Object Camera, the Faint Object Spectrograph, the Goddard High Resolution Spectrometer and the High Speed Photometer. The Fine Guidance Sensors, which maintained tracking lock, were also to be used for some independent observations. When the *Hubble Space Telescope* took its first images, on 20 May 1990, the results were disappointing, and over the next few weeks the results remained disappointing. Astronomers battled to focus the telescope, but no matter how hard they tried the images remained blurred. They eventually deduced that the mirror had a design fault. It suffered from spherical aberration: it did not bring the light rays falling on its edge and those falling on its centre to exactly the same focus. (After decades of work, and development costs of $1.5 billion, this came as quite a shock.) The aberration was particularly detrimental to the performance of the two cameras. NASA decided to replace the High Speed Photometer instrument with a package of corrective optics, and astronauts on board the shuttle Endeavour installed the package in December 1993. Since then, the images from the *Hubble Space Telescope* have stunned people with their beauty.

The *Hubble Space Telescope* is potentially the best instrument we have for making precise astrometric measurements. In principle, we can use it to measure the parallax of stars that are several hundreds of light years away. In practice, there are problems with the stability of the platform that limit its use in astrometry. Besides, the universe is large and filled with interesting objects to study. A programme of astrometric measurements of nearby stars must fight against a host of other programmes for observing time. If we want to measure the distance to nearby stars, what we need is a dedicated astrometry mission.

4.9.2   The *Hipparcos* astrometry mission

The French astronomer Pierre Lacroute (1906–1993) was the first to suggest that pinpoint parallax observations could best be made from a telescope on board an orbiting satellite. The idea was not to take images of the stars; rather, Lacroute imagined an instrument that would precisely measure the angles between stars, scanning the whole sky as the satellite slowly turned. He made this suggestion in 1966. Astronomers from member countries of the European Space Agency (ESA) studied his plan for several years, improving upon the design and adding many new features. In 1980, ESA selected Lacroute's enhanced proposal to be part of its scientific programme. They called the mission *Hipparcos*. The name stands for *hi*gh *p*recision *pa*rallax *co*llecting *s*atellite, but of course it also evokes the name of Hipparchus, who was the first to detect the parallax of a heavenly body. *Hipparcos* became the first space mission dedicated to measuring the positions of the stars.

ESA launched the satellite from Kourou, French Guiana on 8 August 1989 using an *Ariane* rocket. Like the *Hubble Space Telescope*, the mission suffered an initial setback. *Ariane* left the satellite in a transfer orbit, from which it was to be boosted into a geostationary orbit. However, the failure of an apogee booster rocket left it in a highly elongated orbit. This could have been disastrous, but some quick thinking saved the mission. The project scientists revised their operational procedures, and tracked the satellite with three ground stations (Odenwald in Germany, Perth in Australia and Goldstone in the USA) rather than the single ground station as originally planned. The main problem with the highly elliptical orbit was that it took the satellite through radiation belts. This reduced the useful observing time; worse still, the radiation damaged certain critical systems on the satellite. Communication with the on-board computer became difficult, and the mission ended on 15 August 1993.

Despite its initial problems, the *Hipparcos* mission was a success. The original goals of the mission were to measure the positions, parallaxes and proper motions of 100 000 stars with a precision of 2 mas. In addition, the *Tycho* experiment on board the satellite was to gather information on more than 400 000 stars with a lesser precision. In all respects, *Hipparcos* exceeded these targets. The *Hipparcos* catalogue contains positional information on 117 955 stars, with an average precision of better than 1 mas — much better than anticipated. The *Tycho* catalogue gives slightly less precise information on 1 058 332 stars. *Hipparcos* found 10 000 new double stars, 8000 new variable stars, and several candidate planetary systems. (The catalogues became available in June 1997, almost four years after the end of the mission. The delay was not due to laziness on the part of the mission scientists. Since *Hipparcos* observed each star on about one hundred different occasions, the shear volume of data threatened to inundate the teams. The satellite returned more than 1000 gigabytes of information. It was one of the largest data analysis problems in the history of astronomy. The catalogues are available to the public on CD-ROM — so readers who have access to a PC and a CD drive have the chance to make discoveries for themselves!)

So how far out can *Hipparcos* measure distances? If we demand a relative accuracy of 10% in our distance measurements, then *Hipparcos* reaches out to a distance of about 90 pc (about 300 ly). In terms of the number of stars, it determined the distance to 22 396 stars to 10% relative accuracy or better*. If we study groupings of similar stars, and consider the group as a whole rather than those of individual stars, then we can look out even farther into the universe. The implications of these *Hipparcos* results become clear on later rungs of the distance ladder.

For the first time, we have a reliable map of our region of space. Remember, though, that *Hipparcos* studied the parallaxes and proper motions of over 100 000 stars — five times more stars than those for which it obtained reliable distances. The *Tycho* catalogue

---

*It determined the distance of 442 stars to a relative accuracy of better than 1%, and 7388 to better than 5%.

TABLE 4.1: The 35 nearest stars. In nearly all cases, the distances of the 25 stars brighter than apparent magnitude 12.2 come from the *Hipparcos* catalogue (1997). The distances of the stars dimmer than apparent magnitude 12.2 come from the fourth edition of the Yale *General Catalogue of Trigonometric Parallaxes* (1995). The exception is the star GJ 1061. Until recently, astronomers thought that this star was farther away than it really is; see Henry *et al.* (1997). The stellar magnitudes are those given by RECONS (Research Consortium on Nearby Stars). It is entirely possible that more nearby dim stars remain to be found. According to RECONS, which is a team devoted to discovering stars nearer than 10 pc, there are about 130 star systems within a 10 pc radius that are missing from the present catalogues. They base their conclusion on the observed density of stars within a 5 pc radius. If this density is constant out to 10 pc, which seems likely, then there must be stars out there that we do not see. These stars are probably dim red dwarfs. These account for about 70% of all stars but they are inherently faint, with luminosities in the range 10^{-2}–$10^{-4} L_\odot$. As of January 1999, RECONS knows of 323 stars within 10 pc of Earth. These 323 stars occur in 234 separate systems, and include 169 single stars, 47 double stars, 13 triples, 4 quadruples and 1 quintuple. (Appendix A explains the names appearing in the table; the common names for well-known stars are used, but most are catalogue names.)

Name	Distance (light years)	Distance (parsecs)	Apparent magnitude	Absolute magnitude
Sun	—	—	−26.72	4.85
Proxima Centauri	4.22	1.29	11.09	15.53
α Centauri A	4.39	1.35	0.01	4.36
α Centauri B	4.39	1.35	1.34	5.69
Barnard's star	5.94	1.82	9.53	13.21
Wolf 359	7.78	2.39	13.44	16.55
BD +36° 2147	8.31	2.55	7.47	10.44
Sirius A	8.60	2.64	−1.43	1.46
Sirius B	8.60	2.64	8.44	11.33
L 726-8 A	8.72	2.68	12.43	15.29
L 726-8 B	8.72	2.68	13.19	16.05
Ross 154	9.69	2.97	10.43	13.06
Ross 248	10.32	3.16	12.29	14.79
ε Eridani	10.49	3.22	3.73	6.19
CD −36° 15693	10.73	3.29	7.34	9.75
Ross 128	10.88	3.34	11.13	13.50
L 789-6 A	11.26	3.45	13.1	15.4
L 789-6 B	11.26	3.45	13.7	16.0
L 789-6 C	11.26	3.45	13.8	16.1
61 Cygni A	11.36	3.49	5.21	7.50
61 Cygni B	11.42	3.50	6.03	8.31
Procyon A	11.42	3.50	0.38	2.66
Procyon B	11.42	3.50	10.7	13.0
BD +59° 1915 A	11.43	3.51	8.90	11.18
BD +59° 1915 B	11.43	3.51	9.69	11.97
BD +43° 44 A	11.63	3.57	8.08	10.32
BD +43° 44 B	11.63	3.57	11.06	13.30
GJ 1111	11.82	3.63	14.78	16.98
ε Indi	11.82	3.63	4.69	6.89
τ Ceti	11.89	3.65	3.49	5.68
GJ 1061	12.07	3.70	13.03	15.21
L 725-32	12.12	3.72	12.02	14.17
BD +5° 1668	12.38	3.80	9.86	11.96
Kapteyn's star	12.77	3.92	8.84	10.87
CD −39° 14192	12.87	3.95	6.67	8.69

contains over a million stars, and later editions of the catalogue will contain data on over *three million* stars. Clearly, the first space astrometry mission has barely started on the task of determining the distances of our stellar neighbours.

4.9.3   Future missions

The success of *Hipparcos* has led ESA and NASA to consider further space astrometry missions. The proposed ESA mission, called the *Global Astrometric Interferometer for Astrophysics (GAIA)*, would measure the parallaxes of one billion stars down to magnitude 20, with a precision 100 times better than that of *Hipparcos*. The aim is to perform astrometric observations at the 10 microarcsec level at magnitude 15. At this level of precision, *GAIA* could evaluate the distances of stars more than 10 000 pc away to an accuracy of 10%. If ESA decides to proceed with the mission, a suitable launch date might be 2009 or 2014.

The proposed NASA mission, called the *Space Interferometry Mission (SIM)*, would achieve a precision at least 250 times better than that of *Hipparcos*. At this level of precision *SIM* could measure even greater distances than *GAIA*. Rather than survey the whole sky, though, the plan is to point the satellite at chosen objects of interest. Closer to home, *SIM* could search for tiny 'wobbles' in the paths of nearby stars. These 'wobbles' would enable astronomers to search for Earth-sized planets orbiting any of the 100 closest stars.

4.9.4   The nearest stars

Table 4.1 lists the nearest stars to Earth. It shows their distances as determined by parallax measurements. (For stars brighter than magnitude 12.2, these come from *Hipparcos*; stars dimmer than this were ignored by *Hipparcos* so we have to rely on ground-based parallaxes for such stars.) The table also gives their apparent and absolute magnitudes.

As the table makes clear, there is a wide variation in absolute magnitude. Most of the nearby stars are dimmer than our Sun by factors ranging from 100 to 10 000. Stars are not standard candles. Note how the stars are widely separated compared to their size: the ten closest stars have an average separation of about 8 ly, but the typical size of a star is about 2 light seconds. Note also the large number of multiple stars systems included in table 4.1.

CHAPTER SUMMARY

- Astrometry is the science of measuring the precise angular distances between stars. Some effects that must be taken into account in astrometry are proper motion, the aberration of light, and atmospheric refraction.

- Earth's annual motion around the Sun causes the stars to exhibit a parallax, π. A nearby star moves in a small parallactic ellipse on the celestial sphere. The semi-major axis of the ellipse is π, and the semi-minor axis of the ellipse is $\pi \sin \beta$, where β is the star's celestial latitude.

- If we measure parallax in seconds of arc (π'') then the distance d to a star is given by

$$\pi_*'' = 206\,265\frac{R}{d}$$

where R is the radius of the Earth's orbit (i.e. 1 AU).

- One parsec is defined as being the distance at which a star would have a parallax of $1''$.

$$1\,\text{pc} = 3.26\,\text{ly} = 206\,265\,\text{AU}.$$

- The nearest star, apart from the Sun, lies at a distance of 1.29 pc.

- We quantify the brightness of stars using a logarithmic magnitude scale. The absolute magnitude, M, of a star is the apparent magnitude, m, it would have at a distance of 10 pc. The absolute magnitude of a star thus tells us something about the star's luminosity. An equation linking absolute magnitude, apparent magnitude and distance, is:

$$m - M = 5\log_{10}\left(\frac{d}{10\,\text{pc}}\right).$$

If we measure distance in parsecs, then $d = 10^{(m-M+5)/5}$.

- The distance modulus $m - M$ is denoted by the symbol μ_0. Astronomers often refer to the distance of an object simply by giving its distance modulus. We must use one of the above equations to convert the distance modulus into parsecs.

- Since the stars are so distant, their parallax angles are tiny. It is therefore important to know the precision with which a parallax measurement has been made. We need to know e_π/π, the ratio of the error in the parallax to the parallax itself. At large distances, the error in the parallax becomes comparable with the parallax itself.

- The best parallax data come from the *Hipparcos* satellite. It measured parallaxes with an average precision of better than 1 mas.

- Future space-based astrometry missions may include *SIM* and *GAIA*.

QUESTIONS AND PROBLEMS

4.1 The speed of everyday objects can be used to illustrate distance. In medieval times, for example, scholars illustrated the vast distance of the fixed stars (taken to be Ptolemy's value of 20 000 ER) by calculating how many years it would take a man walking at a speed of 25 miles per day to reach the stars. Huygens used to convey the distance scale of the solar system by calculating how long it would take a cannon ball travelling at 600 feet per second to reach the planets.
   Devise some similar illustrations of your own for conveying the vast distance to the nearest stars.

4.2 Some of the stars in the *Hipparcos* catalogue have a *negative* parallax. What does this signify?

4.3 Access the *Hipparcos* catalogue (either on-line or from CD-ROM) and find the absolute visual magnitude of the 200 nearest stars in the catalogue. How many of the stars are more luminous than the Sun?

4.4 How would our parallax measurements change if we could make our observations from the surface of Mars? How large would the parsec be for Martian astronomers?

4.5 If you were an astronomer living on Pluto, would any stars possess an annual trigonometric parallax large enough to be detected with the naked eye?

4.6 The starship *Enterprise* is 25 pc from Earth. A homesick Captain Kirk looks out of the observation port for a sight of the Sun. No telescope is available, but Kirk has perfect vision. Can he see the Sun?

4.7 If you were asked to define a new stellar magnitude scale, using modern techniques and instruments, how would you approach the problem?

4.8 Derive an expression for the distance to a star in terms of its distance modulus.

4.9 How many sixth-magnitude stars equal the brightness of a single first-magnitude star? 100

4.10 What is 1 parsec in terms of (a) astronomical units and (b) metres? (a) 206 265 AU (b) 3.085×10^{16} m

4.11 What is the distance modulus of the Sun? $\mu_0 = -31.57$

4.12 Which star, other than the Sun, has the smallest distance modulus? Proxima Centauri, with $\mu_0 = -4.44$

4.13 If we make an 0.01 mag error in measuring the apparent magnitude of a star, what percentage error does this introduce in our distance estimate of that star? (Assume that we know the absolute magnitude of the star exactly.) 0.46%

4.14 You observe two stars with a photometer. You receive energy from star A at a rate of 5.3×10^{-14} W, and energy from star B at a rate of 3.9×10^{-14} W. What is $m_A - m_B$, i.e. the difference in apparent magnitude between stars A and B? $m_A - m_B = -0.33$

FURTHER READING

GENERAL

Cook A (1998) *Edmond Halley* (Oxford University Press: Oxford)
— The definitive biography of Halley.

MacRobert A M (1996) The stellar magnitude system. *Sky & Telescope* **91 (1)** 42–44
— A short, nicely illustrated explanation of the magnitude system.

Moore P (1996) *Brilliant Stars* (Cassell: London)
— A fascinating survey of the 21 brightest stars in our sky, by the doyen of British amateur astronomers.

Murray C A (1988) The distances to the stars. *Observatory* **108** 199–217
— A detailed history of attempts to measure parallax — from Halley to *Hipparcos*.

Roy A E and Clarke D (1988) *Astronomy — Principles and Practice* (Institute of Physics: Bristol)
— An excellent all-round introduction to astronomy for first-year undergraduates.

Shapley H and Howarth H E (1929) *A Source Book in Astronomy* (McGraw-Hill: New York)
— Newton on the distance of Sirius, Halley's discovery of proper motion, Bradley's discovery of aberration and nutation, Bessel and Struve on parallax — all in their own words.

Westfall R. (1980) *Never at Rest* (Cambridge University Press: Cambridge)
— The best biography of Newton.

Williams M E W (1979) Flamsteed's alleged measurement of an annual parallax for the Pole Star. *J. Hist. Astron.* **10** 102–116
— How Flamsteed fooled himself into thinking he had discovered the parallax of Polaris.

http://www.obspm.fr/planets
— There are now more extrasolar planets known than there are planets in the solar system! The Extrasolar Planets Encyclopaedia, maintained by Jean Schneider, will keep you up to date.

ASTROMETRY AND GROUND-BASED PARALLAXES

Dahn C C (1992) Stars, distances and parallaxes. In *The Astronomy and Astrophysics Encyclopaedia.* ed.
S. P. Maran. (Van Nostrand Rheinhold, New York)
— Describes the detailed work required to measure a ground-based parallax.

Henry T J *et al.* (1997) The solar neighbourhood. IV. Discovery of the twentieth nearest star system. *Astron. J.*
114 388–395
— The authors recently discovered that GJ 1061, a dim red dwarf, was much closer than previously thought.
There may be about 130 more star systems in the solar neighbourhood remaining to be discovered.

Kovalevsky J (1995) *Modern Astrometry.* (Springer: Berlin)
— A detailed account of all aspects of astrometry.

Kovalevsky J (1998) The new astrometry. *Rep. Prog. Phys.* **61** 77–115
— A nice review of modern astrometry, including *Hipparcos* and its possible successors.

Smith H Jr and Eichhorn H (1996) On the estimation of distances from trigonometric parallaxes. *MNRAS* **281**
211–218
— Estimating a stellar distance from an observed trigonometric parallax is not as simple as might be thought!

van Altena W F (1983) Astrometry. *Ann. Rev. Astron. Astrophys.* **21** 131–164
— An excellent history of astrometry.

van Altena W F, Lee J T and Hoffleit D E (1995) *General Catalogue of Trigonometric Parallaxes* 4th edn. (Yale
University Observatory: Yale)
— The best of the ground-based parallax observations.

HIPPARCOS

Arenou F *et al.* (1995) Zero-point and external errors on *Hipparcos* parallaxes. *Astron. Astrophys.* **304** 52–60
— An interesting discussion of why the *Hipparcos* parallaxes are both accurate and precise.

Høg E *et al.* (1997) The *Tycho* catalogue. *Astron. Astrophys.* **323** L57–L60
— A brief review of the principal observational characteristics of the *Tycho* catalogue.

Kovalevsky J (1998) First results from *Hipparcos. Ann. Rev. Astron. Astrophys.* **36** 99–129
— An overview of how *Hipparcos* results are already being applied to many questions in astronomy.

Kovalevsky J *et al.* (1995) Construction of the intermediate *Hipparcos* astrometric catalogue. *Astron. Astrophys.*
304 34–43
— Of interest if you wish to learn of the detailed data analysis needed to produce the *Hipparcos* catalogue.

Perryman M A C *et al.* (1997) The *Hipparcos* catalogue. *Astron. Astrophys.* **323** L49–L52
— A brief review of the principal observational characteristics of the *Hipparcos* catalogue.

http://astro.estec.esa.esa.nl:80/SA-general/Projects/Hipparcos/hipparcos.html
— The *Hipparcos* Space Astrometry Mission home page. Definitely a page to bookmark!

http://astro.estec.esa.esa.nl:80/SA-general/Projects/GAIA/gaia.html
— The *GAIA* home page. Look here to find information on the proposed follow-up mission to *Hipparcos.*

THE LUTZ–KELKER BIAS

Hanson R B (1979) A practical method to improve luminosity calibrations from trigonometric parallaxes. *MN-
RAS* **186** 875–896
— A more realistic way of treating the Lutz–Kelker bias.

Lutz T E and Kelker D H (1973) On the use of trigonometric parallaxes for the calibration of luminosity systems:
theory. *Pub. Astron. Soc. Pacific* **85** 573–578
— This paper describes the statistical basis of the Lutz–Kelker bias.

5

Mezzanine: the nature of stars

Just over a century ago, astronomy consisted of two disciplines. There was *astrometry* — the measurement of celestial angles, and in particular the angles between stars — and its theoretical counterpart, *celestial mechanics*. Both disciplines treated stars simply as points of light. But what *are* the stars? Are there different types of star, or are all stars like the Sun? What makes them shine? Do they contain the same chemical elements that we find here on Earth, or are they made of some strange 'star stuff'? Since the *nearest* stars are tens of trillions of miles away, do we stand any chance of learning something about their nature? The French positivist philosopher Isidore Auguste Marie François Xavier Comte (1798–1857) thought not. Comte believed there were absolute limits to mankind's scientific knowledge. In 1842, for instance, writing in his *Course of Positive Philosophy*, he claimed that 'Man will never encompass in their conceptions the whole of the stars'. Two years later he explained why it would forever be impossible for us to know the chemical composition of the stars:

> 'The stars are only accessible to us by a distant visual exploration. This in-
> evitable restriction therefore not only prevents us from speculating about life
> on all these great bodies, but also forbids the superior inorganic speculations
> relative to their chemical or even their physical natures.'

Comte was wrong. Soon after he made his pronouncement scientists developed a tool that revealed the chemical make-up of stars, the motion of stars through space, and much else besides. A third discipline emerged: *astrophysics*. (Astrophysics — the study of stars as physical objects rather than as mysterious points of light — was deemed to be of more interest than astrometry and celestial mechanics, and both of these subjects went into relative decline. With the success of *Hipparcos* the disciplines have regained some of their glamour, but most of the interest in the *Hipparcos* results stems from their implications for astrophysics.) No matter how precisely we measure the positions of stars, the fact remains that we can see innumerable stars whose parallaxes are too tiny to measure. They are simply too far away. If we want to step farther out we must abandon the direct geometric method of parallax and instead use indirect techniques based on astrophysics.

A description of the main tool of astrophysics, and what it tells us about stars, requires a detour. The detour is a long one, but at the end of it we will be in a position to take the next step on the distance ladder.

5.1   SPECTROSCOPY

While Newton was busy with his work on gravity and mechanics and calculus, he somehow found time to develop the science of optics. In a celebrated experiment he passed
a beam of white sunlight through a clear glass prism to make a continuous spectrum of
colours on a screen. By making a small hole in the screen he could create a beam of
coloured light. When he passed a beam of coloured light through a second prism, he
found that the beam spread out slightly but stayed the same colour. On the other hand,
when he passed light from the entire spectrum through a second prism, held upside down
with respect to the first, the spectrum turned back into a beam of white light.

The explanation for this behaviour is that white light is a mixture of many different
colours. Each colour corresponds to a different wavelength. Visible light that has a 'long'
wavelength (7×10^{-7} m, say, or 700 nm) appears red. Light that has a 'short' wavelength
(about 400 nm) appears violet. Light of intermediate wavelength appears as one of the
other colours of the spectrum. If many different wavelengths are present in a beam of
light, our eyes register the combined effect as white. When a beam of white light passes
through a prism, the prism bends, or *refracts*, each wavelength by a different amount. A
prism bends short wavelengths more than long wavelengths, and so a *continuous spectrum*
forms. See figure 5.1.

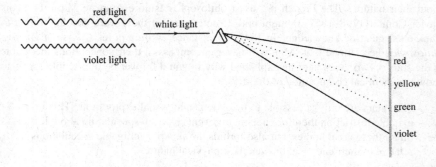

FIGURE 5.1: White light is a mixture of all colours. Each colour has a different wavelength: red light, for
instance, has a longer wavelength than violet light. A prism bends different wavelengths by different amounts.

In 1802, the English physicist William Hyde Wollaston (1766–1828) repeated Newton's experiment. He noticed that the continuous spectrum formed by sunlight was not
quite continuous. Seven dark lines crossed the spectrum. Unfortunately, he took no further steps to investigate his observation. (It is surprising that Newton, of all people, did
not notice these lines in the spectrum: in some of his experiments they should have been
visible. Perhaps imperfections in his prisms smeared out the lines. Or perhaps his assistant, who ran some of the experiments because Newton's own eyes were not keen enough,
saw the lines but did not bother to report them.)

In the early years of the nineteenth century, the best manufacturer of high-quality
prisms was Fraunhofer. In 1814, he decided to test the quality of his prisms by their ability

to form spectra. He pointed a telescope at the Sun and passed a beam of sunlight through a slit and then through a prism. The prism refracted the light onto a screen, each wavelength being refracted through its characteristic angle. Fraunhofer's apparatus thus formed an image of the slit in a particular colour at a particular position on the screen. The millions of images of the slit overlapped to form the familiar continuous rainbow-like spectrum. Except that Fraunhofer, like Wollaston before him, noticed that the spectrum was *not* completely continuous: there were dark lines in the spectrum. These *spectral lines* were missing wavelengths in sunlight. Unlike Wollaston, who saw only seven lines, Fraunhofer saw hundreds of lines, and eventually mapped the position and intensity of nearly 600 of them. (Modern astronomers have found tens of thousands of them.) Fraunhofer, though, could provide no explanation for these mysterious lines in the solar spectrum.

During the 1850s, the German chemist Robert Wilhelm Bunsen (1811–1899) developed his eponymous burner. The burner produced so little light of its own that he could use it to heat a vapour to incandescence and be sure that any light emitted was from the vapour alone. His colleagues used the Bunsen burner to confirm a curious fact: if the vapours of different chemical elements are heated to high temperatures, they emit light of different colours. For instance, hot sodium vapour emits a strong yellow light; mercury vapour emits a green light; potassium vapour emits a dim purple light. The question then arises: what happens if this light, rather than sunlight, is passed through a prism?

The first man to do the experiment was Bunsen's friend, the German physicist Gustav Robert Kirchhoff (1824–1887). Kirchhoff passed the light through a slit to form a tight beam and then through a prism to form a spectrum. He studied the spectrum with a telescope. (Nowadays we call such a device a *spectroscope*. In a modern observatory the spectroscope is almost as important as the telescope.) Kirchhoff found that, unlike the solar spectrum, which was almost continuous, only a few bright lines appeared on the screen. He called them *emission lines*. After further work he noticed that each element always produced emission lines of the same colour in the same place on the screen, and that no two elements produced lines in exactly the same place. Each element therefore had a unique 'fingerprint': its *emission spectrum*.

When Kirchhoff carried out the same experiment with light from a glowing solid — any solid — he obtained a continuous spectrum. Glowing solids, in other words, emit light at all wavelengths. Then, in 1859, he devised a crucial experiment. First, he heated a solid to form continuous-spectrum light. He then passed this light through sodium vapour, which he kept at a temperature lower than that of the solid. Finally, he passed the light through a spectroscope. When he examined the resulting spectrum he saw that it contained a dark line, just like the solar spectrum, at *precisely* the wavelength where sodium produced its unique emission line. The same effect occurred for the vapours of other elements. Kirchhoff announced that each element emits and absorbs light only at certain fixed wavelengths unique to itself. The *absorbtion lines* were, in some sense, just the reverse of the emission lines.

Kirchhoff then understood the origin of the missing wavelengths in the Sun's spectrum. He argued that the very hot interior of the Sun emits a continuous spectrum. As this light passes through the Sun's cooler outer layers, chemical elements in those layers absorb light at their characteristic wavelengths. He realised immediately that he had a powerful tool for detecting chemical elements in the laboratory. And by comparing the

Edward Charles Pickering

absorbtion lines in the solar spectrum with absorbtion lines found in his laboratory experiments, he could study the chemical composition of the Sun. Had Comte lived for just two more years he would have seen his pessimism about the limits of science proved spectacularly wrong.

Kirchhoff quickly showed that the Sun contains the same chemical elements that make up the Earth — though in very different proportions and under very different conditions, of course. By 1861, he had detected sodium, calcium, magnesium, chromium, nickel, barium, copper and zinc in the Sun. (Kirchhoff's banker was unimpressed with his client's ability to find elements in the Sun. 'Of what use is gold in the Sun if I cannot bring it down to Earth?' he is reputed to have asked. Some time later Kirchhoff's work earned him a medal and a prize in gold sovereigns from Britain. He gave the sovereigns to his banker and said 'Here is gold from the Sun'.)

Kirchhoff and Bunsen even discovered two entirely new chemical elements with their spectroscope. In 1861, they discovered *caesium* (named from the Latin word for 'sky blue', due to its brilliant blue spectral lines). The following year they discovered *rubidium* (named from the Latin word for 'red', since it has prominent red lines in its spectrum). The importance of the spectroscope was demonstrated even more impressively in 1878, when the English astronomer Joseph Norman Lockyer (1836–1920) suggested that some dark lines in the solar spectrum, which corresponded to no known substance, were the absorption lines of a hitherto unknown element. He called this element *helium*, after Helios, the Greek god of the Sun. It was not until 1895 that the Scottish chemist William Ramsay (1852–1916) discovered terrestrial helium.

The English astronomer William Huggins (1824–1910) was the first to appreciate the profound importance of the spectroscope in astronomy. He can be considered the founder of modern astrophysics. He took the spectrum of just about any astronomical object whose light he could pass through a telescope and then a prism. Huggins was also one of the first to use spectrography in astronomical research and, by using long time exposures,

William Huggins

he obtained spectra of objects that were far too faint for the naked eye to see. In 1863, he showed that the spectral lines in starlight were the same as the spectral lines found on Earth — he identified hydrogen, sodium, iron, magnesium and calcium. William Herschel had earlier proved that the motion of the Earth, the Sun and the stars is governed by the same law of gravitation. Huggins had now proved that the Earth, the Sun and the stars all contain the same few chemical elements.

5.2   SPECTRAL CLASSES

In 1867, the Italian astronomer Pietro Angelo Secchi (1818–1878) studied all the stellar spectra available to him — about 4000 in total — and concluded that they fall into four *spectral classes*. The first of Secchi's four classes contained blue stars and white stars, like Vega and Sirius. Spectra in this class all had prominent hydrogen lines and very weak metallic lines. (The word 'metallic' when used by astronomers refers to any element other than hydrogen and helium.) His second class contained yellow and orange–yellow stars, like Capella and the Sun. Spectra in this class had weaker hydrogen lines and more prominent metallic lines. His third class contained orange–red stars, like Betelgeuse and Antares. A typical spectrum in this class had few hydrogen lines but an abundance of metallic and molecular lines. Secchi's fourth class contained red stars, like R Cygni and Barnard's star. These spectra all had prominent carbon lines.

During the 1880s, the director of the Harvard College Observatory, the American astronomer Edward Charles Pickering (1846–1919), began a programme of photographing the spectra of thousands of stars. On the basis of his work he introduced a spectral classification scheme that was more refined than Secchi's. Pickering classified the spectra according to the strength of the hydrogen lines. He labelled his classes by letter, beginning with A (which corresponded to the first of Secchi's spectral classes) through to Q, omitting the letter J.

Henry Draper

Funding for much of Pickering's programme came from a bequest by Anna Palmer Draper, who wished to establish a memorial to her husband, the American astronomer Henry Draper (1837–1882). The bequest enabled Pickering to employ a team of assistants, led by Annie Jump Cannon (1863–1941), who herself analysed and classified the spectra of 225 300 stars! She published her work, between 1918–1924, as the nine-volume *Henry Draper Catalogue*. Initially, she followed Pickering and classified spectra alphabetically, A through Q, according to the strength of the hydrogen lines, but later rationalised her classification, throwing out some letters and adding others. Astronomers also realised that it made more sense to order the classes according to the surface temperature of the star rather than by hydrogen line strength. They kept the original letter designations, even though they were no longer in alphabetical order. Thus the Harvard classification scheme of stellar spectra, which is still widely used, follows a rather odd sequence*:

O B A F G K M (R N S).

Type O stars are blue. Type M stars are red, like the fourth of Secchi's categories. R, N, S stars are subgroups of the M stars. Each class is further divided into ten subclasses, distinguished by the numbers 0 to 9. A G1 star, for instance, is hotter than a G9 star.

Why do spectral classes exist at all? The answer is not that different stars have wildly different chemical compositions. The reason is that, as mentioned above, different stars have different surface temperatures. To understand how astronomers can measure the surface temperature of a star, and why different surface temperatures should give rise to the spectral classes, we have to understand in some detail the atomic processes that cause spectral lines.

---

*Unfortunately, despite the work of Cannon and others, astronomy has remained a male-dominated discipline. This explains the usual mnemonic for remembering the order of the spectral classes: 'Oh Be A Fine Girl, Kiss Me (Right Now Sweetheart)'.

Annie Jump Cannon

5.3   THE CAUSE OF SPECTRAL LINES

A *blackbody* is an object that absorbs all the radiation that falls onto it. In the real world no
object can be a perfect blackbody, since a real object always reflects some of the radiation
that hits it. But we can approximate a blackbody by putting a tiny hole in an otherwise
enclosed metal box. Suppose that some radiation enters the box through the hole. Once
inside, the radiation bounces around and some of it is absorbed each time it hits one of
the inner walls. The radiation hits the walls so many times before it finds the hole again
that any radiation entering via the hole is unlikely ever to leave the box. The hole is thus
a blackbody. See figure 5.2.

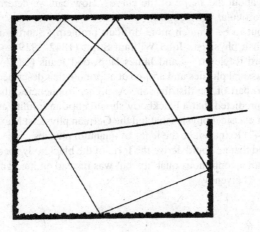

FIGURE 5.2: We can make a blackbody by punching a small hole in a metal box. Any radiation that enters the
hole is almost certain to be absorbed, and so the hole acts like a blackbody. If we heat the box the walls will
radiate; radiation leaving the box through the hole is thus blackbody radiation.

A perfect absorber of radiation is also a perfect emitter of radiation. If we heat the box, the metal walls will start to radiate. Some of that radiation will leave through the hole, and we can then study it. This is *blackbody radiation*. The radiation from many objects is, to a good approximation, blackbody radiation. The surface of a star, for instance, is a good blackbody.

Blackbody radiation is very simple: its features depend only upon the temperature, T. For instance, the total energy R emitted each second from a unit area of blackbody was shown by the Austrian physicist Josef Stefan (1835–1893) to be

$$R = \sigma T^4. \tag{5.1}$$

(The constant σ is the Stefan–Boltzmann constant, named after the Austrian physicist Ludwig Edward Boltzmann (1844–1906), who showed that Stefan's empirical relation follows directly from the theory of blackbodies. The Stefan–Boltzmann constant has the value $5.67 \times 10^{-8} \, \mathrm{W \, m^{-2} \, K^{-4}}$.) The distribution of radiation as a function of frequency also depends only upon the temperature. Figure 5.3 shows the characteristic blackbody spectrum at three different temperatures. Note how the peak of the distribution shifts to shorter wavelength (higher frequency) as the temperature increases. The German physicist Wilhelm Carl Werner Otto Fritz Franz Wien (1864–1928) showed that blackbody radiation peaks at a wavelength given by

$$\lambda_{max} T = \text{constant} \tag{5.2}$$

where the constant takes the value $2.898 \times 10^{-3} \, \mathrm{m \, K}$.

From (5.1) and (5.2) we can describe some of the features of the curves in figure 5.3: for any given temperature we can predict the position of the peak and the area under the curve. But what about the *shape* of the curve? How can we determine the blackbody radiation at any particular wavelength?

This turned out to be a much more difficult problem. Shortly after the turn of the century, the English physicists John William Strutt (1842–1919) — who is invariably referred to as Lord Rayleigh — and James Hopwood Jeans (1877–1946) analysed the problem using classical physics and arrived at a formula that described the low-frequency (long-wavelength) part of the distribution. At higher frequencies, though, their formula was hopeless: it predicted that a blackbody should emit an infinite amount of energy. It was this failure of classical physics that led the German physicist Max Karl Ernst Ludwig Planck (1858–1947) to introduce the ideas of quantum physics.

Planck showed that he could derive the form of the blackbody spectrum if he assumed that energy was not a continuous quantity, but was instead emitted and absorbed by photons with energy, E, given by

$$E = hf = \frac{hc}{\lambda} \tag{5.3}$$

where f is the frequency, λ the wavelength and h ($= 6.6262 \times 10^{-34} \, \mathrm{J \, s}$) is the Planck constant. The photon thus has particle-like properties (for instance, it has a definite discrete energy) and wave-like properties (for instance, it has a wavelength).

The energy density $\rho(\lambda)$, between wavelength λ and $\lambda + \mathrm{d}\lambda$, is represented by

$$\rho(\lambda)\,\mathrm{d}\lambda = \frac{8\pi hc}{\lambda^5}\,\frac{\mathrm{d}\lambda}{e^{hc/kT\lambda} - 1} \tag{5.4}$$

where k is the Boltzmann constant (which has the value $1.38 \times 10^{-23}\,\mathrm{J\,K^{-1}}$). Both (5.1) and (5.2) can be derived from (5.4). The Planck distribution is the fundamental equation of a blackbody.

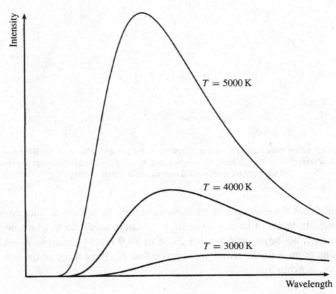

FIGURE 5.3: The higher the temperature, the more radiation a blackbody emits. As temperature increases the peak shifts to shorter wavelength (higher frequency). The shape of a blackbody spectrum always has this characteristic 'hump'.

One of the key advances of twentieth century physics was the realisation that quantum effects are not confined to the emission and absorbtion of photons in blackbody radiation. On the smallest distance scales the universe obeys the laws of quantum physics, not classical physics. We can understand the atom only if we use quantum ideas.

In the early years of this century, the Danish physicist Niels Hendrik David Bohr (1885–1962) developed a simplified but nevertheless quite useful model of the atom. The Bohr atom consists of one or more *electrons* orbiting a *nucleus*. Electrons are light, negatively charged particles. The nucleus, which contains most of the mass of an atom, consists of positively charged *protons* and electrically neutral *neutrons*. The number of protons in the nucleus determines the type of atom. If a nucleus contains one proton then it is the nucleus of a hydrogen atom; the nucleus of a helium atom always has two protons, that of a lithium atom has three, and so on through the various elements. Uranium is the most massive of the naturally occurring elements. The nucleus of a uranium atom contains 92 protons. Since atoms overall are electrically neutral, the positive charge of the nucleus must exactly balance the negative charge of the electrons. An atom of hydrogen, for instance, has one electron: the negative charge of the single electron exactly balances the positive charge of the single proton. An atom of helium contains two electrons, whose negative charges balance the positive charges of the two protons in the nucleus. An atom of lithium contains three electrons and, using the same argument, an atom of uranium contains 92 electrons. See figure 5.4.

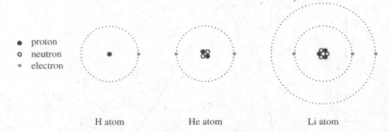

FIGURE 5.4: Atoms of the three simplest elements, hydrogen, helium and lithium, in the Bohr model. Most of the mass of each atom resides in the nucleus. The negative charge of the electrons balances the positive charge of the protons, so overall the atom is electrically neutral.

Bohr argued that the electrons in an atom exist only in certain distinct *energy states* or *energy levels*. In terms of basic energy units, an atom might have 1 unit or 3 units or 27 units of energy, but never, say, 0.62 or 29.76 or 81.9 units of energy. In other words, the energy of an atom is *quantised*. Bohr showed that E_n, the energy of the nth level of a hydrogen atom, is given by

$$E_n = -\frac{13.6\,\text{eV}}{n^2}. \tag{5.5}$$

(The unit eV in (5.5) is an electron volt. It is a convenient unit for expressing atomic energies. One electron volt is the energy acquired by an electron when accelerated through a potential difference of one volt. In terms of SI units, $1\,\text{eV} = 1.6 \times 10^{-19}\,\text{J}$.) When an electron in an atom moves from a state of high energy to a state of low energy the atom emits a photon with an energy that is *precisely* the difference in energy between the two states. So the frequency of the light emitted in a transition depends on the energy difference between the two states. For example, the familiar red laser beam, visible at any supermarket checkout, occurs when trillions of atoms make a transition between the same energy states. The energy difference between those states corresponds to a deep red colour — and *only* to that colour.

The transition of electrons from high-energy states to low-energy states is how emission lines form. See figure 5.5. In the same way, an atom can absorb light only of particular frequencies. These frequencies correspond to energies that are just right to raise an electron from one state to another. This is how absorbtion lines form.

An atom of hydrogen is the simplest possible atom: just one proton and one electron. Physicists can calculate the allowed energy states of the hydrogen atom, and it turns out

Example 5.1 Frequency of a photon

Calculate the frequency of the photon emitted when a hydrogen atom transits from the $n = 4$ to the $n = 1$ state.

Solution. From (5.5),

$$E_4 - E_1 = -13.6\,\text{eV}\left(\frac{1}{4^2} - \frac{1}{1^2}\right) = \frac{15 \times 13.6\,\text{eV}}{16} = 12.75\,\text{eV}.$$

From (5.3),

$$f = E/h = 12.75 \times 1.6 \times 10^{-19}\,\text{J}/6.6262 \times 10^{-34}\,\text{J s}$$
$$\approx 3 \times 10^{15}\,\text{Hz}.$$

This frequency, which corresponds to a wavelength of about 100 nm, is in the ultraviolet region.

that the spectral pattern is quite plain. Helium, the next simplest atom, has a spectral pattern that is much more complicated. Iron, which has 26 electrons, has thousands of prominent spectral lines in the visible part of its spectrum. The more electrons an atom contains, the more intricate is the pattern of allowed energy states and thus the more complicated the spectrum becomes. (The complexity of the Sun's spectrum, which baffled Fraunhofer, arises chiefly from the presence of iron in its outer layers.)

But complex atoms are not the only objects that produce complex spectra. Suppose energy is added to a sample of hydrogen gas by heating it. The atoms in the gas share out the energy between them. When a particular atom absorbs its share of the available energy, the electron in that atom jumps to a higher energy state. As the temperature continues to rise, the electron occupies states of higher and higher energy. Eventually, so much energy is imparted to the electron that it leaves its atom. The same thing happens with other elements as well as hydrogen. If enough energy is added electrons can be stripped away from atoms. When an electron leaves its atom, so that the atom has less than its normal complement of electrons, the atom is *ionised**. The allowed energy states of an atom with one electron missing are very different to the energy states of the non-ionised atom, so ionised and non-ionised atoms have different spectral lines. An ionised atom with two electrons missing has a completely new set of allowed states, and thus produces a still different set of spectral lines. An ionised atom with three electrons missing produces yet another set of spectral lines, and so on. Table 5.1 shows the energy in eV needed to liberate the first electron (single ionisation) and the second electron (double ionisation) of some common atoms.

Atoms of different elements hold on to their electrons with different strengths, so the

---

*An ionised hydrogen atom is still hydrogen, and an ionised calcium atom is still calcium. It is the number of protons in the nucleus that defines the type of atom, and this number does not change during the process of ionisation.

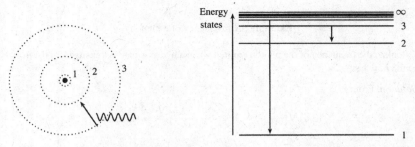

FIGURE 5.5: Energy states in a hydrogen atom. The ground state — labelled 1 — is the state of lowest energy. The left-hand diagram represents an atomic electron jumping from the second excited state (labelled 3) to the first excited state (labelled 2). As it does so, the atom emits a photon. The energy of the photon is exactly equal to the difference in energy between the two states. The right-hand diagram represents the different energy states of a hydrogen atom. If an electron jumps from a highly excited state to the ground state (from 4 to 1, say) the atom emits an energetic photon. In fact, the photon is in the ultraviolet range of the spectrum. If an electron jumps from one excited state to another (from 3 to 2, say) the emitted photon has less energy. In all cases, though, the photon can only have certain well-defined energies. This is why spectral lines form.

TABLE 5.1: The single and double ionisation energies of some common atoms.

Atom	Single ionisation energy (eV)	Double ionisation energy (eV)
H	13.6	—
He	24.6	54.4
C	11.3	24.4
N	14.5	29.6
O	13.6	35.1
Na	5.1	47.3
K	4.3	31.8
Ca	6.1	11.9
Fe	7.9	16.2

energy needed to ionise an atom varies from element to element. A temperature that can ionise a sodium atom, for instance, will not ionise an oxygen atom. Furthermore, it always requires more energy, and thus a higher temperature, to free a second electron from an atom than was required to free the first. To liberate a third electron requires a still higher temperature, and so on.

We can now understand the origin of spectral classes. They have less to do with the chemical composition of stars than with the surface temperature of stars. Stars at different temperatures have their chemical elements in different ionisation states and therefore produce different spectra.

Stars in spectral class O are the hottest stars. They have a surface temperature in the range 20 000–35 000 K. This is so hot that their spectra show lines from multiply ionised atoms; for example, they show lines from oxygen with three electrons missing. The Sun is an average star. It has a surface temperature of about 5700 K, and is therefore of spectral class G2. Stars in spectral class M are relatively cold: they have a surface temperature of around 3000 K.

Spectroscopy thus yielded a second piece of information about stars. Not only did it reveal the chemical composition of stars, it also revealed their surface temperature.

5.4   THE HERTZSPRUNG–RUSSELL DIAGRAM

Boxing correspondents always give two facts about a boxer: his weight and his height. Police try to give the same information when they issue a description of a suspect. Weight and height are important because we often use the two properties to classify an individual within a group.

Now, suppose a graph of weight against height is produced for a random sample of 100 adults, with each person in the sample represented by a point on the graph. The result would be something like the graph shown in figure 5.6. Most points lie within a narrow band — a band that we might call the 'main sequence' of physical properties of humans. The 'main sequence' in figure 5.6 is similar to the sequence that would be produced from any random sample of people in your town or village. It simply means that there is a connection between height and weight in the general population: the taller the person, the heavier that person is likely to be. This is not a rule that applies to individuals. An individual can be short and fat, or tall and skinny. But the connection holds true for people *on average*. In the early years of this century, the American astronomer Henry Norris Russell (1877–1957) and the Danish astronomer Ejnar Hertzsprung (1873–1967) wondered if there were any properties, analogous to height and weight in humans, that they could use to classify stars.

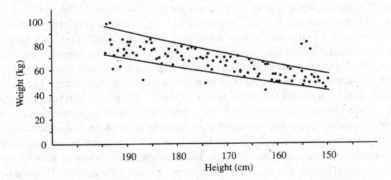

FIGURE 5.6: A graph of weight against height for a random sample of 100 adults. Following the perversity of astronomers, I plot height as increasing from right to left.

Astronomers at the turn of the century were not exactly inundated with information regarding the stars, but they did have information on two important properties. First, they knew the surface temperature of a star, determined from its spectral type. Second, they

Henry Norris Russell

could calculate the absolute magnitude of those stars for which they had a parallax-based distance. (Alternatively, they could study clusters, which contain stars that are essentially all at the same distance. In this case there is no distance effect to cloud the issue, and apparent magnitude is equivalent to absolute magnitude — at least up to some overall constant.) It is hard to think of a third property that is as important as the temperature and intrinsic brightness of a star. So what happens if a graph is produced of absolute magnitude against spectral type or temperature for a random selection of stars?

When Hertzsprung and Russell, independently, made such a graph they had no idea of what they might find, though they hoped that the graph might provide a clue as to the nature of stars. In this they succeeded. Figure 5.7 shows the Hertzsprung–Russell (HR) diagram for the 60 nearest stars and the 20 brightest stars. Each dot on the diagram represents a star of a particular spectral type and a particular luminosity. Figure 5.8 is more impressive. It shows the HR diagram of the 18 860 single *Hipparcos* stars with relative parallax errors of less than 10%. (It is common to plot magnitude against colour index rather than spectral type; figure 5.8 is an example of a *colour–magnitude diagram*.) As can be seen, the stars occupy only certain parts of an HR diagram: not all combinations of temperature and luminosity are allowed. Furthermore, the majority of stars lie along a *main sequence*. There is a connection, on average, between the intrinsic brightness of a star and its surface temperature. In general, the hotter the star, the brighter it is.

Some stars lie off the main sequence. These are strange objects, the like of which we never see on a graph of weight against height for people. Three-ton dwarfs and two-ounce giants just do not exist. We will return to these unusual stars later. For the moment it is more important to ask: why does the main sequence exist? How does one main sequence star differ from another? What property of stars ensures that dim, cool, red stars are at one end of the main sequence, while bright, hot, blue stars are at the other end?

Ejnar Hertzsprung

The answer is that stars at different positions on the main sequence differ primarily in their mass. The bright stars at the top of the main sequence are extremely massive. The faint stars at the bottom of the main sequence have very little mass. (The term 'very little' is relative, of course; these stars are about 30 000 times more massive than the Earth.) The main sequence, then, is a progression of stars arranged in order of increasing mass. The mass of a main sequence star determines its temperature, its brightness and its position on the HR diagram. (The main sequence appears as a band rather than as a thin line, because as a star ages its chemical composition changes. This change can affect the luminosity and spectral class of the star, and thus its position on the main sequence.)

The main sequence is important because it includes most of the stars that we see. If we can understand the main sequence stars, we understand about 80% of the stars in the sky. As we shall see in the next chapter, results from *Hipparcos* have increased our store of knowledge about the HR diagram — but also raised fresh puzzles.

5.5   THE BIRTH, LIFE AND DEATH OF A MAIN SEQUENCE STAR

We now come to the most important question that we can ask about main sequence stars: what makes them shine? The answer came not from astronomers, but from nuclear physicists. In 1938, the German–American physicist Hans Albrecht Bethe (1906–) was the first to calculate the details of the fusion mechanism that powers stars, though our modern understanding of stellar energy sources came about in the 1950s from the work of the English physicist Fred Hoyle (1915–) and his colleagues.

5.5.1   Birth

A star like the Sun begins its life as a *protostar* — a spherical cloud of cold hydrogen gas that is perhaps 200 AU in diameter. Gravity causes the cloud to contract, and as its density increases, the temperature also increases. When the cloud is about 0.5 AU in diameter the temperature at the centre is 10^5 K — hot enough to completely ionise the gas. It is no longer a cloud of hydrogen gas: it is a mixture of electrons and nuclei.

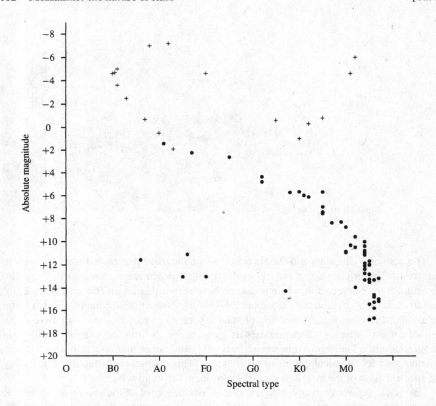

FIGURE 5.7: The HR diagram for the stars within 5 pc (black dots) and the 20 brightest stars (crosses). The absolute magnitude is plotted against the spectral type. (Surface temperature increases from right to left.)

The contraction continues, and this makes the cloud even hotter: the particles inside the cloud move faster and faster. Sometimes two protons might smash into each other at high speeds. When this happens the two protons simply bounce off each other, because the energy of electrical repulsion is greater than their kinetic energy. And then, about 20 million years after the contraction starts, a critical event occurs. The temperature at the centre of the protostar reaches 10^7 K. At this temperature some protons near the centre of the protostar have enough energy to overcome the barrier of electrical repulsion. If these protons collide they stick together. They take part in a *nuclear fusion reaction*.

Once fusion starts, the protostar is well on the way to becoming a main sequence star. Five billion years ago a process similar to this formed the Sun.

5.5.2   Life

Figure 5.9 shows what happens when two protons collide at the high temperatures found near the centre of a protostar. Before the collision there are two separate hydrogen nuclei.

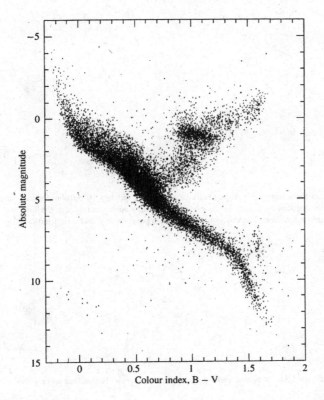

FIGURE 5.8: The HR diagram for *Hipparcos* stars with relative parallax errors of less than 10%.

After the collision there is just one heavier deuteron* nucleus. This particular fusion reaction therefore turns one of the protons into a neutron. In doing so it creates two new subatomic particles, called a *positron* (e^+) and a *neutrino* (ν).

We can represent the reaction by the equation

$$p + p \longrightarrow {}^2H + e^+ + \nu. \tag{5.6}$$

Even at high temperatures the chance of two protons fusing to form a deuteron is tiny, but once a deuteron nucleus has formed, it quickly snaps up another proton. The proton and deuteron fuse to form a three-particle nucleus containing two protons and a neutron:

$$p + {}^2H \longrightarrow {}^3He + \gamma. \tag{5.7}$$

Figure 5.10 illustrates this fusion reaction. Since this nucleus has two protons it is a type of helium, and it has the symbol ${}^3He$. Hydrogen thus turns into helium. The transmutation

---

*The deuteron nucleus contains one neutron and one proton. The number of protons determines the element, so the deuteron nucleus is still a type of hydrogen nucleus. The symbol for the deuteron is ${}^2H$, which indicates that it is a form of hydrogen.

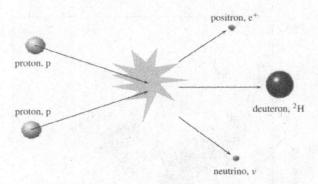

FIGURE 5.9: When the temperature at the centre of a protostar reaches 10 million degrees, fusion reactions begin. For the protostar that eventually became the Sun, the first such reaction was the fusion of two protons into a deuteron. The reaction also creates a positron (e^+) and a neutrino (ν).

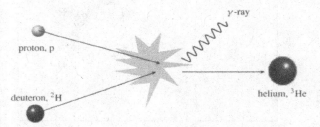

FIGURE 5.10: A proton and a deuteron fuse to form a ^{3}He nucleus and a γ-ray.

of elements, that age-old dream of alchemists, takes place in the centre of stars. Note the release of energy in the form of a γ-ray.

Suppose that reactions (5.6) and (5.7) both happen twice. Two light helium nuclei are now free to collide, and if they do so they fuse. Figure 5.11 shows this reaction. This time they fuse to form an 'ordinary' helium nucleus — a nucleus with two protons and two neutrons, denoted by ^{4}He — with two protons as a by-product:

$$^3\text{He} + {}^3\text{He} \longrightarrow {}^4\text{He} + \text{p} + \text{p}. \tag{5.8}$$

The two protons are free to collide with other protons, and start the process all over again.

The net result of this set of fusion reactions is that four protons turn into one helium nucleus. As a by-product, the reactions also form two positrons, two neutrinos and two γ-rays. Symbolically, we can represent what happens as follows:

$$4\text{p} \longrightarrow {}^4\text{He} + 2e^+ + 2\nu + 2\gamma. \tag{5.9}$$

The neutrinos escape immediately and begin a journey through space that, for most of them, will last for eternity. The two positrons, on the other hand, soon meet two electrons. The positrons and electrons annihilate and form more γ-rays.

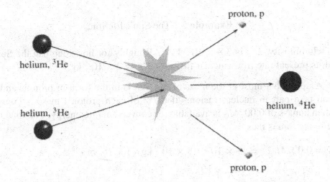

FIGURE 5.11: Two ^{3}He nuclei fuse to form a ^{4}He nucleus and two protons.

The key point to note is that the mass of four separate protons is greater than the mass of a helium nucleus. The mass of the particles on the right-hand side of the above equation is about 0.7% less than the mass of the four protons on the left-hand side. The mass does not simply disappear. It turns into energy in accord with the famous equation $E = mc^2$, which the German physicist Albert Einstein (1879–1955) obtained when he developed his special theory of relativity. The symbol c here represents the speed of light. Since this is a large quantity, the square of the speed of light is very large indeed. A small mass loss in a nuclear fusion reaction creates a vast amount of energy.

The energy release from the onset of nuclear fusion is enough to halt the gravitational collapse of a star. It is a balancing act. Gravity still tries to crush the star while nuclear reactions try to blow the star apart. About 17 million years after nuclear fusion begins, and about 50 million years after the first collapse of the protostar, the star arrives at its resting place on the main sequence. As we have seen, high-mass stars live at the top of the sequence, and low-mass stars at the bottom. But all stars on the main sequence have one thing in common: they 'burn' hydrogen. (The reaction chain described above is the principal set of reactions taking place in the Sun, and all main sequence stars less massive than the Sun. For main sequence stars more massive than the Sun, a different set of fusion reactions takes place. Although the details are different, the principle is the same: light elements turn into heavy elements and in the process some mass turns into energy.)

Every second that it stays on the main sequence, a star like the Sun 'burns' about 4.5 million tons of hydrogen. This seems like a lot of mass to lose but, as the worked example shows, the Sun contains so much hydrogen that it will continue to shine steadily on the main sequence for several billion years to come.

A star less massive than the Sun stays longer on the main sequence. Although such a star has less material to burn, the temperatures at its centre are not so high. Therefore, protons collide less violently and overcome their mutual electrical repulsion less often. In short, the nuclear reaction rate decreases. The star shines, but not brightly. A dim star will still be there, doling out its energy in miserly fashion, long after the Sun has died.

A star more massive than the Sun spends less time on the main sequence. This may

Example 5.2 The Sun's lifetime

The Sun's luminosity, $L_\odot$, is 3.8×10^{26} J s^{-1}. Roughly for how long can the Sun produce energy at its present rate from nuclear fusion? ($M_\odot = 2 \times 10^{30}$ kg.)

Solution. Assume that most of the mass of the Sun is in the form of protons and that each proton can take part in nuclear fusion. If 0.7% of each proton's mass is converted into energy, then a mass of $0.007M_\odot$ is available for conversion into energy. The total energy, E, available to the Sun is thus

$$E = 0.007 M_\odot c^2 = (7 \times 10^{-3})(2 \times 10^{30} \text{ kg})(3 \times 10^8 \text{ m s}^{-1})^2$$
$$= 1.26 \times 10^{45} \text{ J}.$$

Divide this energy by the luminosity $L_\odot$ to derive the lifetime t:

$$t = E/L_\odot = \frac{1.26 \times 10^{45} \text{ J}}{3.8 \times 10^{26} \text{ J s}^{-1}} = 3.3 \times 10^{18} \text{ s} \approx 10^{11} \text{ yr}.$$

Probably only about 10% of the Sun's mass is available to take part in nuclear reactions, so a better estimate of the lifetime of the Sun would be 10^{10} years.

seem paradoxical, since it has more nuclear fuel to burn, but massive stars are profligate. The increased mass generates higher central temperatures, so protons collide more violently and fusion is more likely to take place. The reaction rate is very sensitive to temperature: a doubling of the temperature increases the reaction rate by a factor of 30 000. A massive star has more mass to burn, but it burns that mass at a furious rate. It is like a motorist who, realising that fuel is low, steps on the gas to try to reach the next garage. This is unwise, since fuel consumption increases with speed. The motorist is better advised to drive slowly, as then the car is more likely to reach the garage. A bright high-mass star at the top of the main sequence can stay there for less than a million years. This is less than an eyeblink in cosmic terms. A high-mass star is the stellar equivalent of a mayfly that is born, lives and dies on just one afternoon.

Table 5.2 shows the typical surface temperature, mass, luminosity, radius and lifetime on the main sequence for stars of different spectral types.

5.5.3   Death

What happens to a star when it has used up its supply of hydrogen? It starts to move off the main sequence, of course, since main sequence stars are those that 'burn' hydrogen. The details, though, depend upon the initial mass of the star. In the case of a very low-mass star the question is rather academic, since its lifetime on the main sequence is much longer than the present age of the universe. You might say that the question is just as irrelevant in the case of the Sun, since none of us will be around when it moves off the main sequence, some 5–8 billion years from now. On the other hand, the question has some interest because the Sun is *our* star. Furthermore, a distant solar-mass star that

TABLE 5.2: The typical surface temperature, mass, luminosity and radius of stars of different spectral types. Also shown is their lifetime on the main sequence.

Spectral class	Surface temperature (K)	Mass (Sun = 1)	Luminosity (Sun = 1)	Radius (Sun = 1)	Lifetime on main sequence (billions of years)
O5	45 000	60	800 000	12	0.001 or less
B5	15 400	6	830	4	0.07
A5	8100	2	40	1.7	0.5
F5	6500	1.3	17	1.3	0.8
G2	5780	1	1	1	10
G5	5200	0.92	0.79	0.92	12
K5	4600	0.67	0.15	0.72	45
M5	3200	0.21	0.011	0.27	200

formed many billion years ago may be moving off the main sequence at this minute. We have the chance to observe the process.

So what will happen to the Sun? As the Sun 'burns' its hydrogen the helium 'ash' left behind will accumulate at the centre, where most of the fusion reactions take place. At first, the temperature in this helium core will be too low to fuse helium into heavier elements. If there are no fusion reactions there can be no energy release, and if there is no energy release there is no counterbalance to the ever-present force of gravity. The core will contract under its own weight, and so it will become hotter — in the same way that the protostar itself became hotter as it contracted to form the Sun. This tiny helium core will heat the surrounding shell of hydrogen so much that the hydrogen will begin to burn. The energy output from this shell will be enormous, and it will cause two things to happen. First, the Sun will balloon in size: its diameter will increase so that it will engulf Mercury, and possibly even Venus; Earth will become uninhabitable. Second, the Sun will become extremely bright — perhaps 100 times brighter than it is at present. The Sun's outer layers will be quite cool, and will consist of a very tenuous envelope of hydrogen; every square mile of the Sun's surface will radiate much less light than it does now. The Sun will be bright and yet have a low surface temperature simply because there will be *so much* surface area when it reaches this giant phase.

A cool surface means that the Sun will be of spectral class M: it will appear red. In other words, it will be a *red giant*. Stars of spectral class M occupy the rightmost part of the HR diagram, while luminous stars occupy the topmost part. Red giants, therefore, inhabit the top–right part of the HR diagram. Figures 5.7 and 5.8 show that some stars do indeed occupy the top–right corner of the HR diagram. In that region we find bright stars such as Antares and Betelgeuse. (The diameter of Betelgeuse is nearly 350 times that of the present-day Sun, so it has $350^2 = 120000$ times more surface area. So although a unit surface area of Betelgeuse is much dimmer than the same area of the Sun's surface, the vast size of Betelgeuse makes it about 12 000 times more luminous than the Sun. Betelgeuse is by no means the largest of these stars: ϵ Aurigae, for instance, has a diameter 2000 times that of the Sun.)

The mass of the helium core will continue to increase, causing the core temperature to rise. When the temperature reaches 10^8 K, helium will start to 'burn', and helium nuclei will fuse to form carbon. When this happens the Sun is close to the end of its energy-producing life. The temperature in the Sun's core can never become high enough for the fusion of carbon nuclei to take place, so it has no more fuel to burn. At the end of its red giant phase, radiation pressure from the core will puff the outer layers of the Sun into space, and a beautiful *planetary nebula* will form. The hot core it leaves behind will be tiny, so although it will glow white hot its total luminosity will be low. It will be a *white dwarf*. White dwarfs, since they are hot, are of spectral type B or A, and are thus on the left part of the HR diagram. White dwarfs are dim, and therefore on the bottom part of the diagram. A glance at figures 5.7 and 5.8 shows that some stars do indeed occupy the bottom–left corner of the HR diagram.

The nearest white dwarf is Sirius B, a companion star to Sirius. In 1844, Bessel postulated the existence of an unseen companion in order to explain irregularities in the proper motion of Sirius. The American optician Alvan Graham Clark (1832–1897) detected the companion in 1862. Sirius B has a surface temperature of almost 10^5 K, so it is much hotter than the Sun, but it is so tiny (about the same size as the planet Uranus) that the glare of Sirius drowns out its brilliant white light.

A white dwarf is very dense. A sugarcube-sized piece of material from Sirius B would weigh more than a family saloon car if brought to the Earth's surface. Gravity thus crushes down on a white dwarf with tremendous force. What stops the white dwarf from contracting to even higher densities? The Indian-born astronomer Subrahmanyan Chandrasekhar (1910–1995) pondered this question in July 1930, while on a voyage from India to England where he was to study for his PhD. The young Chandra reasoned that *electron degeneracy pressure* (a quantum-mechanical repulsion between electrons that occurs when they get very close to one another) stops the star from contracting even further. However, degenerate electrons can only support the white dwarf if the star is less than 1.4 times the mass of the Sun. A mass of $1.4 M_\odot$ marks the *Chandrasekhar limit*. No white dwarf can exist with a mass greater than $1.4 M_\odot$.

During its lifetime a star can shed a lot of mass. It is the fate of a main sequence star less massive than about $3 M_\odot$ to lose enough mass to become a white dwarf, which over aeons will cool to become an almost invisible black dwarf. But what is the fate of a more massive star, one that cannot shed enough mass to reach the Chandrasekhar limit?

A high-mass star ends its life in a supernova explosion. As we shall see on page 227, when we discuss supernovae in more detail, these violent death throes blow much of the star's mass off into space. The fate of the mass left behind is interesting. If the stellar remnant has a mass greater than $1.4 M_\odot$, electron degeneracy pressure cannot withstand the force of gravity. In such extreme conditions electrons are forced into the atomic nucleus, where they combine with the positively charged protons to become neutrons. The whole star then consists of neutrons. It is a *neutron star*. Neutron stars are unimaginably dense objects. If all the matter of the Earth were converted into the material of a neutron star it would fit into a large football stadium. If the Sun were similarly converted it would have a radius no larger than an average town.

Neutron stars have other unusual properties besides their high density. For example, they rotate very quickly. Angular momentum is always conserved, so when the radius of a rotating star decreases its rate of rotation increases. A main sequence star might take 30 days to rotate on its axis; as a tiny neutron star it completes one revolution in a few seconds. In addition, intense magnetic fields surround a neutron star. As the star rotates, dragging the magnetic fields with it, those fields accelerate charged particles to almost the speed of light. These accelerated particles produce a periodic beam of electromagnetic radiation, rather like a lighthouse. Such a spinning neutron star is a *pulsar* — a pulsating star. In 1967, the English radio astronomers Susan Jocelyn Bell (1943–) and Anthony Hewish (1924–) discovered the first pulsar. It turns out to be a remarkably regular timekeeper, emitting radio 'beeps' with a period of exactly 1.337 301 09 s.

In a neutron star it is neutron degeneracy pressure that keeps gravity at bay. The nuclear force is much stronger than the electromagnetic force, which is why a neutron star can be so much more dense than a white dwarf. But just as there is a limiting mass for white dwarfs, so is there a limiting mass for neutron stars. The value of the maximum mass of a neutron star is not accurately known, but the limit is probably around $3M_\odot$. And for stars of type O, no matter how violent the explosions at the end of their life, it seems that the remnant will be more massive than $3.2M_\odot$. In this case even the strong nuclear force cannot hold out against gravity, which finally wins, and the process of contraction continues. At this point no known force can halt the contraction. At a certain radius, called the *Schwarzschild radius* after the German mathematician Karl Schwarzschild (1873– 1916), the escape velocity of the body becomes equal to the speed of light. Beyond this radius nothing, not even light, can escape. The star is now a *black hole*. The name is appropriate. Objects can fall into the hole but nothing can escape. Gravity squeezes the mass out of existence.

This ends our detour. We have seen how astrophysicists have provided us with some understanding of the nature of stars. We no longer have to regard them as featureless points of light — we know something of how they form, why they shine, and what happens to them when they die. We will need to flesh out some of these ideas in later rungs of the distance ladder. We have enough information now, though, to take the next step — and calculate the distance to stars for which we cannot detect a trigonometric parallax.

CHAPTER SUMMARY

- Every chemical element has a 'fingerprint' — its set of spectral lines.

- We may classify stars by means of their spectral lines. The Harvard spectral classification scheme places the stars into the following groups: O B A F G K M (R N S). Type O stars are blue; type M stars are red.

- The spectral classes arise because of the different surface temperatures of stars, which cause atoms to be in different ionisation states. Type O stars are hot; type M stars are cool. We can determine the surface temperature of a star from its blackbody spectrum.

- Spectral lines exist because of the existence of discrete energy levels in an atom. When an electron jumps from one energy state to another, the atom absorbs or emits a photon that carries precisely the energy difference between states. If that energy difference is E, the photon has a frequency f given by $f = E/h$.

- The HR diagram is a plot of stellar luminosity against spectral type. Most stars lie along a main sequence, which represents the hydrogen-burning phase of a star's life. Other areas of the HR diagram are populated by red giants and white dwarfs.

- A solar-mass star generates energy through the proton–proton chain reaction, i.e.

$$4p \longrightarrow {}^4\text{He} + 2e^+ + 2\nu + 2\gamma.$$

QUESTIONS AND PROBLEMS

5.1 In most circumstances, the faintest stars that we can see with the naked eye have a magnitude of +6. Plot a graph of the maximum distance that we can see a main sequence star against its spectral type.

5.2 Write an account of the nuclear reactions that take place in stars more massive than the Sun.

5.3 What happens to the γ-rays that are emitted in the proton–proton chain in the Sun?

5.4 What star colours can you see with your naked eye? Estimate the surface temperature of these stars.

5.5 You observe two main sequence stars, one blue and one red, to have the same apparent magnitude. Which one is closer to Earth?   The red star

FURTHER READING

GENERAL

Asimov I (1977) *The Collapsing Universe* (Hutchinson: London)
   — Asimov was the greatest of all science popularisers. This book contains an account of stellar gravitational collapse, and the formation of white dwarfs, neutron stars and black holes, in his uniquely lucid style.

Begelman M C and Rees M J (1997) *Gravity's Fatal Attraction* (Freeman: New York)
   — One in the *Scientific American Library* series. A beautifully illustrated explanation of black holes.

Kaler J B (1997) *Cosmic Clouds* (Freeman: New York)
   — The birth, death and recycling of stars in the Galaxy. Another of the *Scientific American Library* series.

ASTROPHYSICS TEXTBOOKS

Aller L H (1991) *Atoms, Stars and Nebulae* (Cambridge University Press: Cambridge)
   — A nice introduction to modern astrophysics, using little mathematics.

Cohen M (1988) *In Darkness Born* (Cambridge University Press: Cambridge)
   — A non-mathematical account of the birth of stars.

Harwitt M (1988) *Astrophysical Concepts* (Berlin: Springer)
   — A thorough coverage of all aspects of astrophysics, aimed at the beginning graduate student.

Jaschek C and Jaschek M (1987) *The Classification of Stars* (Cambridge University Press: Cambridge)
   — This text covers all aspects of stellar taxonomy.

Kaler J B (1989) *Stars and their Spectra* (Cambridge University Press: Cambridge)
   — A thorough account of stellar spectra, pitched at an accessible level.

Kitchin C R (1995) *Optical Astronomical Spectroscopy* (Institute of Physics: Bristol)
   — A modern account of spectroscopy for undergraduate students.

Rolfs C E and Rodney W S (1988) *Cauldrons in the Cosmos* (University of Chicago Press: Chicago)
   — A reference book on nuclear astrophysics. (Warning: this is for the very advanced student.)

Tayler R J (1994) *The Stars: Their Evolution and Structure* (Cambridge University Press: Cambridge)
   — An excellent undergraduate text on the physics of stars.

6

Fourth step: more distant stars

We can use four new techniques to develop this step of the distance ladder and move beyond the limits of trigonometric parallax. First, we have the methods of secular and statistical parallax. Second, we have the moving cluster method. Third, we have the method of main sequence fitting. Fourth, we have the method of spectroscopic parallax. The last two methods are purely spectroscopic in nature. The first two are more complex, and require a mix of spectroscopic data and precise astrometric observations. But before considering these techniques, we first need to understand how astronomers measure the velocity of a star.

6.1 THE DOPPLER EFFECT

In 1868, Huggins took the spectrum of Sirius and found that, although the expected spectral lines were there, they were all shifted by the same small amount toward the red end of the spectrum. Huggins argued that this shift in the spectral lines was an example of the *Doppler effect.*

People had first become aware of the Doppler effect several years earlier, with the rise of the railways. Observers on a station platform noticed that the pitch of a train whistle changed when the train sped by. The whistle seemed to have a higher pitch when the train approached than when it was stationary. As the train receded, the whistle had a lower pitch. Train whistles are less common nowadays, but one often hears the same effect with the sirens of emergency vehicles. As a police car races past us, the pitch of its siren drops quite noticeably.

In 1842, the Austrian physicist Christian Johann Doppler (1803–53) explained the effect. According to Doppler, the whistle sent out a constant number of sound waves each second. In other words, it had a constant pitch, or frequency. As a train approached, each successive wave had less far to travel to reach an observer. The waves 'piled up' on each other and so the observed frequency increased. As a train receded, each successive wave had farther to travel to reach an observer. The waves 'spread out' and so the observed frequency decreased. Figure 6.1 illustrates the principle.

Doppler examined the problem mathematically. Suppose a source moves directly away from an observer with velocity v_s and emits sound waves of frequency f_s. The speed of sound in air is v, and the observer is stationary with respect to the air. The

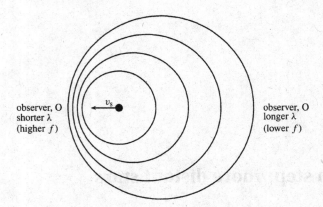

FIGURE 6.1: The source emits waves with a constant frequency but, because the source is moving, an observer on the left sees a shorter wavelength (i.e. higher frequency). An observer on the right, from whom the source recedes, sees a longer wavelength (i.e. lower frequency).

distance between successive wave crests increases from

$$\lambda_s = \frac{v}{f_s}$$

to

$$\lambda_{obs} = \frac{v + v_s}{f_s}.$$

The observed frequency is given by $f_{obs} = v/\lambda_{obs}$, i.e.

$$f_{obs} = f_s\left(\frac{v}{v + v_s}\right) = \frac{f_s}{1 + (v_s/v)}. \tag{6.1}$$

If the source moves towards the observer, v_s is negative in (6.1). The difference, $\Delta\lambda$, between the observed wavelength and the source wavelength is called the *Doppler shift*. It is positive ($\lambda_{obs} > \lambda_s$) if the source recedes from the observer, and negative ($\lambda_{obs} < \lambda_s$) if the source approaches the observer. One usually expresses the shift in terms of its fraction, z, of the source wavelength:

$$z = \frac{\Delta\lambda}{\lambda_s} = \frac{\lambda_{obs} - \lambda_s}{\lambda_s} = \frac{1}{\lambda_s}\left(\frac{v}{f_s} + \frac{v_s}{f_s} - \frac{v}{f_s}\right) = \frac{v_s}{\lambda_s f_s}$$

i.e.

$$z = \frac{\Delta\lambda}{\lambda_s} = \frac{v_s}{v}. \tag{6.2}$$

Using a similar argument we can derive a formula for the Doppler shift when the observer moves directly towards or away from a source that is stationary relative to the air. In the general one-dimensional case we can suppose that both source and observer move along the same line, with speeds relative to the air of v_s and v_{obs}, respectively. The source emits sound with frequency f_s and the speed of sound in air is v. Then the observed frequency is

$$f_{obs} = f_s \left(\frac{1 - (v_{obs}/v)}{1 - (v_s/v)} \right). \tag{6.3}$$

A little algebra shows that (6.3) contains (6.1) as a special case. We can obtain the special case of a stationary source and a moving observer just as easily.

The Dutch scientist Christoph Buys-Ballot (1817–1890) put Doppler's mathematics to the test in a rather unconventional way. Buys-Ballot persuaded some musicians to stand in the open carriage of a train, and had them play notes on wind instruments. Musicians with perfect pitch stood on a platform and recorded the frequency of each note as the train sped by. The description of the musicians agreed perfectly with Doppler's formula.

The Doppler effect occurs with all types of wave — electromagnetic as well as acoustic. (Indeed, Doppler first mentioned the effect in terms of light waves. He incorrectly attributed the colours of stars to their radial motion relative to Earth.) The derivation given above does not apply to electromagnetic radiation, though, because the underlying physics is very different. Unlike sound waves, which require a medium like air through which to propagate, electromagnetic waves can propagate through a vacuum. When electromagnetism is involved, there is no way to distinguish between a source that moves relative to a stationary observer and an observer that moves relative to a stationary source. It makes sense only to talk of the relative motion of source and observer. The Doppler effect must therefore be modified when the speeds involved approach that of light. Any textbook on special relativity will include the derivation of the Doppler effect for light. Below, I simply state the results.

Suppose that a source of electromagnetic radiation emits waves of frequency f_s. Furthermore, suppose that the observer and the source move away from each other with relative velocity v; the term β is this velocity expressed as a fraction of the speed of light, i.e. $\beta = v/c$. Then the observed frequency is given by

$$f_{obs} = f_s \sqrt{\frac{1 - \beta}{1 + \beta}}. \tag{6.4}$$

In terms of wavelength, we have

$$\lambda_{obs} = \lambda_s \sqrt{\frac{1 + \beta}{1 - \beta}}. \tag{6.5}$$

If the observer and the source move toward each other, simply interchange the signs in the numerator and denominator of the radical in (6.4) and (6.5). These equations have a pleasing symmetry lacking in the non-relativistic case. The symmetry arises because in this case there is no preferred reference frame; only relative motions are important.

When velocities are large, these relativistic formulae *must* be used when calculating the size of the Doppler effect. But in the non-relativistic regime $\beta \ll 1$ (i.e. $v \ll c$) these formulae reduce to the familiar acoustical Doppler equations. In particular, if $v \ll c$ the Doppler shift is given by

$$z = \frac{\Delta\lambda}{\lambda} = \frac{v}{c}. \tag{6.6}$$

The *radial velocity* (v_r) of a star is its relative motion directly along an imaginary line joining the star to an observer here on Earth. If a star approaches, the wavelength of its light decreases. It is *blueshifted*. If a star recedes, the wavelength increases. It is *redshifted*. Since light travels about a million times faster than sound, the Doppler effect for light is correspondingly much smaller. Nevertheless, the spectroscope is an acutely sensitive tool. It is quite capable of detecting tiny shifts in the position of spectral lines. So when Huggins measured the redshift in the spectral lines of Sirius, he was able to use Doppler's formula to deduce that Sirius has a radial velocity of 46 km s^{-1} away from us. We now have better data, but Huggins was correct to within an order of magnitude.

Example 6.1 Doppler shift

In the laboratory, the Hα line has a wavelength of 656.28 nm. A star is observed to have a Hα line at 656.26 nm. What is the radial velocity of the star?

Solution. The fractional change in wavelength, $\Delta\lambda/\lambda$, is small:

$$\frac{\Delta\lambda}{\lambda} = \frac{0.02}{656.28} = 3.047 \times 10^{-5}.$$

We can therefore safely use the non-relativistic equation (6.6):

$$z = \frac{\Delta\lambda}{\lambda} = \frac{v}{c} \longrightarrow v = 3.047 \times 10^{-5} \times 3 \times 10^8 \, \mathrm{m\,s^{-1}}$$
$$\longrightarrow v = 9.14 \, \mathrm{km\,s^{-1}}.$$

Since the radiation from the star is blueshifted, its velocity is 9.14 km s^{-1} *towards* us.

Of course, stars do not move in a straight line directly towards us or away from us. They move in three dimensions with some *space velocity* (v). Therefore, as well as having a radial velocity, stars also have a *transverse velocity* (v_t): a velocity at right angles to our line of sight. A transverse velocity causes a star to shift its position on the celestial sphere. In other words, it causes a star to have a proper motion.

The radial and transverse velocities are just the radial and transverse components of the space velocity, as figure 6.2 shows. If the star moves at some angle θ to our line of sight, then

$$v_r = v \cos\theta \qquad v_t = v \sin\theta. \tag{6.7}$$

If we know the distance to a star from its parallax, and if we can measure its proper motion, then we can calculate its transverse velocity. The transverse velocity (in km s^{-1})

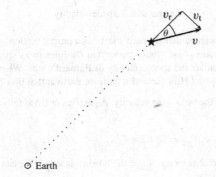

FIGURE 6.2: A star moves through space with a velocity v. We can split this velocity into two components. The radial velocity, v_r, carries the star directly towards or away from us. We derive the radial velocity from Doppler shift measurements. The transverse velocity, v_t, shows up as the proper motion of the star. If we know the distance to the star then, by measuring the proper motion, we can deduce the transverse velocity. Conversely, if we know the transverse velocity and the proper motion we can deduce the distance to the star.

is simply the distance d (in km) multiplied by the proper motion μ (in rad s^{-1}):

$$v_t \, (\text{km s}^{-1}) = d \, (\text{km}) \times \mu \, (\text{rad s}^{-1}). \tag{6.8}$$

The units in which we have expressed (6.8) are rather inconvenient. We have seen how stellar distances are best measured in parsecs; furthermore, since proper motions are so tiny, they are best expressed in seconds of arc per year. (The *largest* proper motion belongs to Barnard's star, and that is only 10.3″ per year.) It therefore makes sense to rewrite (6.8) using the following conversion factors:

$$1 \, \text{rad} \equiv 2.063 \times 10^5 \, '' \qquad 1 \, \text{yr} \equiv 3.156 \times 10^7 \, \text{s} \qquad 1 \, \text{pc} \equiv 3.086 \times 10^{13} \, \text{km}.$$

We obtain

$$\begin{aligned}
v_t \, (\text{km s}^{-1}) &= [d \, (\text{km}) \, \mu \, (\text{rad s}^{-1})] \times [2.063 \times 10^5 \, \text{arcsec/rad}] \\
&\quad \times [3.156 \times 10^7 \, \text{s/yr}] \times [3.086 \times 10^{13} \, \text{km/pc}] \\
&= 4.74 \mu d \tag{6.9}
\end{aligned}$$

where d is in parsecs and μ is in seconds of arc per year.

We now have enough information to calculate the space velocity of a star. Doppler measurements directly provide us with the radial velocity of a star; parallax and proper motion measurements enable us to calculate its transverse velocity through (6.9). The space velocity is obtained simply by combining the two velocities through the relation

$$v^2 = v_r^2 + v_t^2.$$

Example 6.2 Space velocity

Astrometric observations show that Barnard's star has a proper motion of 10.3″ per year and a parallax of 0.55″. Spectroscopic studies show that the lines in its spectrum are shifted to the red by 0.036%. Calculate the space velocity of Barnard's star. When will the star make its closest approach to Earth? How far will it be from Earth at that time?

Solution. First, calculate the transverse velocity. A parallax of 0.55″ tells us that $d = 1.82\,\text{pc}$. So, from (6.9), we have

$$v_t = 4.74 \times 1.82 \times 10.3 = 88.8\,\text{km s}^{-1}.$$

Second, calculate the radial velocity. Since the redshift is small, we can safely use the non-relativistic Doppler formula:

$$z = \frac{v_r}{c} = 3.6 \times 10^{-4} \longrightarrow v_r = 3.6 \times 10^{-4} \times 3 \times 10^5\,\text{km s}^{-1}$$
$$= 108\,\text{km s}^{-1}.$$

Finally, to calculate the space velocity we must combine v_r and v_t using Pythagoras:

$$v = \sqrt{v_r^2 + v_t^2} = 140\,\text{km s}^{-1}.$$

In the diagram below, E represents Earth, B represents the position of Barnard's star now, and C is Barnard's star at closest approach. We know that EB $= 1.82\,\text{pc}$. Furthermore, $\theta = \tan^{-1}(v_t/v_r) = 39.4°$.

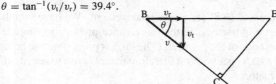

From the diagram, BC $=$ EB $\cos\theta = 1.4\,\text{pc}$. If Barnard's star travels at $140\,\text{km s}^{-1}$, it will take roughly 100 centuries to cover this distance. At closest approach:

EC $=$ EB $\sin\theta = 1.16\,\text{pc}$.

In 1997, astronomers calculated the space velocity of the dim red dwarf GL 710. Measurements from *Hipparcos* indicate that the star is now 63 ly away, and moving towards us at about $14\,\text{km s}^{-1}$. In about one million years, GL 710 will be the closest star to us. It will pass within about one light year, much closer than the α Centauri system is at present. In future skies, this rather unassuming star will shine as brightly as the red giant Antares. More importantly, it may dislodge some comets from the Oort Cloud and cause them to fall towards Earth. Some astronomers believe that something similar happened 65 million years ago: a close encounter with some unidentified star initiated a cometary bombardment of the inner solar system, the result being the extinction of the dinosaurs and much of the life on Earth.

6.2 SECULAR AND STATISTICAL PARALLAXES

The reach of trigonometric parallax depends upon the baseline. If the baseline at our disposal is just an inch or two, as is the case with our eyes, then we can measure distances on an everyday scale. If the baseline is a few thousand miles, as is the case with astronomical observatories based in different countries, then we can measure distances within the solar system. If the baseline is the diameter of the Earth's orbit, as is the case when astronomers make observations separated by six months, then we can measure distances to the nearby stars. If we wish to use parallax to measure even greater distances we clearly need a longer baseline. But how can we proceed beyond the 2 AU baseline provided by the Earth's motion around the Sun?

As we shall see when we take the next step on the distance ladder, the nearby stars are part of a much larger structure called the Galaxy. The details of this structure are not important for the present discussion; what is important is to realise that most of the stars within about 100 pc of the Sun move with more or less a common velocity. This common velocity is the orbital velocity about the centre of our Galaxy. In addition to this shared orbital velocity, each star has a *peculiar velocity* that arises from the combined gravitational tugs of every other star. By carefully measuring the space velocities of a large well-defined group of nearby stars, and determining the average motion for the group, we can define a *local standard of rest* (LSR). This is a reference frame in which the average velocity of nearby stars, relative to the Sun, is zero. The LSR orbits the centre of the Galaxy, making one revolution every 230 million years or so.

Every nearby star, including the Sun, has a peculiar velocity relative to the LSR. Astronomers decompose such a velocity into three mutually perpendicular components, called the U, V and W velocities. The U velocity of a star is its velocity away from the centre of the Galaxy; V is its velocity in the direction of the rotation of the Galaxy; and W is its upwards velocity perpendicular to the plane of the Galaxy. See figure 6.3

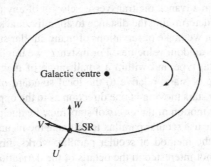

FIGURE 6.3: The U, V, W velocity components.

The precise *Hipparcos* measurements of parallaxes and proper motions have improved our knowledge of the Sun's motion relative to the local standard of rest. (In particular, the Sun's V component of velocity has been shown to be much smaller than the commonly accepted value.) According to the latest analyses, the Sun has a velocity, relative to the local standard of rest, of

$$U_\odot = -10.00 \pm 0.36 \, \text{km s}^{-1}$$
$$V_\odot = 5.23 \pm 0.62 \, \text{km s}^{-1} \quad\quad\quad (6.10)$$
$$W_\odot = 7.17 \pm 0.38 \, \text{km s}^{-1}.$$

If we combine these U, V, W components into a single space velocity, we find that the Sun moves at a speed of about $13.4 \pm 0.7 \, \text{km s}^{-1}$ in the direction of the constellation of Hercules. The point towards which it moves is called the *solar apex*. (Note that Herschel made a good estimate of the position of the solar apex before our modern understanding of the Galaxy came about.)

The point of all this is that the Sun moves rather quickly. In one year it moves towards the apex by about 2.83 AU, relative to the LSR, and it drags the Earth and the rest of the solar system with it. The Sun's motion thus generates a longer baseline with which to measure parallax. If we accumulate measurements for 25 years, say, the Sun will have moved through space by more than 70 AU. About half of this movement will be reflected in the proper motion of stars, which gives us a baseline of more than 35 AU with which to make distance determinations. This is much better than the 2 AU afforded by the Earth's orbit. The parallax of a star due to the Sun's motion is called its *secular* parallax.

Unfortunately, we cannot go straight ahead and use the secular parallax of a star to determine its distance. The problem is that the stars themselves move relative to the local standard of rest, and they do so with random space velocities. The stars resemble a swarm of bees: the swarm moves with a well-defined velocity, but the individual bees buzz about haphazardly. Any measured shift in the position of a star is therefore derived from a combination of the Sun's motion and the star's motion. If we knew the star's transverse velocity then, from measurements of its proper motion and from our knowledge of the Sun's motion, we would have enough information to determine its distance. The difficulty is that we cannot know in advance the transverse velocity of any particular star. Secular parallax is of no help in determining the distance to an individual star.

Suppose, though, that we make observations of many similar stars distributed all over the sky and moving with random velocities. For instance, we might choose to investigate stars of the same spectral type and within a small range of magnitudes. The *average* velocity of such a group of stars, relative to the local standard of rest, should be zero. There should be as many stars moving in one direction as in the opposite direction. So the observed *average* proper motion of the *group* of stars must be due purely to the motion of the Sun. We can therefore use secular parallax to estimate the mean distance to the group.

To understand how the method of secular parallax works, first consider figure 6.4. (Those readers who are not interested in the details of the derivation should skip the next few paragraphs.) The figure shows a star with coordinates (α, δ) and the solar apex with coordinates (α_A, δ_A). The angular distance between the star and the apex is λ. We denote by ψ the angle between the great-circle arcs that join the star to the north celestial pole and to the apex. Some basic spherical trigonometry tells us that

$$\cos \lambda = \sin \delta \sin \delta_A + \cos \delta \cos \delta_A \cos(\alpha - \alpha_A) \quad\quad\quad (6.11)$$
$$\sin \lambda \cos \psi = \cos \delta \sin \delta_A - \sin \delta \cos \delta_A \cos(\alpha - \alpha_A). \quad\quad\quad (6.12)$$

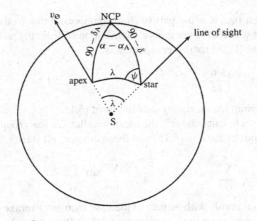

FIGURE 6.4: The angles λ and ψ.

Since we know δ_A and α_A, we can calculate λ and ψ for each star in the group.

The next step is to resolve the star's proper motion into two components. We could resolve it into equatorial components $(\mu_\alpha \cos \delta, \mu_\delta)$, but it is preferable to use a slightly different decomposition. The upsilon (υ) component is the star's proper motion along the great circle connecting the star and the apex. The υ component is defined to be positive in the direction directly away from the apex. Note that this component depends in part upon the motion of the Sun. The tau (τ) component is the proper motion perpendicular to the υ component. It does not depend upon the solar motion. The equations for the two components are:

$$\upsilon = \mu_\alpha \cos \delta \sin \psi - \mu_\delta \cos \psi \tag{6.13}$$

$$\tau = \mu_\delta \sin \psi + \mu_\alpha \cos \delta \cos \psi. \tag{6.14}$$

Now consider figure 6.5, which is derived from a consideration of figure 6.4.

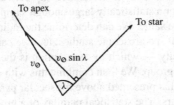

FIGURE 6.5: The component of the Sun's velocity across the line of sight to a star is $v_\odot \sin \lambda$.

It shows that the component of the Sun's velocity across the line of sight to the star is $v_\odot \sin \lambda$. From (6.9) we deduce that this transverse velocity shows up as a proper motion $v_\odot \sin \lambda / 4.74d \equiv \pi'' v_\odot \sin \lambda / 4.74$, where π'' is the star's parallax in seconds of arc. Call this proper motion $\upsilon_\odot$: it is the part of the υ component due to the solar motion. The υ component that we observe contains a contribution υ_* from the motion of the star itself, relative to the LSR. So the total υ component is:

$$\upsilon = \upsilon_* + \pi'' \frac{\upsilon_\odot \sin \lambda}{4.74}. \tag{6.15}$$

Each star in the group has an equation of the form (6.15).

The final step is to calculate $\bar{\pi}''$, the mean parallax for the group of stars. To do this, multiply throughout by $\sin \lambda$ in (6.15) and then sum over all stars:

$$\sum_{i=1}^{n} \upsilon_i \sin \lambda_i = \sum_{i=1}^{n} \upsilon_{*i} \sin \lambda_i + \frac{\bar{\pi}'' \upsilon_\odot}{4.74} \sum_{i=1}^{n} \sin^2 \lambda_i. \tag{6.16}$$

If the stars move randomly with respect to the LSR, then any average over υ_* will be zero. Thus the first summation on the right-hand side of (6.16) is zero. Now divide throughout by n, the total number of stars in the group, and denote averages by angle brackets (i.e. $\langle \ \rangle$ denotes $\frac{1}{n} \sum$). We obtain

$$\bar{\pi}'' = \frac{4.74 \langle \upsilon \sin \lambda \rangle}{\upsilon_\odot \langle \sin^2 \lambda \rangle}. \tag{6.17}$$

Since we know $\upsilon_\odot$, and we can perform the required averages, we can use (6.17) to determine the mean parallax for the group of stars. Note, however, that it is vital to choose the group of stars with great care. We must choose stars with the same spectrum and luminosity class, and which lie in a restricted range of apparent magnitude, in order to be confident that they all lie at a similar distance. The mean distance of the group is then likely to be meaningful. If we are lazy, and choose a random sample of stars to work with, it is not clear what significance we can attach to the results.

We can use a related method to determine the average distance to such a group of stars. This is the method of *statistical* parallax, so called because it involves a statistical averaging procedure. In this method we argue that the average radial component of velocity of a group of stars is equal to the average transverse component of velocity. Of course, for any individual star there is no relationship between the radial and transverse components of velocity. For a statistically large enough sample of stars, though, the average velocities should be the same. We can determine the radial component of velocity for each star in the group from spectroscopic studies, and we can thus calculate the average radial component of velocity — which by assumption is the average transverse component of velocity — for the group. We then combine this with data on proper motions, and use an analysis similar to that presented above for secular parallax, to deduce the average distance to the group of stars. The statistical parallax of a group of stars is given by

$$\bar{\pi}'' = \frac{4.74 \langle |\tau| \rangle}{\langle |\upsilon_r + \upsilon_\odot \cos \lambda| \rangle} \tag{6.18}$$

where $|\tau|$ is the modulus of the τ component, υ_r is the radial component of the observed space velocity of the star, and as usual $\upsilon_\odot$ is the solar velocity.

Example 6.3 Secular parallaxes

Using the data given in the table below, calculate the secular parallax of this group of 13 stars. Take the apex to be at $\alpha_A = 18\,h$, $\delta_A = 30°$ and the solar velocity to be $13.4\,km\,s^{-1}$. Is the value you obtain likely to be accurate? (The data come from the *Hipparcos* catalogue. All coordinates are for epoch 2000.0.)

Hipparcos number	α (hr min sec)	δ (° ′ ″)	μ_α (mas/yr)	μ_δ (mas/yr)	Spectral type	M_V
21273	04 33 50.86	+14 50 40.2	103.69	−25.94	A8 V	4.70
23783	05 06 40.66	+51 35 53.3	−29.31	−173.95	F0 V	5.00
28103	05 56 24.32	−14 10 04.9	−42.23	139.02	F1 V	3.71
33202	06 54 38.59	+13 10 40.9	70.19	−77.79	F0 V	4.65
34834	07 12 33.74	−46 45 34.4	−135.60	106.79	F0 IV	4.49
75411	15 24 29.54	+37 22 37.1	−147.68	84.69	F0 V	4.30
78180	15 57 47.59	+54 44 58.2	−150.18	106.47	F0 IV	4.96
101612	20 35 34.77	−60 34 52.7	70.15	−185.58	F1 III	4.75
104887	21 14 47.35	+38 02 39.6	195.74	410.02	F0 IV	3.74
108036	21 53 17.58	−13 33 06.5	313.03	13.67	F3 IV	5.08
108431	21 57 55.03	−54 59 33.2	43.00	−3.67	F0 V	4.40
109857	22 15 01.80	+57 02 36.5	476.45	50.00	F0 IV	4.10
114570	23 12 32.92	+49 24 21.5	89.70	95.57	F0 V	4.53

Solution. The procedure for calculating the secular parallax of a group of stars like this is straightforward but tedious. It is best to write a simple computer program to do the chores.

The procedure is as follows. First, determine λ and ψ for each star by substituting the appropriate values of α, δ in (6.11) and (6.12). Second, for each star substitute your calculated values of λ, ψ and the tabulated values of μ_α, μ_δ into (6.13). This gives you the υ component for each star. Finally, sum all the values of $\upsilon \sin \lambda$ and $\sin^2 \lambda$, and calculate the secular parallax using (6.17).

You should find that $\bar{\pi}'' = 0.0426''$ ($= 42.6\,mas$). Compare this with the mean trigonometric parallax of these stars of $0.0372''$ ($= 37.2\,mas$).

More accurate values exist for the coordinates of the solar apex; you could repeat the calculation with better values. The stars should perhaps have been chosen with more care: they range from spectral type A8 to F3, and the luminosity class varies from III to V. (For a discussion of luminosity class, see later in the chapter.) It would be better to choose a more homogeneous group of stars — a group with all F0 V stars or all G2 V stars, say. Just as importantly, it would be better to work with a larger sample of stars. (Note, though, that Hertzsprung used this method on a sample of just 13 Cepheids to calibrate the period–luminosity relation!)

Which method is best: secular or statistical parallax? The answer depends on the average radial velocity of the group of stars under consideration. If the solar velocity is greater than the average radial velocity of the group it is better to use the method of secular parallax, because the solar motion has a greater observable effect than the space velocities of the stars. On the other hand, if the solar velocity is less than the average radial velocity it is better to use the method of statistical parallax. In this case it is the stellar space velocities that provide the largest observable effects.

For carefully chosen groups of stars, the methods of statistical and secular parallax reach out to distances of about 500 pc (about 1600 ly) with reasonable accuracy — especially if we use the *Hipparcos* data on proper motions. This is much farther than trigonometric parallax takes us. The trade-off is that we must give up hope of measuring accurate distances to individual stars.

6.3   THE MOVING CLUSTER METHOD

The clouds of gas from which stars condense sometimes give birth not to a single star but to several tens or hundreds of stars. An *open cluster* forms when many stars condense at the same time from the same cloud. The stars do not appear at precisely the same moment; there may be a gap of a few million years between the formation of the first and last stars in a cluster. Nevertheless, a few million years is brief when compared to the main sequence lifetime of all but the most massive stars. For practical purposes, an open cluster is a group of stars with a common birthday*.

At first sight it may seem that all the stars in a given constellation form an open cluster. A moment's thought tells us that this is not so. Stars are distributed through the three dimensions of space, so although *some* stars in a constellation may be cluster stars, others may just happen to lie in the field of view. Distinguishing between cluster stars and unrelated field stars thus poses a problem. It is a particular problem if the cluster is nearby, since it will occupy a large area of sky and field stars are then certain to complicate the picture. So how can we tell if a particular star belongs to a cluster?

To discriminate between cluster stars and field stars we can use their different proper motions. Since the stars in a cluster are physically related to each other, and stay together because of their mutual gravitational attraction, they move through space as a unit. They move in more or less the same direction and at the same speed. Stars that are not part of the cluster do not share this common motion. If we can observe the proper motions of the cluster stars, we call the group a *moving cluster*. For instance, five of the seven stars in the Plough (the Big Dipper), which is part of the constellation Ursa Major, belong to the same moving cluster. The other two stars are not part of the cluster. See figure 6.6.

If one plots the proper motions of all the stars in a moving cluster they will appear to converge on, or diverge from, a single point. To see why, imagine a flock of birds flying over your head and into the distance. Even though the birds fly along parallel paths, perspective makes the flock seem to converge on a single point. A similar effect happens if you drive along a road flanked with telephone poles. The poles seem to diverge from a point on the horizon, and move apart as you approach them. In the first case, the convergent point represents the direction of motion of the flock of birds. In the second case it represents the direction of motion of the car. Figure 6.7 shows a schematic diagram of the convergence of a moving cluster.

If an open cluster has a large angular size, the proper motions of the cluster stars will point in very different directions. We can thus determine the convergent point quite

---

*The best known open cluster is the Pleiades. The chief stars of the Pleiades are hot, white stars that probably formed only recently — about 60 million years ago.

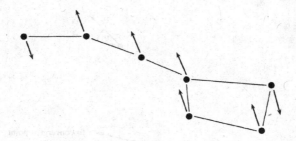

FIGURE 6.6: From observations of proper motions we know that the middle stars of the Plough belong to the same cluster. The two end stars are unrelated to the other five (or to each other).

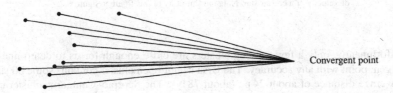

Convergent point

FIGURE 6.7: Due to perspective, the stars in a moving cluster all appear to converge on (or diverge from) a point in the sky.

accurately — much more accurately than from measurements of the proper motion of a single cluster star. Once we know the convergent point, we know the precise direction of motion of each star in the cluster. (In practice, it is unlikely that the space motions of the cluster stars will all be parallel. Some of the stars may have a random motion relative to the cluster motion, and the cluster as a whole may be rotating, or perhaps contracting under gravitational forces.)

Now consider figure 6.8. It shows a cluster star moving with space velocity, v, toward the convergent point. The angle between the star and the convergent point is θ. We can write the transverse velocity, v_t, in terms of the radial velocity, v_r, thus:

$$v_t = v_r \tan \theta.$$

But we know θ, and we can measure v_r from Doppler studies, so we can calculate the transverse velocity of the star. The transverse velocity of a star gives rise to its proper motion, μ. If we measure μ in seconds of arc per year, v_r in $km\,s^{-1}$ and the distance d to the star in parsecs, then from (6.9) we can write

$$d = \frac{v_r \tan \theta}{4.74 \mu}. \tag{6.19}$$

By measuring the distances of many cluster stars and averaging the results we can, in theory, obtain an accurate value for the distance to the centre of the cluster.

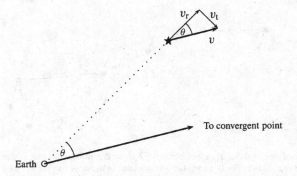

FIGURE 6.8: Knowledge of the convergent point enables us to calculate the radial and transverse components of velocity of a cluster star. Note the similarity of this figure to figure 6.2.

Unfortunately, only a few moving clusters are close enough for us to determine the convergent point with any accuracy. The Ursa Major open cluster contains some 60 stars, at an average distance of about 24 pc (about 78 ly). The Scorpio–Centaurus cluster contains almost 100 stars, at an average distance of about 170 pc (about 550 ly). By far the most important open cluster, though, is a ∨-shaped wedge of about 300 stars covering almost 6° of sky in the constellation of Taurus. This is the beautiful Hyades cluster. (The name comes from Greek words meaning 'rain makers' since, for the Greeks, the rising of the Sun in this cluster marked the beginning of the rainy season.) The proper motions of the Hyades stars converge to a point on the sky near Betelgeuse.

The American astronomer Lewis Boss (1846–1912) first applied the moving cluster method to the Hyades in 1908. Since then astronomers have refined his techniques, and gradually improved the accuracy of his method. They have also tried a number of other ways of determining the distance to the Hyades. Figure 6.9 shows the best estimates since 1930. The graph shows that these best estimates are converging.

There are have been two recent independent estimates of the distance to the Hyades. The first comes from a refinement of the moving cluster method. It drops the physically unrealistic condition that all cluster stars have exactly the same velocity. Astronomers now have models that allow a range of velocities for the Hyades stars. With this refinement, the distance to the Hyades cluster centre is 45.75 ± 1.25 pc (149 ± 4 ly). This agrees well with the second estimate: a direct distance measurement made by *Hipparcos*. Astronomers used the satellite to study 218 stars in the Hyades, and determined a distance of 46.34 ± 0.27 pc (151.1 ± 0.88 ly) to the centre of the cluster.

Determining the distance to the Hyades is one of the most important steps on the entire distance ladder. The cluster contains stars of different spectral types, so it defines the position of much of the main sequence in the HR diagram. Once we know the distance to the Hyades, we know the absolute magnitude of all these different main sequence stars. (We could give the luminosity of main sequence stars in table 5.2 because we knew the distance to some clusters.) And once we calibrate the main sequence in this way, we have a powerful method of estimating distance. This is the method of *main sequence fitting*.

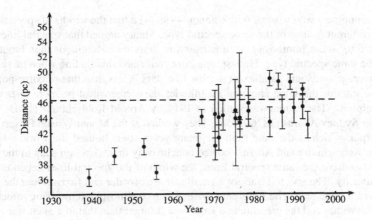

FIGURE 6.9: Some estimates of the distance to the centre of the Hyades. The disks denote the best estimate; the vertical lines attached to each disk represent the uncertainty in the measurement. The broken horizontal line refers to the current best estimate, derived from *Hipparcos* data. The distance to the Hyades almost certainly lies within ± 0.27 pc of the distance represented by this line.

6.4   MAIN SEQUENCE FITTING

As we look out deeper into space, proper motions become too small to measure accurately. Eventually, they vanish completely. Therefore the moving cluster method works only for the closest open clusters. We can still see the stars in far-off clusters, though, and we can still take their spectra. We can thus plot the HR diagram for a remote cluster. Since the cluster's distance is unknown, such a diagram is a plot of *apparent* magnitude against spectral type. If we compare this diagram with the calibrated HR diagram for the Hyades we should find that the two diagrams are identical except for the vertical scale: one scale (that for the Hyades) will be in terms of absolute magnitude; the other (that for the remote cluster) will be in terms of apparent magnitude. (We assume that the main sequences of similar clusters are the same. There might be differences, for example, if the clouds from which the clusters condensed had very different chemical compositions.) This provides us with the clue we need to find the distance to an open cluster. The difference between the absolute and apparent magnitudes of an object fixes its distance through (4.8). In this case the 'object' is the entire main sequence of an open cluster. The method of main sequence fitting thus consists of laying the HR diagram of the distant cluster on top of the HR diagram of the Hyades, and sliding it vertically until the two main sequences overlap. The difference in the magnitude scales determines the distance modulus of the cluster.

There are two obvious queries about this procedure. The first is: how can we be sure that a particular star is on the main sequence? If we see a distant red star, for instance, how can we determine if it is a dim main sequence star that is relatively nearby, or if it is a luminous red giant that is far away?

The first indication of how this might be answered came in 1897, when the American astronomer Antonia Caetana De Paiva Pereira Maury (1866–1952) — one of the many female astronomers who worked with Cannon — showed that the widths of spectral lines could be different in stars of the same spectral type. Maury argued that spectral linewidth was related to stellar luminosity, with narrow-line stars more luminous than broad-line stars of the same spectral type. Hertzsprung later confirmed this finding when he showed that the average statistical parallax of narrow-line stars is less than that of corresponding broad-line stars of the same apparent magnitude: the former must be more distant and more luminous. The matter was settled in 1914 by Arnold Kohlschütter (1883–1969) and Walter Sydney Adams (1876–1956). They worked at the Mount Wilson Observatory in California, which at the time (and for many years after) housed the world's largest telescope. Kohlschütter and Adams showed conclusively that, although stars in the same spectral class have the same spectral lines, the width of the spectral lines depends upon their luminosity. The spectral lines of a small star are broader and fuzzier than the spectral lines of a giant star. The difference arises because atmospheric pressure affects the spectral linewidth, and the pressure in a small star is larger than that in a giant star.

When astronomers describe the spectrum of a star they not only give the spectral class to which it belongs; they also give its *luminosity class.* In the MK system, developed by the American astronomers William Wilson Morgan (1906–1994) and Philip Childs Keenan (1908–), stars belong to one of the luminosity classes I to VI. Stars in class Ia are extreme supergiants, with luminosities typically in excess of one million times that of the Sun. Luminous supergiants are about 200 000 times as luminous as the Sun; they form luminosity class Ia. Stars in class Ib are about 40 000 times as luminous as the Sun. Class II stars are the bright giants, and have luminosities between 1000 and 10 000 times that of the Sun. Normal giants are in class III; they typically have luminosities that are 100 times greater than the Sun. The subgiants, with luminosities in the range 10 to 100 times that of the Sun, are in class IV. Dwarf stars, otherwise known as main sequence stars, are in luminosity class V. Finally, the subdwarfs are in luminosity class VI. White dwarfs have their own special classification — wd. See table 6.1 for some typical examples. (The table includes an entry for luminosity class 0, which contains hypergiants.)

TABLE 6.1: Stellar luminosity classes. (There are no well-known examples of subdwarfs, which are all faint.)

Class	Stellar type	Example
0	Extremely luminous supergiant	S Doradi
Ia	Luminous supergiant	Rigel
Ib	Less luminous supergiant	Antares
II	Bright giant	Polaris
III	Giant	Arcturus
IV	Subgiant	Procyon
V	Main sequence (dwarf)	Sun
VI	Subdwarf	—
wd	White dwarf	Sirius B

The Sun, therefore, is not just a G2 star. It is a G2 V star. By classifying a distant star's spectrum within the MK system we can locate its position on the HR diagram. Our first query regarding main sequence fitting can thus be answered satisfactorily.

The second query is: how much confidence can we place in the calibrated main sequence? We now know the distance to the Hyades cluster very accurately, and so we know the absolute magnitude of the Hyades stars. But the Hyades is quite an old cluster. Results from *Hipparcos* suggest that it formed between 575 million and 675 million years ago, which is enough time for massive stars to have evolved away from the main sequence. The Hyades cluster therefore only provides us with information on the lower part of the main sequence. To establish the upper part of the main sequence we need to use the main sequence of nearby younger clusters, like the Pleiades. If we could do this, the calibrated main sequence would extend all the way up to the bright O and B stars.

Various workers have used *Hipparcos* data to deduce a distance to the Pleiades of 116.3 ± 3.3 pc (380 ± 10.8 ly). This is about 15% closer than previously thought, so if the result is correct the Pleiades stars are dimmer than once believed. *Hipparcos* astronomers have obtained distances to ten more clusters to a precision of 10%, and a further ten clusters with a precision of between 10–20%. They found something strange. The different clusters seem to define three different main sequences. The main sequence defined in part by the Hyades is about 0.5 magnitudes brighter than the main sequence defined by the Pleiades; it is 0.7 magnitudes brighter than the main sequence defined by the open cluster NGC 2516. At present, no one understands the cause of this variation.

Before *Hipparcos*, astronomers had used main sequence fitting to estimate distances out to about 7 kpc. (A kiloparsec, or kpc, is one thousand parsecs, so 7 kpc is about 23 000 ly). This is the largest distance we have yet discussed. Even before *Hipparcos*, though, astronomers knew that the method of main sequence fitting depended critically upon the calibrated HR diagram. In particular, it depended upon knowing the distance to the Hyades. Any distance derived using the method of main sequence fitting depended upon the distance estimate of the Hyades. When that estimate decreased or increased, as it sometimes did, the whole universe shivered in sympathy. It is worth bearing in mind figure 6.9: in just 50 years the best estimate for the Hyades distance increased by about one third — and the size of the universe altered accordingly. The *Hipparcos* distance estimate for the Hyades is likely to be definitive, but the discovery of the variation in main sequence luminosity once again casts doubt on the technique of main sequence fitting. (Alternatively, the differences between direct *Hipparcos* measurements to some open clusters and the distances derived from main sequence fitting may suggest that the *Hipparcos* parallaxes are systematically in error.) Until we understand the differences between the cluster main sequences, any distance derived by main sequence fitting must be viewed with some caution.

6.5   SPECTROSCOPIC PARALLAX

Why bother with main sequence fitting at all? Once we have the spectral and luminosity classifications of an individual star we can calculate its absolute magnitude and therefore its distance. This is the method of *spectroscopic* parallax.

Main sequence fitting is just a special case of spectroscopic parallax. The reason to prefer main sequence fitting, of course, is that spectroscopic parallax is more accurate when applied to clusters. We can pick out a main sequence quite easily from a sample of,

Example 6.4 Spectroscopic parallax

We measure an apparent magnitude of 12 for a main sequence A5 star. How distant is it?

Solution. From figure 5.7 we see that an A5 star has $M \approx +2.0$. We measure $m = 12$. The distance modulus is thus $m - M = 12 - 2 = 10$. We use this in (4.8) to find the distance:

$$5 \log_{10} \left(\frac{d}{10 \, \text{pc}} \right) = 10 \quad \text{or} \quad \log_{10} \left(\frac{d}{10 \, \text{pc}} \right) = 2$$

thus $d = 1000 \, \text{pc}$. There will of course be a large uncertainty attached to this answer.

say, 100 stars, and cluster main sequences show much less variation than individual stars. A single star might be 'peculiar' in some way, or it may be partially obscured from view for some reason, and then the method fails. Nevertheless, if we can determine the spectral and luminosity classes of a star, we have a means of crudely estimating its distance. And since the spectroscope is so sensitive, the method works for *very* distant stars.

CHAPTER SUMMARY

- At this step on the distance ladder we can no longer directly measure the distance to a star. Trigonometric parallax — our only direct method of determining distance — does not work at this range (at least with our present generation of instruments).

- We can measure the mean distance to an homogeneous group of stars using the method of secular parallax. This method makes use of the solar velocity ($v_\odot = 13.4 \pm 0.7 \, \text{km s}^{-1}$ towards the solar apex) relative to the local standard of rest.

- The mean secular parallax, $\bar{\pi}''$, of a group of stars is given by

$$\bar{\pi}'' = \frac{4.74 \langle v \sin \lambda \rangle}{v_\odot \langle \sin^2 \lambda \rangle}$$

where λ is the angle between a star and the apex, and v is the component of a star's proper motion directly away from the apex. The angle brackets denote that an average must be taken over all the stars in the group.

- We can also use the method of statistical parallax to estimate the mean distance to a group of stars. In this approach we argue that the average radial velocity of the stars is equal to their average transverse velocity. The statistical parallax, $\bar{\pi}''$, of the group is

$$\bar{\pi}'' = \frac{4.74 \langle |\tau| \rangle}{\langle |v_r + v_\odot \cos \lambda| \rangle}$$

where τ is the component of proper motion perpendicular to v, and v_r is the radial component of the observed space velocity of a star. The vertical bars denote a modulus, and the angle brackets denote an average.

- We can use the methods of secular and statistical parallaxes only because we are able to measure the radial velocities of the stars. We do this by measuring the Doppler shift of spectral lines. For a radial velocity v much less than the speed of light c, the Doppler shift z is given by

$$z = \frac{\Delta\lambda}{\lambda} = \frac{v}{c}$$

where $\Delta\lambda$ is the shift in wavelength. If a star approaches, its light is blueshifted. If a star recedes, its light is redshifted.

- We can measure the distance to a nearby open cluster if we can determine the position of the cluster's convergent point. If the angle between the convergent point and a cluster star is θ, then the distance to the star (in pc) is given by

$$d = \frac{v_r \tan\theta}{4.74\mu}$$

where v_r is its radial velocity (in $km\,s^{-1}$) and μ is its proper motion (in seconds of arc per year). The average of several such cluster stars gives the distance to the centre of the cluster.

- The most important open cluster is the Hyades, because we can use it to calibrate the absolute magnitude of the lower part of the main sequence. We can then use the main sequence as a kind of standard candle. By fitting the main sequence of a remote cluster to that of the Hyades we obtain the distance modulus of the cluster. This is the method of main sequence fitting.

- Main sequence fitting is a special case of spectroscopic parallax. Spectroscopic parallax only works if we know the luminosity classification of a star as well as its spectral class.

QUESTIONS AND PROBLEMS

6.1 What happens to the Hα line in a star due to the Earth's motion around the Sun?

6.2 Write an account of the Wilson–Bappu effect as a distance indicator.

6.3 The radial velocity (in $km\,s^{-1}$) of each of the stars in the table on page 121 is, respectively, $+39.6$, -5.4, -2.4, 0.0, -0.8, -12.5, -11.0, -20.0, -21.1, -21.5, $+15.0$, -0.6, $+12.5$. (With the exception of HIP 33202, the data come from the *General Catalogue of Stellar Radial Velocities*.) Calculate the statistical parallax of this group of stars, and comment upon the likely accuracy of your result. In this case would it be better to use the v or the τ component of proper motion?

6.4 Rigel and Vega are very nearly of the same spectral class (they are B8 and A0 respectively). In what ways might their spectra be different?

6.5 Write an account of OB associations. How do they differ from open clusters? How can we determine their distance?

6.6 Use your library to find examples of HR diagrams of some open clusters, and also find an HR diagram of the Hyades plotted in terms of absolute magnitude. Use the method of main sequence fitting to estimate the distances of the open clusters.

6.7 A star is observed to have a trigonometric parallax of 0.01″ and a proper motion of 0.1″ per year. What is its tangential speed?   47.4 km s^{-1}

6.8 The radial velocity of a moving cluster is 35 km s^{-1}. If the centre of the cluster has a proper motion of 0.07″ per year, and it is at an angle of 26° to the convergent point, how far away is it?   52 pc

6.9 A main sequence star of spectral class A0 is observed to have an apparent magnitude of +5. Estimate its distance (assuming no interstellar absorption).   100 pc

6.10 You observe two stars to have the same apparent magnitude. One star is of spectral class F0 II; the other is of spectral class F0 III. Which star is closer?   The star of spectral class F0 III

6.11 In the laboratory the sodium D line has a wavelength of 589.616 nm. A star is observed to have the line at 589.518 nm. What is the radial velocity of the star?   50 km s^{-1} toward us

6.12 For a group of G2 V stars it is found that the mean value of the τ component of their proper motions is 15 mas and the mean radial velocity is 10 km s^{-1}. What is the mean parallax of the group of stars?   7.11 mas

FURTHER READING

GENERAL
French A P (1966) *Special Relativity* (Nelson: London)
— Contains a particularly clear derivation of the Doppler effect for light.

Hoskin M (1980) Herschel's determination of the solar apex. *J. Hist. Astron.* **11** 153–163
— How William Herschel managed to derive a value for the solar apex that is extraordinarily close to the best modern value.

Mihalas D and Binney J (1981) *Galactic Astronomy: Structure and Kinematics* (Freeman: San Francisco)
— An in-depth treatment of the local standard of rest. The book is a standard text for galactic astronomy, which is the next step of our distance ladder.

STELLAR KINEMATICS
Dehnen W and Binney J (1998) Local stellar kinematics from *Hipparcos* data. *MNRAS* **298** 387–394
— The Sun's motion with respect to the local standard of rest, as determined by *Hipparcos* data.

Layden A C *et al.* (1996) The absolute magnitude and kinematics of RR Lyrae stars via statistical parallaxes. *Astron. J.* **112** 2110–2131
— An example of the use of the statistical parallax method. The authors use the method to derive the distance, and thus magnitude, of a particularly important type of star: RR Lyrae variables. These stars are a useful distance indicator for the next two steps of the distance ladder.

Wilson T D *et al.* (1991) Absolute magnitudes and kinematic properties of Cepheids. *Astrophys. J.* **378** 708–717
— The authors perform a maximum likelihood statistical parallax analysis of 90 classical Cepheids. The luminosity of Cepheids is vitally important on later rungs of the distance ladder. At $\log_{10} P = 0.8$ they obtain $\langle M_V \rangle = -3.46 \pm 0.33$.

OPEN CLUSTERS
Cooke W J and Eichhorn H (1997) A new and comprehensive determination of the distance to member stars of the Hyades. *MNRAS* **288** 319–222
— The authors deduce a distance of 45.8 ± 1.25 pc to the cluster's centre.

O'Dell M A, Hendry M A and Collier Cameron A (1994) New distance measurements to the Pleiades and α Persei clusters *MNRAS* **268** 181–193
— The authors use a new technique to measure the distance to the Pleiades and α Persei clusters, and compare their results with the method of main sequence fitting.

7

Fifth step: the Galaxy

Scattered light from our homes, our cars and our industries drowns out light from all but the brightest stars. If you live in a large city you are lucky if you can see a few tens of stars, even in a cloudless sky. Go where the night sky is really dark, though, and the view is different. You can make out about three thousand stars with your naked eye. If the sky is clear you will also see a band of milky-white light stretching across the sky. The Greeks called this band of light 'kiklos galaxias', from their word 'gala' meaning 'milk'. In Latin it was 'via lactea', which means 'road of milk'. We call it the *Milky Way*.

In 1610, Galileo became the first man to look at the Milky Way through a telescope. He saw that it was a vast collection of stars. A few of them were bright, but the number of dimmer stars was 'quite beyond calculation'. Galileo concluded that the foggy appearance of the Milky Way was due to its great distance. The individual stars were so far away they merged into a misty circle of light.

In 1750, the English philosopher Thomas Wright (1711–1786) published a book entitled *An Original Theory or New Hypothesis of the Universe*. In it, he proposed that the Milky Way is a flat slab or disk of stars; at the centre of this system is the Sun, around which the other stars rotate. Wright's book was theological in tone, and was not taken seriously by contemporary astronomers. Nevertheless, reviews of the book appeared in several European periodicals. The German philosopher Immanuel Kant (1724–1804) read such an account in a Hamburg newspaper. Kant was a fine mathematician, and he reformulated Wright's ideas in more scientific style. In 1755, he published his own improved version of the idea that the system of stars has a definite disk-like structure.

Wright and Kant were correct. The stars are not scattered equally throughout infinite space, but instead form a system with a definite structure. There then follows the obvious question: just how big is this system? The answer to that question proved elusive, because the astronomers were *inside* the system they were trying to measure. It was a classic case of being unable to see the wood for the trees. They finally solved the problem a few decades ago, but the story begins over two centuries ago with William Herschel.

7.1 STAR COUNTING

For Herschel, the chief aim of astronomy was to obtain 'a knowledge of the construction of the heavens'. In other words, he wanted to know the position of every star in three-dimensional space: not just their apparent position on the celestial sphere, which he could measure, but also their distance from us. Herschel would have been delighted with the *Hipparcos* mission, and without doubt he would have been an enthusiastic supporter of the *GAIA* and *SIM* missions.

The difficulty with his plan was that he could not measure the distance even to one star, let alone the many millions that were visible in his telescopes. (As we saw on page 71, it was not until 1838, sixteen years after Herschel's death, that Bessel first made an accurate measurement of a stellar distance.) Of course, Herschel was well aware of the difficulty. To solve the problem he proposed to make a statistical analysis of large aggregates of stars, rather than study each star individually.

Herschel made three assumptions, which he hoped would enable him to measure (or 'gauge') the extent of the stellar system.

First, he assumed that wherever stars exist they are more or less equally distributed throughout space. In other words, equal volumes of space contain roughly equal numbers of stars.

Second, he assumed that stars have more or less the same intrinsic brightness and that no starlight is absorbed in its passage through space. Under this assumption, if star A is four times fainter than star B, then A is twice as distant as B. This is just the usual inverse-square law. For practical work, Herschel needed a way of comparing the brightness of different stars. He did this by viewing each star through telescopes of different sizes. To see why this works, suppose the same star is observed with two reflecting telescopes of the same design but with mirrors of different sizes. The light transmitted by the telescope to the eye is proportional to the area of the collecting mirror, and thus to the square of the radius of the mirror. So the brightness of a star viewed through a telescope is proportional to the inverse square of its distance, and to the square of the radius of the mirror. The 'distance effect' cancels the 'radius effect'.

Example 7.1 If 'bright means near' and 'dim means far'...

You observe star A through a telescope with a mirror of radius 5 cm. You observe star B through a telescope with a mirror of radius 15 cm. Both stars appear to be the same brightness. How much farther away is star A than star B?

Solution. Suppose star A is at a distance d_A and star B is at a distance d_B. We observe A through a telescope with mirror radius r_A and B through a telescope of mirror radius r_B. If A and B have the same intrinsic luminosity, and if there is no absorbtion of light on its journey to us, then apparent brightness is proportional to r^2/d^2. So:

$$\frac{r_A^2}{d_A^2} = \frac{r_B^2}{d_B^2} \longrightarrow \frac{d_A}{d_B} = \frac{r_A}{r_B} = \frac{1}{3}.$$

So B is three times as distant as A, but *only if the assumptions hold*.

Third, he assumed that his telescopes could penetrate to the edge of the system of stars. He used first the '20-foot' telescope and later the infamous '40-foot' telescope —

the biggest telescopes anyone had ever made. So if any astronomer of his day could see to the edge of the stellar system it would have been Herschel.

In 1784, armed with his assumptions, he began his programme of 'star gauging': put simply, he counted the stars. Even a tireless observer like Herschel could not count all the stars in the sky. He therefore divided the sky into circular patches, each patch covering an area equal in size to one quarter of the full Moon, and then counted the stars in each patch and recorded their magnitudes. In 1785, he published the results of star counts made in 683 regions distributed all over the sky. He later added a further 400 regions of sky.

Herschel was sure that this programme would provide information on the size and shape of the stellar system. For instance, suppose that the Sun were at the centre of a spherical distribution of stars. In this case, he would count equal numbers of stars in each direction because he would be looking through equal volumes of star-filled space. On the other hand, suppose the distribution of stars was not spherical. Suppose, as Wright and Kant argued, that it was a flat slab with the Sun close to the centre. In this case he would count many stars if he looked along the plane of the slab, but only a few stars if he looked out of the plane.

Mathematically, Herschel's argument was as follows. The space density of stars $D(r, l, b)$ — where r is the distance from the observer, and l and b are galactic latitude and longitude — is, by the first assumption, constant. We can call this density D, and drop any functional dependence on r, l, b. Now, suppose that we study an area on the celestial sphere that subtends a solid angle Ω sr*. At a distance r from us this solid angle subtends an area $A = \Omega r^2$. The element of volume dV between r and $r + dr$ is thus given by $dV = \Omega r^2 \, dr$. The number of stars $n(r)$ in this volume is just

$$n(r) = D \, dV = \Omega D r^2 \, dr.$$

To obtain $N(r)$, the total number of stars contained in the solid angle Ω out to a distance r, we simply integrate over all the volume elements:

$$N(r) = \int_0^r n(r) \, dr$$

$$= \Omega D \int_0^r r^2 \, dr$$

$$= \tfrac{1}{3} \Omega D r^3. \tag{7.1}$$

Herschel planned to measure the apparent magnitude, m, of stars rather than their distance. But from (4.8) we know that there is a link between apparent magnitude, absolute magnitude and distance. We can rewrite (4.8) as

$$r = 10^{0.2m+k} \tag{7.2}$$

where k is a constant; the constant arises from the second assumption, namely that all stars have the same absolute magnitude M. Substituting (7.2) into (7.1) we obtain

$$\log N(m) = 0.6m + K \tag{7.3}$$

---

*The steradian (sr) is the dimensionless unit of solid angle. A complete spherical surface subtends 4π sr. Thus 1 sr is equal to 3283 square degrees of arc.

where K is a constant that depends on the absolute magnitude M of stars, the space density D of stars, and the solid angle Ω. Even without knowing the value of K, (7.3) tells us that there should be $10^{0.6} = 3.98$ times as many stars with apparent magnitude $m + 1$ than with apparent magnitude m. According to Herschel's counts the ratio of second-magnitude stars to first-magnitude stars was 3.4, which agreed well with the prediction of (7.3). His counts were in much worse agreement for faint stars, though. We shall see later why this was the case.

Herschel found that the different regions of sky contain very different numbers of stars. In some of the circular patches that he studied he could see only one star; in other patches he could see 600 stars. His star counts agreed with Galileo's early observations. Perpendicular to the plane of the Milky Way there were few stars per unit area of sky. The number of stars rose steadily as one approached the Milky Way, and they reached a maximum in the plane of the Milky Way. If his assumptions were correct they implied that, perpendicular to the plane of the Milky Way, he was looking through the forest of stars out into the emptiness beyond. In the plane of the Milky Way he was looking through a much larger thickness of stars, which were so numerous that they merged to form a luminous band encircling the sky.

Herschel thus suggested that the stars form a finite system in the shape of a 'grindstone' or a lens. Since the Milky Way seemed to encircle the sky, and to be equally bright on all sides, Herschel placed the Sun somewhere near the centre of the system. (This is exactly the same suggestion made by Wright and Kant. There is no indication that Herschel knew of the work of the other men, and it seems likely that he made the suggestion independently. Furthermore, whereas Wright and Kant made a speculative suggestion, Herschel made his claim on the basis of hard observational data.)

Astronomers soon accepted the existence of a stellar system in the shape of a lens. It became known as the Milky Way. The term 'Milky Way' can thus refer either to the system of stars or to the band of light that crosses the night sky. (To avoid confusion, the stellar system is now often called the *Galaxy* — with a capital letter — or sometimes the *Milky Way Galaxy*. In this book I will use the word 'Galaxy' to mean the system of stars to which our Sun belongs. The term 'Milky Way' will refer only to the band of light that encircles the sky.)

Upon analysing his star counts, Herschel estimated that the long diameter of the lens was about 800 times the average distance between stars and the short diameter was about 150 times the average distance between stars. Of course, he had no idea of the average distance between stars; he did not know the distance to even one star. In the absence of any better figure, though, he could take the average star separation to be the distance between the Sun and Sirius, for which he had Newton's estimate (see page 66). This made the long diameter of the Galaxy to be 80 million AU, and the short diameter to be 15 million AU — or, in modern parlance, about 1250 ly by 235 ly. See figure 7.1.

This was the first estimate of the size of the Galaxy. Although philosophers such as Kant speculated that other galaxies or 'island universes' might exist, as far as anyone knew the Galaxy contained all the stars in the universe. Herschel's estimate could thus serve as an estimate of the size of the universe. But how good was his estimate? Herschel's entire method of star gauging depended upon three assumptions: equal stellar brightness, uniform stellar distribution and the ability of his telescopes to penetrate the system of

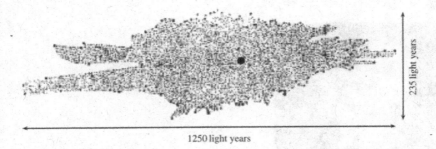

235 light years

1250 light years

FIGURE 7.1: A cross section of Herschel's model of the Galaxy. Note the position of the Sun, marked with a black circle, near the centre. The model resembled the disk-like structure of Wright and Kant, but Herschel based his model on hard observation rather than speculation.

stars. His estimate was only as good as these assumptions.

Herschel knew from the outset that his assumptions might be wrong. His own work on double stars invalidated the assumption of equal stellar brightness. Remember that he found binary systems in which the two stars, which were both clearly at the same distance from Earth, had very different apparent magnitudes. The assumption of a uniform distribution of stars was also shaky. For instance, Herschel discovered many star clusters in which hundreds of faint stars appeared together in a small part of the sky. Unless in each case he was looking along a 'spoke' of stars, which seemed unlikely, he had to deduce that clusters contain more stars than other regions of the sky. As for his assumption that telescopes could plumb the depths of the Galaxy: by 1817 he believed that the Galaxy was fathomless even with the 40-foot telescope.

By the end of his life, Herschel admitted that his model of the Galaxy was untenable. He also realised that, while it was possible to criticise the assumptions upon which his model was based, it seemed impossible to 'construct the heavens' without them. During the hundred years after his death, anyone wishing to investigate the size and shape of the Galaxy used his method of star counting. It seemed the only way forward*.

This statistical approach to stellar astronomy culminated in the work of two men: the German astronomer Hugo von Seeliger (1849–1924) and the Dutch astronomer Jacobus Cornelius Kapteyn (1851–1922). Both men had access to the detailed star catalogues that began to appear in the nineteenth century. These catalogues held information on the position and apparent magnitude of hundreds of thousands of objects. By far the most comprehensive of them was the *Bonner Durchmusterung des Nordlichen Himmels* (the *Bonn Survey of the Northern Skies*), which was compiled by the German astronomer Friedrich Wilhelm August Argelander (1799–1875) and first published in 1861. It contained information on 457 848 stars. Seeliger and Kapteyn sought to understand the distribution of stars in space using data mainly from the *Bonn Survey*, but also from the other star catalogues.

---

*Although Herschel's model of the Galaxy was influential, far more influential was his *approach* to the problem. In many ways it ushered in the modern approach to observational astronomy.

Jacobus Cornelius Kapteyn

Seeliger was a highly accomplished mathematician, and he brought sophisticated new mathematical techniques to bear on the problem. In 1898, he managed to find a way around the assumption of equal stellar brightness, which by now was known to be completely invalid.

Seeliger argued that *all* stellar brightnesses are possible, but that in a given volume of space the *spread* of brightnesses is the same. In other words, two equal volumes of space on average contain the same number of magnitude 1 stars, the same number of magnitude 2 stars, and so on. The exact form that this spread of brightnesses took was unknown. It seemed likely, though, that in a given volume of space there would be more dim stars than bright stars. So there would be more magnitude 3 stars than magnitude 2 stars, more magnitude 2 stars than magnitude 1 stars, and so on. The number of stars in each magnitude is known as the *stellar luminosity function*, which we denote by $\phi(M)$. He reformulated Herschel's problem by assuming a particular form for the luminosity function and investigating different approximations for the density function $D(r, l, b)$. The resulting problem was hard, but Seeliger solved it. For the next two decades he refined his theory, and in 1920 he published his definitive model of the Galaxy. He believed it to be an ellipsoidal system of stars some 33 000 ly long in the plane of the Milky Way and some 6000 ly thick (about 10 kpc long and 1.8 kpc thick).

Seeliger's work was very mathematical and he published solely in German journals, so his work was not immediately influential within the international astronomical community. Kapteyn, on the other hand, was hugely influential. Kapteyn had contacts at observatories around the world, and he had access to more data regarding the positions, magnitudes and proper motions of stars than any other astronomer. He spent many years analysing these data in an attempt to derive the distance to similar groups of stars. In 1901, he published a model of the Galaxy that he called a 'first approximation', but that other astronomers regarded as an important step in our understanding of the size of the Galaxy. Kapteyn continued to collect and analyse data. He published his definitive model of the Galaxy in 1920, in the same year that Seeliger published his.

In many ways the two models were similar. In both cases the Sun was at the centre of an ellipsoidal distribution of stars, and the density of stars decreased with increasing distance from the Sun. Perpendicular to the plane of the Galaxy, for instance, the density decreased to half of the central density after only 150 pc. The main way that the two models differed was in their dimensions. According to Kapteyn the Galaxy was larger than Seeliger thought: about 60 000 ly long in the plane of the Milky Way and about 8000 ly thick (18 kpc by 2.4 kpc). This model became known as the *Kapteyn universe*. See figure 7.2.

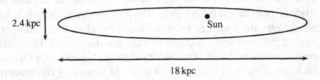

FIGURE 7.2: A cross section of the Kapteyn universe. Kapteyn believed that the stars were most densely packed at the centre of the system. The density dropped smoothly as one moved out from the centre. The Sun was close to the centre of the ellipse, but not *exactly* at the centre.

Both Seeliger and Kapteyn assumed there was no dust or gas in interstellar space. If dust dimmed the light from stars, then their size estimates would need radical revision.

7.2   CEPHEID VARIABLES: A FLICKERING YARDSTICK

The work of Seeliger and Kapteyn was as far as one could go by simply counting stars. If astronomers were to verify or disprove Kapteyn's universe they would need an entirely new approach to the problem of measuring the size and shape of the Galaxy. More than anything else they needed a method of measuring distances over tens of thousands of parsecs. The key to measuring galactic distances lay in a discovery made by a young man in 1784, the same year that Herschel began his star counts.

John Goodricke (1764–1786) was a deaf-mute, as a result of a fever contracted in infancy. His liberal parents were determined that his disability would not bar their son from an education, and at the age of eight Goodricke was sent to a special school in Edinburgh. At the age of 14 he was able to attend the respected Dissenting school, War-rington Academy, an institution where the famous chemist Joseph Priestley (1733–1804) had recently taught. At the Academy Goodricke excelled in mathematics.

Goodricke went on to work with the English astronomer Edward Pigott (1735–1825), who counted many notable scientists, including Herschel, amongst his colleagues. The partnership between Goodricke and Pigott is one of the most interesting in the history of astronomy. The Goodricke family home in York was close to Pigott's observatory. When Goodricke was 17 years old he visited his neighbour's observatory, and quickly became fascinated by the fine instruments on display. They began a partnership that was essentially that of teacher and pupil (Pigott was already an accomplished astronomer, and almost 30 years older than Goodricke), but it soon developed into a close working relationship.

Pigott was interested in *variable* stars: stars that change in brightness. The German astronomer David Fabricius (1564–1617) discovered the first variable star*. In 1596 he observed *o* Ceti to be a third-magnitude star; by 1597 it had faded from view. He thought it might be a nova, but astronomers soon detected it again and eventually found that *o* Ceti followed an 11-month cycle of brightness. They named it Mira: 'the miraculous one'. It became the prototype for a class of star called *Mira variables*, which change in brightness over long periods — anything between 100 to 700 days.

Throughout 1784, while Herschel counted stars, Pigott maintained a programme of observing all suspected variables. On the night of 10 September he noticed that the star η Aquilae was brighter than the star θ Serpentis. The previous year, though, it had been the other way around: θ Serpentis had been brighter than η Aquilae. From comparisons with nearby stars it seemed to Pigott that η Aquilae was the star that varied. He soon calculated its period of variation to be one week. Two months later, in a paper of 5 December 1784, he had refined the value of the period to 7 days 4 hours 38 minutes. (The modern value for the period of this star is 7 days 4 hours 14 minutes 33 seconds.) Pigott further noted that the star took much less time to increase in brightness than it did to decrease.

Goodricke shared his teacher's interest in variable stars. He studied fewer stars than Pigott, but those stars he worked with he studied in greater detail. One star that particularly interested him was δ Cephei. On 23 October 1784, Goodricke wrote that 'I am now *almost* convinced that δ Cephei varies'. To prove that it varied, Goodricke, in the first ten months of 1785, observed δ Cephei on 100 nights (unlike η Aquilae, it was far enough north to study all year round). He patiently watched its rapid rise to maximum brightness (a process that took 1 day 14 hours 30 minutes) and its slower fall to minimum brightness (a process that took 3 days 18 hours 17 minutes). The entire cycle thus repeated itself every 5 days 8 hours 47 minutes. During this time the magnitude of δ Cephei varied by 0.86, which meant that at maximum brightness it was about twice as luminous as at minimum brightness. Soon after making these observations, Goodricke died of an illness caused by exposure to the night air. He was just 21.

The brightness fluctuations of δ Cephei were very similar to those of η Aquilae and to several other stars that astronomers subsequently observed in detail. In each case the cycle repeated itself over a period of a few days, the period was very stable, and the rise to maximum brightness was much quicker than the fall to minimum brightness. Because δ Cephei was the first such object to be studied in detail, astronomers assigned the name 'Cepheid' to this class of variable star. Figure 7.3 shows the light curve of a typical Cepheid variable.

Goodricke suggested that a star such as Algol varies in brightness because a dimmer unseen companion star regularly eclipses it. This is exactly what happens in the case of Algol. The 'eclipsing binary' explanation does not work for a Cepheid, though, because of the different way in which it brightens and dims. A detailed explanation of the Cepheid phenomenon had to wait until quite recent times.

---

*There are indications that Mira, the star studied by Fabricius, was known by the Babylonians to be variable.

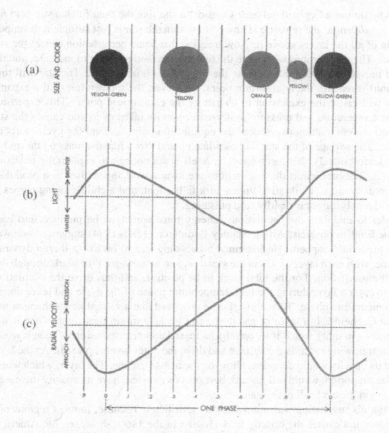

FIGURE 7.3: The light curve of a typical Cepheid. Part (a) shows that Cepheids vary in colour and size (the relative sizes have been exaggerated; at its largest, the radius of a Cepheid is typically only 20% larger than at minimum). Part (b) shows the variation in brightness. Note how the Cepheid increases in brightness much more quickly than it decreases. Part (c) shows the radial velocity of the star's atmosphere.

7.2.1 Pulsating stars

A Cepheid variable star pulsates. The pulsation is nothing like the rotation of pulsars. Rather, Cepheids move radially in and out — periodically shrinking and expanding like a vast breathing lung. We deduce this from the variable Doppler shifts exhibited by spectral lines from the star. This behaviour was first noticed by the Russian astronomer Aristarkh Apollonovich Belopolsky (1854–1934), who showed that the spectral lines from δ Cephei shift to and fro with the same 5 day 9 hr period as the brightness fluctuations. From the size of the Doppler shifts we can construct *radial velocity curves*, and from these we can deduce the change in radius of the star. The radius of δ Cephei, for instance, varies between 23 and 26 solar radii.

But why should a Cepheid pulsate? Consider a star like the Sun. Such a star is in *hydrostatic equilibrium:* the pressure at any point within the star is just sufficient to support the weight of all the layers above it. Now imagine that some perturbation causes the star to contract. The contractions will cause the temperature and pressure to increase, and the increased pressure will overcompensate the inward crush of gravity. The star will thus expand until it reaches the equilibrium point. However, the momentum of the expanding gases will cause the expansion to overshoot the equilibrium point. This expansion causes the temperature and pressure to decrease; gravity takes over, and causes the star to contract; the contraction overshoots the equilibrium point ... and the cycle starts all over again. The surface of the star will oscillate up and down like the mass on the end of a spring. Unfortunately, this mechanism by itself is not enough to explain the pulsation of Cepheids. The problem is that oscillations are invariably damped. Just as a pendulum clock eventually stops, a vibrating tuning fork falls silent, and a child's swing comes to rest, so will a pulsating star rapidly stop pulsating.

In order to maintain the oscillations, energy must somehow be pumped into each cycle. The English physicist Arthur Stanley Eddington (1882–1944) suggested one way in which this could happen. He imagined a pulsating star to be like a thermodynamic heat engine, with each layer of gas in the star doing an amount $p \, dV$ of work through one cycle of the oscillation. For the total work to be positive, and thus drive the oscillation, heat had to enter a layer during the high-temperature phase of the cycle and leave during the low-temperature phase. To achieve this he proposed that a valve-like mechanism was at work in Cepheids. If a layer of gas in the star became more 'heat-tight' when it was compressed — in other words if its *opacity* increased upon compression — then it would dam the heat flowing towards the surface and drive the outer layers upwards. As the layer expanded its opacity would decrease, allowing the heat to escape. The layer, which would then not be supported, would fall inward, become compressed, have its opacity increased, and thus begin the cycle all over again.

Eddington's suggestion was not accepted immediately because, in most regions of a star, compression causes the opacity to *decrease*. In the 1960s, however, the American astronomers John Paul Cox (1926–1984) and Norman Hodgson Baker (1931–) along with the German astronomer Rudolf Kippenhahn (1926–) calculated that Eddington's valve could operate in the *He partial ionisation zone*, a region of a star where the gaseous material is partially ionised. When the zone is compressed, some of the energy that would normally raise the temperature of the gas instead produces further ionisation. Because the temperature rise is smaller than usual, the increased density upon compression gives rise to an increased opacity — exactly what Eddington's valve mechanism requires.

A Cepheid, then, pulsates for the following reason: the compressed helium ionisation zone traps the heat produced by the star and pushes the star's surface outwards. As the star expands, the temperature in the zone drops and the He^+ ions recombine with electrons. This helium gas is more transparent to radiation and so the trapped energy escapes. The star's surface falls inward under the pull of gravity, the helium is once more compressed, and the cycle starts again.

The location of the helium ionisation zone within the star depends upon the star's temperature. If the star is hot (hotter than about 7500 K) the zone lies close to the surface of the star. In this case the oscillations do not build up because there is insufficient mass

Arthur Stanley Eddington

for the driving mechanism to work effectively. On the other hand, if the star is cold (colder than about 4500 K) the zone lies close to the centre of the star. In this case the oscillations do not build up because efficient convection in the outer layers of the star prevents the storage of energy required by the driving mechanism. So oscillations occur only over a relatively narrow range of temperature. This in turn means that pulsating stars can occupy only a restricted, well-defined region of the HR diagram. This region, located between the main sequence and the red giant branch, is called the *instability strip*.

We now have a good idea of what a Cepheid is. It is a high-mass star that has begun to evolve off the main sequence. As it evolves it will cross through the top end of the instability strip, and it may cross the strip several times. When it enters the instability strip it will start to pulsate — it will become a Cepheid. When it leaves the strip it will stop pulsating.

Before we move on to the most important aspect of Cepheids, there is one further feature of these stars worth noting. The radial oscillation of a Cepheid is essentially a standing acoustic wave in the interior of the star; the centre of the star is a node and the surface of the star is an antinode. This should sound familiar to anyone who has taken a basic physics course: the same thing happens to the air in an organ pipe. The closed end of the pipe, where air does not move, is a node; the open end of the pipe, where the air is maximally displaced, is an antinode. In the fundamental mode of an organ pipe (and of a Cepheid) all the material taking part in the standing wave moves in the same direction. But just as an organ pipe can resonate with overtones, a Cepheid is not restricted to oscillating in the fundamental mode. A first-overtone Cepheid, for instance, has a single node between the centre and the surface; the stellar material moves in opposite directions on either side of the node. Kepler talked of the 'music of the spheres' in relation to the orbits of the planets. The phrase is even more appropriate when applied to Cepheids.

7.2.2   The period–luminosity relation

Cepheids are one of the most important types of star in the sky because they are our best standard candle. They provide us with a 'flickering yardstick' with which to measure distances in the Galaxy and beyond.

Henrietta Swan Leavitt

The possibility of using Cepheid variables as a yardstick was discovered by chance by the American astronomer Henrietta Swan Leavitt (1868–1921) of the Harvard College Observatory. In the early years of this century, Leavitt began studying photographs of the Large and Small Magellanic Clouds (the LMC and SMC). To the naked eye, the Magellanic Clouds are patches of milky-white light, rather like disconnected pieces of the Milky Way. Like the Milky Way itself, their cloudy appearance is due simply to their great distance from us. In reality they contain vast numbers of stars. A good telescope, such as the one Leavitt used, resolves the Clouds into their constituent stars. Leavitt was therefore able to search the Clouds for Cepheids.

In 1907, as part of a paper on variable stars, she published data on 16 Cepheids in the Small Magellanic Cloud, including their period and average brightness. When preparing the data for publication she decided to rank the Cepheids in order of increasing period. After doing this she noticed that they were then also in order of increasing average brightness. In other words, bright Cepheids had longer periods than dim Cepheids.

Because the Small Magellanic Cloud is so far away, Leavitt was not too wrong in saying that all stars in the Cloud were at the same distance from Earth. It mattered little whether a star was at the far edge or the near edge of the Cloud: the extra distance was tiny compared with their overall remoteness. (Suppose you stand in Times Square, New York. You can make the equally valid statement that every street light shining in the City of London is the same distance from you. A street light might be on the western edge or the eastern edge of the City, but the difference is irrelevant when compared with the distance of 3470 miles between London and New York.) Leavitt therefore had found not just a relationship between period and average apparent magnitude. She had found a relationship between period and average *absolute* magnitude!

In modern notation we would say that the relation is of the form

$$\langle M_V \rangle = a + b \log_{10} P \tag{7.4}$$

where P is the period of oscillation in days, and a and b are constants to be determined.

Leavitt confined her discovery of this Cepheid *period–luminosity relation* to one sentence at the end of her paper. Not surprisingly, no astronomer picked up on the importance of the result. In 1912, though, she published a paper focusing just on the period–luminosity relation. She gave the period and magnitude data for 25 Cepheids in the Small Magellanic Cloud. This time the importance of the finding could not be overlooked.

Before we discuss the importance of the period–luminosity relation it is worthwhile trying to understand, in terms of a model of radially pulsating stars, *why* there should be such a relationship.

Consider a simple pendulum. The period of oscillation, P, is given by

$$P = 2\pi \sqrt{\frac{l}{g}} \tag{7.5}$$

where l is the length of the pendulum and g is the acceleration due to gravity. We can obtain a crude estimate of the period of oscillation of a Cepheid by treating it as a pendulum. This is not as outlandish as it might seem: both systems are mechanical oscillators, with the oscillations driven by gravity. The difference is mainly one of scale. For a Cepheid, then, we need to replace l in (7.5) by R, the radius of the star. And we must replace g by GM/R^2 — the acceleration due to gravity at the surface of the star. Making these substitutions we obtain

$$P = 2\pi \sqrt{\frac{R^3}{GM}}. \tag{7.6}$$

The mean density $\bar{\rho}$ of the star is given by

$$\bar{\rho} = \frac{M}{\frac{4}{3}\pi R^3} = \frac{3M}{4\pi R^3}.$$

Substituting this expression into (7.6), and rearranging, we obtain

$$P\sqrt{\bar{\rho}} = Q. \tag{7.7}$$

Equation (7.7) is called the *pulsation equation*; Q is the *pulsation constant* for Cepheids. Although the derivation given above is crude, the period–mean-density relation of (7.7) is found to be true over many orders of magnitude.

The pulsation equation says that $P \propto R^{1.5}$. According to Stefan's law, if we model a star as a blackbody the luminosity L is given by

$$L = 4\pi R^2 \sigma T^4 \tag{7.8}$$

where σ is Stefan's constant and T is the surface temperature. In our earlier discussion we noted that Cepheids occur only in a restricted range of temperature. So, although there is a weak dependence on temperature, in general we can say that $L \propto R^2$. Since both period and luminosity depend upon the radius of the star, there must be a relationship between period and luminosity. In fact, theory suggests that $L \propto P^{1.6}$.

As we have noted several times, one of the great barriers against determining distances in astronomy is our ignorance of the intrinsic brightness of objects. Leavitt's discovery meant astronomers could determine the relative brightness of Cepheids simply by timing their periods. If the period–luminosity relation held for *all* Cepheids, and not just those in the Magellanic Clouds, then astronomers would be able to make a scale model of the Galaxy (or at least those portions of the Galaxy in which they could detect Cepheids). For instance, suppose we observe that Cepheids A and B have the same period, and that A is four times as bright as B. Because they have equal periods they must have equal intrinsic luminosities. So, from the inverse-square law, A must be twice as close as B. We are more likely to observe Cepheids with different periods, of course, but the period–luminosity relation means we can handle these cases just as easily.

A scale model of the Galaxy would be interesting, but how much more interesting it would be if Cepheids gave absolute, rather than relative, distances! Unfortunately, Leavitt could not calibrate the period–luminosity relation because she did not know the distance to the Small Magellanic Cloud. If the Cloud was nearby, the Cepheids would be dim. If it was remote, the Cepheids would be bright. In modern terminology we would say that Leavitt's observations of the Clouds enabled her to determine the slope of the relation (the term b in (7.4)) but not the zero point (the term a).

If we can find the distance to just *one* Cepheid we can calibrate the period–luminosity relation and we would then have a means of determining the distance to *all* Cepheids. The problem is finding the distance to that one Cepheid. Polaris, the closest Cepheid, is unfortunately too far away to show an annual parallax. How, then, can we determine the scale of the Cepheid yardstick?

Astronomers resorted to the method of statistical parallax (described on page 120) to determine the distance to Cepheids. Hertzsprung first used the method on Cepheids in 1913. He concluded that a Cepheid of absolute magnitude -2.3 had a period of 6.6 days. Combined with observations of Cepheids in the Magellanic Clouds, this effectively gave him a period–luminosity relation of the following form:

$$\langle M_V \rangle = -0.6 - 2.1 \log_{10} P.$$

He had only 13 Cepheids to work with, so his estimate was necessarily imprecise. A young American astronomer called Harlow Shapley (1885–1972) therefore decided to repeat Hertzsprung's work.

Shapley became an astronomer by accident. He arrived at the University of Missouri as a young man intent on studying journalism, but found that the buildings that would house the school were not yet finished. He was told to return the following year, which he did — only to find that the buildings were *still* unfinished. Rather than wait a further year he decided to take a different course. He picked up a prospectus and, in his own words, 'got a further humiliation. The very first course offered was A. R. C. H. A. E. O. L. O. G. Y. — and I couldn't pronounce it (although I did know roughly what it was about). I turned over a page and saw A. S. T. R. O. N. O. M. Y. — I could pronounce that — and here I am!'

Harlow Shapley

After graduating, and then gaining a PhD from Princeton, he obtained a job at the Mount Wilson Observatory. His first task was to investigate Cepheids.

Shapley used the statistical parallax method on 11 of the 13 stars that Hertzsprung had used; he discarded κ Pavonis and l Carinae because he thought they might not be Cepheids. His mean parallax for this group of stars was tiny: only 0.0034″. Shapley had confidence in his procedure, though, and he decided that Hertzsprung's calibration of the period–luminosity relation required a small correction. According to Shapley, a Cepheid with a period of 5.96 days had an absolute magnitude of −2.35 ± 0.19. He then used this calibration of the period–luminosity relation to estimate the distances to several much more remote Cepheids with various periods. From these distances, and the observed brightnesses of the Cepheids, he was able to calculate their luminosities. Finally, he plotted the luminosities against the periods. He reasoned that, if his initial distance estimates were correct, he should obtain the same smooth curve that Leavitt obtained. If his statistical method was incorrect there would be no correlation between period and luminosity. When he drew the graph he obtained a smooth curve: his distance estimates seemed to be correct.

Shapley had calibrated the period–luminosity relation (at least to his own satisfaction; many contemporary astronomers thought Shapley guilty of trying to squeeze too much information out of unreliable data). By measuring the period of a Cepheid he could deduce its absolute magnitude, and by comparing its absolute magnitude with its apparent magnitude he would have its distance. Shapley now had a standard candle. Furthermore, it was an ideal standard candle: a Cepheid is intrinsically very bright (Cepheids are supergiants of luminosity class Ib; a typical Cepheid is 10 000 times more luminous than the Sun) and so is easy to see at large distances. And although Cepheids are not particularly common, neither are they exceptionally rare. He could use Cepheids to measure large distances with precision. Shapley immediately set about using them to determine the size of the Galaxy.

7.3   GLOBULAR CLUSTERS

The German astronomer Johan Hevel (1611–1687) discovered the first *globular cluster,* which now goes by the name of M 22. Since Hevel's time, astronomers have catalogued about 160 globular clusters in the Galaxy. We now know that these clusters are groupings of tens or hundreds of thousands of stars packed into a spherical region no larger than about 10 pc (roughly 30 ly) in diameter. Figure 7.4 shows a typical example. These globular clusters are completely unlike open clusters such as the Hyades or the Pleiades.

The stars in a globular cluster are 500 times more densely packed than in an open cluster or in the stellar environment around the Sun. (If you were to stand on a planet near the centre of a globular cluster you would immediately notice that the starlight was brighter than the full Moon here on Earth. Such a planet would never have night.) Another difference between the two types of cluster is that light from a globular is dominated by old stars that formed very early in the history of the Galaxy. The young O and B type stars prevalent in open clusters are completely absent in globulars.

FIGURE 7.4: The globular cluster ω Centauri. Only a few globular clusters are visible to the naked eye; at magnitude +3.6, ω Centauri is the brightest. It is also the most massive ($5.1 \times 10^6 M_\odot$). ω Centauri is relatively close; the most distant cluster, NGC 2419, is so far away that it is difficult to consider it a member of the Galaxy.

Soon after starting work at Mount Wilson, Shapley began a study of globular clusters. He hoped to calculate their distances and thus learn something about how they were distributed within the Galaxy (or to find out if they perhaps lay beyond the realm of the Galaxy). To estimate their distance, he built a type of distance ladder.

First, he found the distance to the three clusters ω Centauri, M 3 and M 5 by detecting Cepheids in them and applying his newly calibrated period–luminosity relation. He was limited to just three distance determinations with this technique because unfortunately not all clusters hold Cepheids. For instance, he counted about 70 000 stars in M 22, but even today we have identified only 32 of them as being variable. Shapley found Cepheids only in the largest, and therefore presumably the closest, globular clusters.

Second, he used the distance of these three clusters to estimate the absolute magnitude of another type of variable star: *RR Lyrae* stars. These are old, low-mass yellow variable stars with typical periods of less than one day. They pulsate in the same way as Cepheids, but they inhabit a lower part of the instability strip. Not one of them is particularly bright. Indeed, no RR Lyrae star is bright enough to be seen with the naked eye. They were not recognised until the 1890s, when the American astronomer Solon Irving Bailey (1854–1931) began a catalogue of short-period variable stars in globular clusters. Bailey noted that all RR Lyrae stars in a given cluster have much the same apparent magnitude, and therefore pretty much the same absolute magnitude, independent of their period. They are less luminous than Cepheids but, since they are much more common than Cepheids in globular clusters (indeed, until 1948, they were officially known as 'cluster variables'), they are an ideal standard candle for measuring the distance to globulars. Using his Cepheid-based calibration of the RR Lyrae luminosity, Shapley measured the distance to another four globular clusters. He now had the distance to seven clusters: three from Cepheids and four from RR Lyrae stars.

Third, he used these distance estimates to estimate the luminosity of the 30 brightest stars in these seven clusters. He discarded the five very brightest stars in each cluster, in case they were foreground stars, and then calculated the mean luminosity of the remaining 25 stars. He found that this mean luminosity was more or less the same in each of the seven clusters. If he assumed that this held true for every cluster he had yet another standard candle; and, since the brightest stars in a cluster outshine any Cepheid, he could use it over greater distances. He used this technique to estimate the distance to a further 21 clusters. He now had a total of 28 globular cluster distances.

Fourth, he used these distances to estimate the actual diameter of each cluster. All 28 clusters had more or less the same intrinsic size, so he now had a standard rod: the apparent diameter of a cluster. In this way he estimated distances to a further 41 clusters, producing a total of 69 globular cluster distances.

Shapley knew that his distance ladder was a rickety structure, and it was built on the quicksand of his Cepheid period–luminosity calibration. But though his estimate of the distance to any single cluster might have a large error associated with it, surely the *average* cluster distance would be quite accurate. In 1917, he published a list of the 69 globular clusters and their distances. He found that a typical globular cluster was perhaps 15 kpc (50 000 ly) distant. This was surprisingly large — about the size of the entire Kapteyn universe. He then went on to make an even more astonishing claim.

It was a well known but puzzling fact to astronomers of the day that all the globular clusters seemed to be in one hemisphere of the sky. The distribution of globular clusters is even more asymmetrical than this simple fact suggests. Over 40% of the globular clusters are to be found in a patch of sky covering less than 3% of the area of the celestial sphere. (The patch of sky containing so many globular clusters lies in the direction of Sagittarius, a constellation in the southern sky.) Shapley boldly postulated that globular clusters are spherically distributed about the massive centre of the Galaxy. According to him, the clusters occupy a small patch of sky simply because we see them from a large distance. In other words, Herschel, Seeliger and Kapteyn were wrong. The Sun was not even close to the centre of the Galaxy.

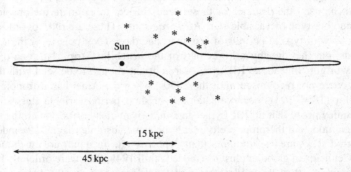

FIGURE 7.5: The size of the Galaxy according to Shapley. The Sun is about one-third of the way towards the edge of the Galaxy and a halo of globular clusters surrounds the central region. Because the Sun is so far away from the centre, the globular clusters are seen to occupy just a small region of sky.

By 1920, Shapley had arrived at a radical model of the Galaxy. It looked like two fried eggs placed back to back. Apart from the shape, it differed from the Kapteyn universe in two main ways. First, it was much bigger. The galactic lens was about 90 kpc (290 000 ly) in diameter, and the central bulge was 9 kpc (29 000 ly) thick. Second, and perhaps more importantly, the Sun was 15 kpc (50 000 ly) from the centre of the Galaxy. See figure 7.5.

7.4   INTERSTELLAR DUST

From Herschel to Shapley, all astronomers who studied the size of the Galaxy assumed that no starlight is lost on its journey to Earth. In 1930, the Swiss–American astronomer Robert Julius Trumpler (1886–1956) showed that this common assumption is false. *Interstellar absorption* by dust dims the light from some stars, and completely obscures the light from others. Not only is a study of our Galaxy a case of being unable to see the wood for the trees, but most of the wood is shrouded in fog.

Trumpler reached his conclusion by studying open star clusters, rather than the globular clusters with which Shapley worked. He used two methods to determine their distances. First, he assumed that open clusters all have the same intrinsic size. Since he believed he knew the distance and therefore true size of one open cluster, namely the Hyades, a measurement of a cluster's apparent angular diameter produced an estimate of its distance. Second, he used the method of main sequence fitting (described on page 125). By 1929, Trumpler had estimated the distance to 80 open star clusters by using both methods. Estimates from the two methods disagreed systematically. To explain the disagreement he argued that space was not transparent: light from distant star clusters was absorbed, making them appear dimmer than they really were.

To illustrate this effect with a concrete example, consider the two open star clusters M 103 and the Hyades.

Let us use the first of Trumpler's methods to find the distance to M 103. The Hyades cluster has an apparent angular diameter of 400'. The cluster M 103 has an apparent angular diameter of 7'. If the two clusters have the same intrinsic size, then M 103 must be 400/7 = 57 times more distant than the Hyades cluster. As we saw on page 124, the current best estimate for the Hyades distance is about 46 pc (150 ly). This places M 103 at a distance of about 2.6 kpc (8500 ly).

Trumpler's second method, that of main sequence fitting, produces a different answer. Trumpler found that the greater distance of M 103 caused its main sequence stars to appear 10.5 magnitudes fainter than those in the Hyades. If space is transparent, a magnitude difference of 10.5 corresponds to a distance ratio of $10^{10.5/5} \approx 125$. So if the Hyades cluster is 46 pc away, M 103 must be about 5.75 kpc away. This is larger than the first estimate by a factor of more than two.

Of course, a discrepancy for a single pair of open clusters might be due simply to a difference in the intrinsic size of the two clusters. Trumpler analysed 80 clusters, though, so variations in intrinsic size should have averaged out. The effect remained. Distances calculated using main sequence fitting were systematically larger than distances calculated from angular diameters.

Trumpler could resolve the discrepancy if he supposed that interstellar dust obscures 0.7 magnitudes for every kiloparsec (about 0.2 magnitudes per 1000 light years).

Example 7.2 Reconciling main sequence fitting and angular diameter distances

Show that interstellar absorption of 0.7 mag pc^{-1} resolves the discrepancy between main sequence fitting distances and angular diameter distances for M 103 and the Hyades.

Solution. If we believe the angular diameter data then M 103, at a distance of 2.6 kpc, is 57 times more remote than the Hyades. Light from M 103 is thus dimmed by 0.7 × 2.6 = 1.8 magnitudes. So in the absence of dust the stars are really only 10.5 − 1.8 = 8.7 magnitudes fainter than the Hyades. This corresponds to a distance ratio of $10^{8.7/5} = 55$: there is thus agreement between the two methods.

Since interstellar absorption can play a significant role in dimming starlight, we must take it into account in our expression for the distance modulus. Equation (4.8) becomes

$$d \text{ (pc)} = 10^{(m-M+5-A_\lambda)/5} \tag{7.9}$$

where A_λ is the total number of magnitudes that are absorbed. The subscript λ indicates that we must specify the wavelength at which we make the observation: the amount of absorption depends upon wavelength. It is clearly vital for accurate distance determination that we know the amount of absorption. There is no space here to give the details of how astronomers measure absorption except to say that, by comparing the spectrum and brightness of stars of the same spectral type at various wavelengths, astronomers have created a model for absorption in different regions of the Galaxy.

Example 7.3 Interstellar absorption and the distance modulus

A G0 V star is observed to have an apparent magnitude of +12. How far away does the star appear to be? It is estimated that interstellar dust grains between us and the star absorb 2 magnitudes. What is the true distance to the star?

Solution. A G0 V star has an apparent magnitude of +5. From (4.8) we see that

$$d\ (\mathrm{pc}) = 10^{(m-M+5)/5} = 10^{(12-5+5)/5} = 250\,\mathrm{pc}.$$

Taking into account interstellar absorption, we see from (7.9) that

$$d\ (\mathrm{pc}) = 10^{(m-M+5-A_\lambda)/5} = 10^{(12-5+5-2)/5} = 100\,\mathrm{pc}.$$

These answers assume that the star itself is not somehow abnormal (which is the usual proviso for using the method of spectroscopic parallax) and also that it does not affect the dust grains in its vicinity (which might cause the estimate of absorption to be in error).

Trumpler gained further support for his conclusions when he pointed out that the more distant the cluster the redder its light appeared, a phenomenon called *interstellar reddening*. Note that this has nothing to do with the Doppler effect; rather it arises from wavelength-dependent scattering of photons off interstellar dust grains. Imagine equal numbers of red and blue photons leaving a distant star and impinging on a dust cloud. Since dust scatters blue photons more efficiently than red photons, fewer blue photons than red photons reach an observer here on Earth. The light is 'de-blued' or, equivalently, reddened. The greater the amount of dust that must be traversed, the greater the reddening.

In his studies of globular clusters Shapley had neglected absorption, with good reason: he saw stars of every colour in globular clusters, which implied that there was no absorption or reddening of starlight en route to Earth. Trumpler explained this by postulating that the dust was confined to a layer of just 120 pc (400 ly) on either side of the plane of the Galaxy. Since open clusters occur only in the plane, they were all equally affected; we can see globular clusters at high galactic latitudes, though, well away from the dust band. This had important implications for both Kapteyn's and Shapley's work, as the American astronomer Joel Stebbins (1878–1966) demonstrated.

Stebbins had one of the longest productive careers in astronomy: he published papers in the academic journals for more than 64 years. His most significant publication was in 1933. He showed that absorbtion caused globular clusters in the plane of the Galaxy to appear four times more distant than they really are. The thinness of the obscuring dust layer meant that globular clusters more than 10° above or below the plane of the Galaxy suffered little obscuration, and these distances needed no correction. The net result of all this, according to Stebbins, was that Shapley had overestimated the size of the Galaxy. By the same token, Kapteyn had underestimated the size of the Galaxy: the obscuring effects of dust meant he was simply unable to see most of the stars in the Galaxy. Consequently, Stebbins arrived at a compromise between the Kapteyn and Shapley universes.

According to Stebbins, the Galaxy was a disc of stars. The disc had a diameter of 28 kpc (90 000 ly) and a thickness of about 920 pc (3000 ly), except for the central bulge, which had a thickness of about 3.7 kpc (12 000 ly). The Sun was about 9 kpc (30 000 ly) from the centre of the Galaxy. See figure 7.6.

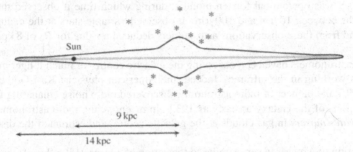

FIGURE 7.6: The size of the Galaxy according to Stebbins. The Sun is about two-thirds of the way towards the edge of the Galaxy. A thin layer of dust crosses the plane of the Galaxy. As in the Shapley model, a halo of globular clusters surrounds the central region.

7.5   THE MODERN VIEW OF THE GALAXY

In recent years our knowledge of the Galaxy has increased dramatically. It turns out to be a far stranger and more exciting place than Herschel, Kapteyn or even Shapley could have imagined. We now know that it shines with the light of 150 billion suns, arranged in a disk of two main spiral arms and several smaller spiral arms. These spiral arms, which are even now giving birth to stars, wrap around a central bulge that contains many violent and so-far mysterious objects — including a massive black hole that lures stars to their deaths. A few hundred globular clusters, containing stars that are almost as old as the Universe, inhabit the outskirts of the Galaxy: the so-called *galactic halo*. If we could see the Galaxy from outside, it would be a truly awe-inspiring sight. (Figure 8.2 on page 172 shows an example of how our Galaxy might appear to distant alien astronomers.)

Although we know much more about the Galaxy, our knowledge of its size — and in particular the distance from the Sun to the centre of the Galaxy* — remains uncertain. The good news is that we now have many more techniques available to estimate R_0, the distance to the centre of the Galaxy.

We can define the centre of the Galaxy in different ways. However, it seems likely that the radio source Sgr A is within one parsec of the dynamical centre of the Galaxy. And recent observations indicate that Sgr A* is probably a black hole, with a mass about three million times that of the Sun, of the type believed to exist at the heart of most galaxies. So for all practical purposes we can take Sgr A* to be at the galactic centre.

7.5.1   Infrared and radio observations

The direct method of estimating the distance to the galactic centre is to look for objects there. Unfortunately, intervening gas and dust blocks our view of the centre, at least at visible wavelengths. Our Sun, for instance, would have an apparent visual magnitude of $+49$ if it were at the galactic centre; even our largest telescopes would be unable to detect it. We therefore need to study the galactic centre at wavelengths that can penetrate the dust and gas: we need to use radio and infrared telescopes.

Some of our best views of the central regions of the Galaxy have come from the *InfraRed Astronomical Satellite* (*IRAS*), an orbiting telescope launched by NASA in 1983. *IRAS* was operational for ten months, during which time it observed the sky at wavelengths between $10\,\mu$m and $100\,\mu$m. It observed variable stars at the centre of the Galaxy, and from these observations astronomers deduced a value for R_0 of 8 kpc (about 26 000 ly).

Radio astronomers have been observing the galactic centre for much longer than their colleagues working in the infrared. Indeed, the American physicist Karl Guthe Jansky (1905–1950), the pioneer of radio astronomy, discovered radio noise emanating from the central regions of the Galaxy as early as 1932. In recent years, radio astronomers have detected *water masers* in gas clouds at the galactic centre, and estimated the distance to them.

The action of a maser is very similar to the action of a laser. (Like 'laser', the word is an acronym: it stands for *m*icrowave *a*mplification by *s*timulated *e*mission of *r*adiation.) An interstellar maser can occur when giant gas clouds containing OH or H_2O molecules absorb large amounts of energy; typically they occur at the periphery of newly formed massive stars. When an H_2O molecule absorbs a photon, it can jump from its ground state to an excited rotational energy state. This process is known as *pumping*. The molecule does not stay in the excited state, of course; it emits a photon and drops to a state of lower energy. Suppose, though, that it does not drop back down directly to the ground state; suppose, as can occur in OH and H_2O molecules, it de-excites to a *metastable* state. A molecule can exist in a metastable state for some time before it returns to the ground state. The process of pumping, therefore, can cause lots of molecules to be in an excited metastable state. If more molecules are in the excited state than are in the ground state, we have a situation called *population inversion*. Now imagine that a microwave photon, with energy equal to the difference between the ground state and the metastable state, crosses a region in which population inversion has occurred. The photon eventually hits a molecule, causing it to drop from the excited state to the ground state and thus stimulating the emission of a photon. The emitted photon has the same energy as the original photon and travels in the same direction: we now have two identical photons. These two photons can stimulate the emission of a further two photons, producing four identical photons. These four photons give rise to eight ... and we quickly have a cascade of microwave photons. This is a maser. Figure 7.7 is a simplified energy-level diagram of a maser.

An individual maser in space is transient, lasting for perhaps a few weeks. While it exists, though, its microwave energy output is intense, focused and at a single frequency. And in regions like the galactic centre new masers are continuously switched on just as old masers fade away. Furthermore, masing spots are small in astronomical terms (typically

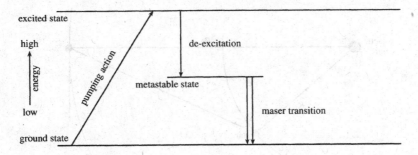

FIGURE 7.7: The energy levels of a simplified three-level maser.

10^8 km or smaller). This combination of brightness and small size means that we can use very long baseline interferometry (VLBI) to make precise astrometric observations of masers. In 1988, a group led by the American astronomer Mark Jonathan Reid (1948–) studied the proper motion of a masing source called Sgr B2 (North), which is close to the galactic centre. They deduced that $R_0 = 7.1 \pm 1.5$ kpc. Even though the answer was imprecise, this was the first *direct* measurement of the distance to the centre of the Galaxy. So far, observations of the H_2O maser sources in Sgr B2 constitute the best direct measure of R_0.

7.5.2   The kinematic distance to the galactic centre

In the previous chapter we discussed how the Sun moves relative to the local standard of rest (LSR). We also mentioned that the LSR orbits the centre of the Galaxy, behaving in some ways like a planet orbiting the Sun. If we can describe the galactic rotation with sufficient accuracy, we can use this information to determine the distance of the LSR from the centre of the Galaxy.

In 1927, Oort developed a theory of the *differential rotation* of the Galaxy, which supposed that the orbital angular speed of a star depends on its distance from the galactic centre. He began by assuming that all motions take place in circular orbits around the galactic centre and that the orbits all lie in the galactic plane. He defined R to be the distance of the star from the centre of the Galaxy*, Θ to be its circular orbital speed and ω to be its angular speed. These same symbols with a subscript 0, i.e. R_0, Θ_0, ω_0, refer to the LSR (more simply, they refer to the Sun once we make corrections for the solar motion relative to the LSR.) The star is at distance d from the Sun, has galactic longitude l, and the angle between the line-of-sight to the star and its orbital velocity is α. Figure 7.8 illustrates the geometry of Oort's model.

We are interested in v_r and v_t, the star's radial and tangential speed. From figure 7.8 we see that

$$v_r = \Theta \cos \alpha - \Theta_0 \sin l.$$

---

*The symbol R is often called the galactocentric radius of the star; so R_0 is the *solar galactocentric radius*.

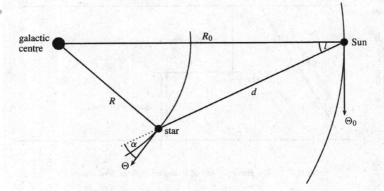

FIGURE 7.8: Oort's picture of galactic differential rotation.

From the triangle formed by the Sun, the star and the galactic centre, we have

$$R \cos \alpha = R_0 \sin l.$$

From the two equations above, and using $\omega = \Theta / R$ and $\omega_0 = \Theta_0 / R_0$, we have

$$v_r = R_0 (\omega - \omega_0) \sin l. \tag{7.10}$$

We obtain an expression for v_t in the same way. We have

$$v_t = \Theta \sin \alpha - \Theta_0 \cos l$$

and

$$d \sin l = R (\cos \alpha \cos l - \sin \alpha \sin l).$$

Solving for $\sin \alpha$ above and substituting it in the expression for v_t, and then using our expressions for $\cos \alpha$, ω and ω_0, we obtain

$$v_t = R_0 (\omega - \omega_0) \cos l - \omega d. \tag{7.11}$$

Equations (7.10) and (7.11) are the Oort equations for circular coplanar orbits around the galactic centre. They are valid for stars and gas clouds that are an arbitrary distance d from Earth. We are interested in the solar neighbourhood, though, so here we take $d \ll R_0$. In this case ω is very nearly equal to ω_0, so we can produce a Taylor series expansion and keep just the first two terms:

$$\omega - \omega_0 \approx \left(\frac{d\omega}{dR} \right)_{R_0} (R - R_0).$$

The expression $(d\omega / dR)_{R_0}$ is just the rate of change of the orbital angular speed with respect to distance, at the solar galactocentric radius. It is a measure of the local velocity shear, which is due to differential galactic rotation.

Finally, we define the *Oort constant A*:

$$A = -\left(\frac{R_0}{2}\right)\left(\frac{d\omega}{dR}\right)_{R_0}$$ (7.12)

to write the radial velocity as

$$v_r = -2A(R - R_0)\sin l.$$ (7.13)

A similar analysis yields

$$v_t = d(A\cos 2l + B)$$ (7.14)

where

$$B = A - \omega_0$$ (7.15)

is the Oort constant *B*. In the decades after Oort's work, astronomers put much effort into finding values for the constants *A* and *B*.

If we know the radial velocities and distances of several stars, we can use (7.13) to make a best fit to the data and thus estimate R_0. The radial velocities are relatively easy to obtain; the problem, as usual, is with distance. To use this method to estimate R_0 we are thus driven to work with Cepheids. Astronomers have used this method for many years, but recently it has undergone a revival. A recent study of the radial velocities of 107 Cepheids combined with a slightly modified version of (7.13) produced a best-fit value of $R_0 = 8.09 \pm 0.30\,\text{kpc}$.

As usual, the *Hipparcos* data have shed light on this problem. The South African-based astronomers Michael William Feast (1926-) and Patricia Ann Whitelock (1951–) studied the *Hipparcos* proper motions (and hence v_t) of 220 galactic Cepheids. The parallaxes of these Cepheids were all tiny (some, indeed, had negative recorded parallaxes) so it was important to treat the data in a statistically appropriate way. Once this was done, they used a modified version of (7.14) to deduce the Oort constants of galactic rotation; they deduced that

$$A = -14.82 \pm 0.84\,\text{km s}^{-1}\text{kpc}^{-1} \qquad B = -12.37 \pm 0.64\,\text{km s}^{-1}\text{kpc}^{-1}.$$

The value of *A* that is deduced from proper motions is essentially independent of the distance scale R_0, whereas the value of *A* deduced from radial velocities is directly proportional to R_0. A comparison of the two methods thus provides an estimate of the distance scale. Feast and Whitelock conclude that the kinematic distance to the centre of the Galaxy is $R_0 = 8.5 \pm 0.5\,\text{kpc}$.

7.5.3 RR Lyrae stars and subdwarfs

We can also follow Shapley's lead, and use the distribution of globular clusters to determine the distance to the galactic centre. If the distribution of globulars is spherical and centred on the massive nucleus of the Galaxy then the mean distance of the globulars will allow us to estimate R_0. Two types of star are particularly suitable for estimating the distance to globulars: RR Lyrae stars and subdwarfs.

One might suppose that it would be impossible to see RR Lyrae stars in globulars near the galactic plane: the dust band would dim their light to invisibility. Fortunately there are 'windows' in the dust through which we can study RR Lyrae stars even far across the Galaxy. So potentially we can use them as standard candles to calculate precise cluster distances. There are two difficulties we must be aware of when we do this.

The first difficulty has to do with the mean absolute magnitude of RR Lyrae stars: unless we know their intrinsic brightness we cannot use them to deduce absolute distances. Shapley was the first to calibrate their magnitude when he calculated a Cepheid distance to three RR Lyrae-containing clusters. He obtained a reasonable answer almost by accident: his calibration of the Cepheid period–luminosity relation was incorrect, but the Cepheids he studied were not the same as nearby Cepheids. By pure chance, his period–luminosity relation was nearly correct for the cluster Cepheids. We shall discuss the calibration of the Cepheid period–luminosity relation in more detail when we take the next step on the distance ladder. At this point, though, it is worth pointing out that a *direct* calibration of the RR Lyrae magnitude is desirable: that way we are not subject to knock-on effects from a possibly incorrect Cepheid calibration.

Unfortunately RR Lyrae stars are too far away for us to use trigonometrical parallax. *Hipparcos* measured only one RR Lyrae star with reasonable precision, and that was RR Lyrae itself. Astronomers have therefore used the method of statistical parallax to determine the distance, and hence absolute magnitude, of groups of RR Lyrae stars. (The *Baade–Wesselink method* can also be used to estimate the distance to RR Lyrae stars, but since it is better suited to a study of Cepheids we leave a discussion of the method to the next chapter.) The method of statistical parallaxes typically produces a value for $M_V(RR)$, the mean absolute magnitude of RR Lyrae stars, of 0.7 ± 0.1. For instance, a group of Japanese astronomers recently used *Hipparcos* data on almost 100 RR Lyrae stars and used a variety of methods to conclude that $M_V(RR) \approx 0.6$–0.7. Such a value for $M_V(RR)$ has been more or less the accepted value for several years. However, if the work of Feast and Whitelock is correct, then the absolute magnitude of RR Lyrae stars may be about 0.3 magnitudes brighter than this. The method of main sequence fitting supports a brighter value for $M_V(RR)$. So the value of $M_V(RR)$ is still under some dispute.

The second difficulty strikes at the heart of the distance-ladder approach: how can we be sure that distant objects are identical to the same type of objects nearby? In this case, how can we be sure than RR Lyrae stars in remote clusters have the same absolute magnitude as the nearby RR Lyrae stars for which statistical parallaxes exist? There is at least one way in which RR Lyrae stars vary: in their *metallicity*.

The metallicity of a star is the ratio of iron to hydrogen — [Fe/H] — in its atmosphere. It is thus an attempt to describe in a quantitative way the composition of a star. Astronomers define metallicity through a comparison with the Sun's composition:

$$\left[\frac{Fe}{H} \right] \equiv \log_{10}\left(\frac{N_{Fe}}{N_H} \right) - \log_{10}\left(\frac{N_{Fe}}{N_H} \right)_{\odot} \tag{7.16}$$

where N_{Fe} is the abundance of iron and N_H is the abundance of hydrogen. A star with the same composition as the Sun has [Fe/H] $= 0$. Young metal-rich stars have positive values of metallicity; [Fe/H] $\approx$ +1 is the largest value measured in the Galaxy. Old metal-poor stars have negative values of metallicity; [Fe/H] $\approx$ −4.5 is the smallest value measured in the Galaxy. The reason for a correlation between age and metallicity is that, over aeons, stars forge heavy elements through the nuclear fusion reactions taking place in their cores. When they die, stars eject these heavy elements into space; as time passes, the interstellar medium becomes enriched with metals. So a star that has formed only recently from these enriched gas clouds is likely to contain more heavy elements than an old first-generation star.

Metallicity defined in this way is just a single number, so it is almost certain to be an oversimplification of a complicated situation. There is no reason to believe that the abundances of *all* heavy elements vary in *exactly* the same way as iron. Nevertheless, the [Fe/H] notation is extremely useful and is commonly employed in the literature.

RR Lyrae stars are old stars, and thus have negative values of metallicity, but they do not all have the *same* value of metallicity. RR Lyrae stars in globular clusters are among the oldest stars and can have extremely low metallicity values. The galactic RR Lyrae stars, which are used in statistical parallax measurements, may have larger metallicities. And if M_V(RR) depends upon metallicity, which seems certain, then we must know the functional dependence before we can use RR Lyrae stars to deduce accurate distances. We must also give an [Fe/H] value whenever we mention M_V(RR). For example, the Japanese group working with *Hipparcos* quoted [Fe/H] = −1.6 when giving M_V(RR) $\approx$ 0.6–0.7. Feast and Whitelock derived a value of M_V(RR) = 0.3 at [Fe/H] = −1.9.

The functional dependence of M_V(RR) on metallicity is not well established, but it seems certain that metal-poor RR Lyrae stars are brighter than metal-rich RR Lyrae stars. For example, in 1992, the American astronomer Allan Rex Sandage (1926–) proposed the following relation:

$$M_V(\text{RR}) = 0.94 + 0.3\left[\frac{\text{Fe}}{\text{H}}\right]. \tag{7.17}$$

This relation, if correct, would confirm the suggestion that RR Lyrae stars are about 0.3 magnitudes brighter than the value derived from statistical parallaxes. This relation is by no means universally accepted, though, and much work remains to be done on finding the absolute magnitude of RR Lyrae stars and its dependence on [Fe/H].

Example 7.4 The distance of an RR Lyrae star

An RR Lyrae star has an apparent magnitude $m = 20$; measurements indicate that interstellar dust has dimmed its light by 4.5 magnitudes. Its [Fe/H] value is −1.47. Assuming (7.17) to be correct, how distant is the star?

Solution. Its apparent magnitude, taking account of absorption, is $m = +15.5$. Its absolute magnitude, from (7.17), is $M_V = (0.3 \times -1.47) + 0.94 = 0.5$. We can now calculate the distance:

$$d(\text{pc}) = 10^{(m-M+5)/5} = 10^{(15.5-0.5+5)/5} = 10^4$$
$$d = 10\,\text{kpc}.$$

Astronomers have made many estimates of the distance to the galactic centre based on RR Lyrae distances to globular clusters. Nearly all estimates have been in the range 7–9 kpc. There is thus reasonable agreement with other methods.

A different way of estimating the distance to globular clusters is to study subdwarfs. The Scottish astronomer Iain Neill Reid (1957–) studied *Hipparcos* parallax data on 15 metal-poor subdwarfs. These stars are all within 60 pc of Earth, so his calculated distances were accurate. He discovered that these stars are more distant, and therefore more luminous, than previously thought. The subdwarfs that Reid studied are all similar to globular cluster stars; indeed, they may once have been members of a cluster that was torn apart as it orbited the Galaxy. So Reid was able to use these stars as a standard candle for probing cluster distances. Since cluster stars are brighter than previously thought, the clusters must be farther away than was previously accepted. Reid found that the distance to metal-poor globular clusters had been underestimated by 2–5%. For extremely metal-poor clusters like M 92, though, the underestimate was 15% or more. Table 7.1 shows some recently measured globular distances, and the new distances calculated by Reid. (The distances derived for M 92-like clusters depend upon the selection of the reference stars, and it may be that the lengthening of the distance scale proposed in the immediate wake of the *Hipparcos* data was excessive. Reid would now drop the distance moduli of such extremely metal-poor clusters by about 0.15 magnitudes. But this still represents an overall increase in the globular cluster distance scale, and seems to confirm that the distance to the galactic centre has been underestimated.)

TABLE 7.1: The distance to some globular clusters. The second column gives recent Cepheid and RR Lyrae values. The third column gives values found by Reid (1997) using *Hipparcos* data. The most recent work suggests that the distance to M 92-like clusters may fall somewhere between the values given here.

Cluster	'Old' distance (kpc)	'New' distance (kpc)
M 5	7.24	7.76
M 13	6.50	7.87
M 15	10.14	11.91
M 68	9.42	11.43
M 92	8.32	9.68
NGC 6752	3.98	4.29

So if we take account of all the various ways of measuring R_0, how far away is the centre of the Galaxy? The answer to that question is still uncertain. In 1963 the International Astronomical Union recommended the value $R_0 = 10$ kpc. In 1985 they adapted the value $R_0 = 8.5$ kpc. In 1993, Mark Reid surveyed all the recent work and concluded that the best value was 8.0 ± 0.5 kpc. There was thus a trend towards a smaller Galaxy over the three decades from 1963. That trend seems to have been reversed: the *Hipparcos* data on subdwarfs and Cepheid proper motions indicate that the Galaxy is slightly larger than Reid suggested. Noting that this seems to contradict the findings of many astronomers regarding the value of $M_V(\text{RR})$, perhaps the best value to take is:

$$R_0 = 8.5 \pm 0.5 \, \text{kpc.} \tag{7.18}$$

Attempts are now being made to obtain the value of R_0 with much more precision: astronomers are using the new Very Large Baseline Array (VLBA) to determine the trigonometric parallax of Sgr A*. This will directly give the distance to the centre of the Galaxy. The proposed *SIM* and *GAIA* missions will also have the capability of measuring trigonometric parallaxes at the galactic centre. Within a few years we should know the distance to galactic centre to within 1% or so.

The full extent of the Galaxy is more difficult to measure. Just as there is no sharp edge to the solar system, there is no well-defined boundary to the Galaxy. Stars and clouds of gas extend outwards, with decreasing density, to vast distances. The overall diameter of the Galaxy is probably of the order of 30 kpc (over 100 000 ly).

CHAPTER SUMMARY

- The stars are part of a system with a definite size and structure: the Galaxy.

- Herschel, Seeliger and Kapteyn attempted to discern the size and shape of the Galaxy by counting the number of stars of each magnitude in different parts of the sky. Such attempts were doomed to failure because interstellar absorption causes most stars in the Galaxy to be hidden from view.

- Cepheid variables are radially pulsating stars; they obey a period–luminosity relation. By calibrating this relation, we can use Cepheids as a yardstick.

- Shapley used Cepheids and other methods to determine the distance to globular clusters. From these distances, and the observed distribution of the clusters, he arrived at a model of the Galaxy in which the Sun was far removed from the galactic centre.

- The current best estimate of the distance of the Sun from the centre of the Galaxy is

$$R_0 = 8.5 \pm 0.5 \, \text{kpc.}$$

- Estimates of R_0 come from:

  (a) a direct measurement of the proper motion of H_2O masers near the galactic centre

  (b) an analysis of the differential rotation of the Galaxy

  (c) distances to globular clusters from measurements on subdwarfs and RR Lyrae stars.

- The disk of the Galaxy is perhaps 30 kpc in diameter and about 700 pc thick (except for the central bulge, which is about 5 kpc thick). The disk contains a great deal of dust and gas; to see through this interstellar material and study the central regions we must either look for 'windows' at visible wavelengths or else use infrared and radio telescopes.

QUESTIONS AND PROBLEMS

7.1 Write an account of how Mira variables may be used as distance indicators.

7.2 Observations of Cepheids at visible wavelengths are plagued by absorption. What are the advantages and disadvantages of observing these stars at infrared wavelengths? Can we still use them as distance indicators?

7.3 We know of 55 Cepheids and 26 RR Lyrae stars that lie within 1 kpc of the Sun. Discuss the sort of investigation that is required to yield parallax-based distance estimates of these objects to better than 1%. If such an investigation were to be successful, what would it mean for our estimate for R_0?

7.4 Can we use the Oort equation to determine d in terms of R_0? Is this method a valid distance indicator for stars and gas clouds in the Galaxy?

7.5 Cepheids A and B, whose light curves are shown in the diagrams, are found in the same open cluster. Which is more luminous?

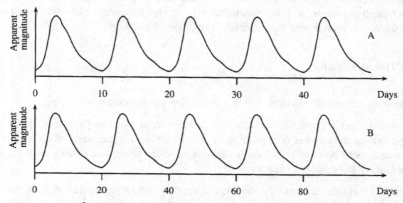

Cepheid B has a longer period than A so, since they are at the same distance, B is brighter.

7.6 Show that $A = \Theta_0/2R_0$, where A is the Oort constant and R_0 is the solar galactocentric radius. What is the value of Θ_0 if $A = -15\,\mathrm{km\,s^{-1}\,kpc^{-1}}$ and $R_0 = 8.5\,\mathrm{kpc}$. If the Sun is 5×10^9 years old, how many circuits of the Galaxy has it completed? $\Theta_0 = 255\,\mathrm{km\,s^{-1}}$; about 24 circuits

7.7 If the Sun were transported to the galactic centre (8.5 kpc away) what would be its apparent magnitude? Assume there is no absorption by dust and gas.   +19.44

7.8 Typically how much more luminous is an RR Lyrae star than the Sun?   About 50 times

7.9 Regarding a Cepheid simply as an oscillating system, estimate a typical Q value for a Cepheid. Compare this with Q values for some other common oscillators. 10^9–10^{10}; this is a high value

FURTHER READING

(Note: see the next chapter for references on the calibration of the Cepheid period–luminosity relation.)

GENERAL
Croswell K (1996) *The Alchemy of the Heavens* (Oxford University Press: Oxford)
— A well-written popular account of the latest research on our Galaxy.

Elitzur M (1992) *Astronomical Masers* (Kluwer: Dordrecht)
— A detailed examination of all aspects of the maser phenomenon in astronomy, including their use as distance indicators. Suitable for graduate students and above.

Elitzur M (1995) Masers in the sky. *Sci. American* **272 (2)** 52–58
— A more popular account of masers.

Mihalas D and Binney J (1981) *Galactic Astronomy: Structure and Kinematics* (Freeman: San Francisco)
— The standard text on galactic astronomy.

Paul E R (1993) *The Milky Way Galaxy and Statistical Cosmology, 1890–1924* (Cambridge University Press: Cambridge)
— The history of the Herschel/von Seeliger/Kapteyn approach to statistical cosmology.

Percy J R (1975) Pulsating stars. *Sci. American* **232(6)** 67–75
— A dated but exceptionally clear, non-mathematical account of stellar pulsations.

Whitney C A (1971) *The Discovery of our Galaxy* (Angus and Robertson: London)
— Of interest for its biographical sketches of the astronomers who shaped our understanding of the Galaxy.

CLASSIC PAPERS

Leavitt H S (1907) 1777 variables in the Magellanic Clouds. *Ann. Harvard College Observatory* **60** 87
— The first indication of a period–luminosity relation for Cepheids.

Leavitt H S (1912) Periods of 25 variable stars in the Small Magellanic Cloud. *Harvard College Observatory Circular* no 173
— The first paper devoted solely to the period–luminosity relation for Cepheids.

Trumpler R J (1930) Preliminary results on the distances, dimensions and space distribution of open star clusters. *Lick Observatory Bull.* **14** 154–188
— Proof of the existence of stellar absorption.

RR LYRAE STARS

Chaboyer B *et al.* (1998) The age of globular clusters in light of *Hipparcos:* resolving the age problem? *Astrophys. J.* **494** 96–110
— The authors review five techniques for setting the distance scale to globular clusters, and show that globulars are more distant than previously thought. They adopt $M_V(RR) = 0.39 \pm 0.08$ at [Fe/H] $= -1.9$.

Sandage A R (1993) The Oosterhoff period effect and the age of the galactic globular cluster system. In: *New Perspectives on Stellar Pulsation and Pulsating Variable Stars* ed. J. M. Nemec and J. M. Matthews. (Cambridge University Press: Cambridge)
— Sandage derives the calibration $M_V(RR) = 0.94 + 0.30$[Fe/H]. This volume also contains several other interesting papers on RR Lyrae stars and Cepheid variables, and the references in these papers give a thorough overview of the recent research literature on pulsating stars.

Smith H A (1995) *RR Lyrae Stars* (Cambridge University Press: Cambridge)
— The definitive account of this class of star; the book is accessible to undergraduates but is mainly aimed at the graduate student and researcher.

Tsujimoto T, Miyamoto M and Yoshii Y (1998) The absolute magnitude of RR Lyrae stars derived from the *Hipparcos* catalogue. *Astrophys. J.* **492** L79–L82
— The statistical parallax of 99 RR Lyrae stars gives $\langle M_V \rangle = 0.69 \pm 0.10$. Applying a Lutz–Kelker correction to HIP 95497, the RR Lyrae star with the most accurately measured parallax, gives $M_V = (0.58-0.68)^{+0.28}_{-0.31}$. Using two more methods, the authors conclude that $M_V \approx 0.6$–0.7.

STELLAR PULSATION

Christy R F (1966) Pulsation theory. *Ann. Rev. Astron. Astrophys.* **4** 353–392
— Useful for the references it contains to the original papers on stellar pulsation theory.

Hansen C J and Kawaler S D (1994) *Stellar Interiors: Physical Principles, Structure and Evolution* (Springer: Berlin)
— The fundamentals of stellar structure, including variable stars, for the beginning graduate student.

DISTANCE TO THE CENTRE OF THE GALAXY

Feast M W (1997) RR Lyraes, galactic and extragalactic distances, and the age of the oldest globular cluster. *MNRAS* **284** 761–766
— The author suggests an alternative treatment of RR Lyrae magnitudes that escapes the usual biases, and concludes that $R_0 = 8.1 \pm 0.4$ kpc.

Feast M W and Whitelock P (1997) Galactic kinematics of Cepheids from *Hipparcos* proper motions. *MNRAS* **291** 683–693
— The authors study the *Hipparcos* proper motion of 220 galactic Cepheids, thereby obtaining improved values for the Oort constants and a value for the distance to the galactic centre of $R_0 = 8.5 \pm 0.5$ kpc.

Paczyński B (1998) Galactocentric distance with the OGLE and *Hipparcos* red clump stars. *Astrophys. J.* **494** L219–L222
— The author uses a particular class of stars to deduce that $R_0 = 8.4 \pm 0.4$ kpc.

Pont F, Mayor M and Burki G (1994) New radial velocities for classical Cepheids. Local galactic rotation revisited. *Astron. Astrophys.* **285** 415–439
— The authors use radial velocity measurements on 107 Cepheids to fit to a model of galactic rotation. They find $R_0 = 8.09 \pm 0.30$ kpc.

Reid I N (1997) Younger and brighter — new distances to globular clusters based on *Hipparcos* parallax measurements of local subdwarfs. *Astron. J.* **114** 161–179
— The author uses *Hipparcos* parallaxes on 15 nearby metal-poor stars to redefine the subdwarf main sequence. The derived distances to globular clusters then increases.

Reid M J *et al.* (1988) *Astrophys. J.* **330** 809–816
— From observations of the masing source Sgr B2 (North), the authors obtain $R_0 = 7.1 \pm 1.5$ kpc.

Reid M J (1993) The distance to the centre of the Galaxy. *Ann. Rev. Astron. Astrophys.* **31** 345–372
— The author discusses all the various ways of estimating R_0.

8

Sixth step: the Local Group

We know the distance scale of the Galaxy to an accuracy of only about 10% or so. This accuracy should improve to perhaps 1% in forthcoming years, but that will still leave us with an unanswered question: what is the distance scale *beyond* the Galaxy?

Even as late as 1925, the very existence of extragalactic objects was controversial. Astronomers belonged to one of two camps: those who thought the Galaxy was big enough to contain all the celestial objects that had been seen, and those who believed in the existence of extragalactic objects lying at distances that dwarfed the distance scale of the Galaxy. On 26 April 1920, the National Academy of Sciences in Washington, DC hosted a debate on 'The Scale of the Universe'. The two participants in the debate were Harlow Shapley and the American astronomer Heber Doust Curtis (1872–1942). Although the debate between them was not newsworthy at the time, and in fact settled nothing, it took on an almost mythical status as the years passed. We now know it as the 'Great Debate'. It is worth briefly reviewing the arguments presented during the Great Debate because the same sorts of argument are presented nowadays when astronomers are uncertain of the distance scale of some class of object. (Indeed, in 1995 the distance scale associated with γ-ray bursters was debated in exactly this manner.)

Once we have established the existence of an extragalactic distance scale we can then proceed to show how our distance indicators — particularly the Cepheids — can calibrate this step of the distance ladder.

8.1 THE GREAT DEBATE

In the early years of this century one of the great unsolved mysteries in science was the nature and distance of the so-called *spiral nebulae*. Were they clouds of gas, as some astronomers maintained? Or were they, as many others believed, 'island universes' — galaxies like our own, but appearing small because of their huge distance from us?

The French astronomer Charles Messier (1730–1817) was the first to take notice of nebulae. Messier was a comet hunter. During his searches he often came across fuzzy patches of light that looked as if they might be cometary nuclei. Sometimes these nebulous objects were indeed comets: they developed the familiar tail and they moved quickly across the sky. Sometimes, though, they were clearly not comets: they were unchanging in appearance and fixed in position. If they were not comets — and they certainly were not stars — they had to be some new type of celestial object. To ensure that he and his fellow comet hunters wasted no more time on such objects, he prepared a catalogue containing the positions of 107 nebulae. The first object in his catalogue became known as M 1, the second as M 2, and so on. (Messier showed no further interest in the nebulae. This was unfortunate, because the objects in his catalogue are of far more scientific interest than any comet he discovered.)

Later Herschel, in his systematic way, decided to make a list of all the non-stellar objects he could find. Over the next 20 years he catalogued over 2500 nebulae. But what were they? His first idea, that they were all objects like our own Galaxy, did not last long. Although he found that some nebulae contain stars (he was the first to resolve individual stars in the globular cluster M 13, for instance) he also found nebulae that did *not* contain stars. For example, he discovered that M 42, the Orion Nebula, is a large glowing cloud. (In an astonishingly prophetic remark, Herschel described M 42 as 'the chaotic material of future suns'. The tenuous mass of gas and dust in M 42 will indeed collapse and give birth to protostars. The nebula already contains some young bright stars, and it is the ionizing radiation from these stars that causes the surrounding gas cloud to glow.) Such a gas cloud is what modern astronomers call a nebula. In fact, M 42 is one of the few true nebulae in Messier's catalogue*.

The most enigmatic objects in Messier's catalogue were not globular clusters like M 13, nor gas clouds like M 42, but the spiral nebulae. The Irish astronomer William Parsons, the third Earl of Rosse (1800–1867), was the first to show that M 31, the Andromeda Nebula, had a spiral shape. Astronomers found other spiral nebulae in Messier's list — objects like M 33 and M 101, for instance. Such definite structure led some astronomers to conclude that these objects could not be chaotic clouds of gas; the spirals had to be groupings of stars like our own Galaxy. On the other hand, no observer could resolve any stars in the spirals. And there was no definite evidence that *anything* lay outside the confines of our Galaxy. So were the spiral nebulae galactic or extragalactic objects? This was the subject of the Great Debate.

8.1.1 Shapley's argument

Shapley was sure that the Galaxy was large enough to contain everything that astronomers had yet observed. (Remember that his original estimate for the galactic diameter was 90 kpc, a huge figure at that time). According to Shapley, spiral nebulae were associated in some way with the Galaxy. Some historians suggest that Shapley had deep-seated psychological reasons for believing in the uniqueness of a Galaxy he had done so much to chart. Maybe so. But he also had clear evidence that seemed to indicate that spiral nebulae were local objects. Astronomers of the time could present several arguments to support this point of view. Below, I describe five of them.

Rotation of a spiral nebula.   Adriaan van Maanen (1884–1946), Shapley's colleague at the Mount Wilson Observatory, announced in 1916 that the spiral nebula M 101 rotated about its centre like a catherine wheel. The period of rotation, according to van Maanen,

---

*The Orion Nebula is an *emission* nebula, since the gas cloud emits light of its own. We can see a *reflection* nebula, on the other hand, only because the gas cloud reflects light. A *dark* nebula, such as the Coal Sack, is a gas cloud that obscures the light from stars behind it.

was 85 000 years. If M 101 were the same size as the Galaxy, and appeared small just because it was distant, then such a rotation period would imply rotation speeds much greater than the speed of light. Since such speeds are impossible, the only conclusion was that M 101 was nearby.

Example 8.1 Rotation of a spiral nebula

You observe a spiral nebula to have an angular diameter of 25′ and a period of rotation of 85 000 years. Estimate the *maximum* distance of the nebula.

Solution. Suppose for simplicity that the nebula rotates as a rigid body. It is unlikely to rotate so quickly that its material approaches relativistic speeds, so suppose that the maximum speed it can attain is $0.1c$. If r_{max} is the radius at which the nebular material moves at $0.1c$, then $r_{max} = 0.1c/\omega$, where $\omega = 2\pi/T$. Substituting

$$T = 85\,000 \text{ years} = 2.68 \times 10^{12} \text{ s}$$

we see that

$$r_{max} = 1.3 \times 10^{19} \, m = 420 \, \text{pc}.$$

Since we know that the nebula subtends an angle of 25′ we can calculate d, its distance from us: $d \approx 115 \text{ kpc}$.

This distance would put the nebula just outside the confines of Shapley's Galaxy. If we demand maximum speeds that are *much* less than c, the nebula must be closer. Furthermore, a radius for the nebula of 420 pc means it is tiny compared to the Galaxy itself. This would make it seem likely that the nebula is a galactic phenomenon rather than a stellar system in its own right.

Conversely, if the nebula is the same size as the Galaxy, the outer reaches of the nebula must be travelling at superluminal speeds — which is impossible.

Novae.   In order to understand another of Shapley's arguments, we must first discuss the phenomenon of novae — 'new stars'. A nova is an uncommon event. Between 1570 and 1890 only a dozen new stars became visible to the naked eye. The most famous of these were Tycho's nova in 1572 and Kepler's nova in 1604. For a brief while they became the brightest stars in the sky, before dimming and disappearing from view. Later novae were less spectacular, but astronomers trained their telescopes and their spectroscopes on them and studied novae in great detail. By the time of the Great Debate, astronomers knew much about the evolution of novae — in large part thanks to the work of amateurs.

Nova T Coronae Borealis, for instance, reached a peak apparent magnitude of +2: bright, but not incredibly so. An Irish amateur astronomer, John Birmingham (1816–1884), first saw this nova on 12 May 1866, just before midnight. Four hours earlier, Johann Friedrich Julius Schmidt (1825–1884), who was then Director of Athens Observatory in Greece, had observed the same patch of sky and was sure that no such star was visible to the naked eye. In other words, the nova had been fainter than magnitude +6. Within the space of four hours, therefore, it must have brightened by several magnitudes.

The nova gradually dimmed, and nine days later it was again invisible to the naked eye. (Nova T Coronae Borealis was the first to be studied with the aid of a spectroscope. Huggins examined its spectrum on 16 May 1866.)

In November 1876, Nova Cygni reached magnitude +3 before dimming over the next two weeks to magnitude +6. By March 1877 the star was completely lost from view, but it reappeared in September of that year. Upon its reappearance, its spectrum had changed completely, and was by then characteristic of a gas cloud being heated from within.

Similar events occurred with Nova T Aurigae. A Scottish amateur astronomer discovered this nova on 1 February 1892, and sent a note to the Astronomer Royal for Scotland, Ralph Copeland (1837–1905). Astronomers scoured their records immediately. On 8 December 1891, and perhaps even later, the star was fainter than magnitude +13. On 10 December 1891, astronomers at Harvard College Observatory had seen it in its nova state. It was then of magnitude +5.4. On 17 December 1891 it had reached its peak magnitude of +4.2. The nova dimmed and disappeared from view on 26 April 1892, but reappeared with magnitude +10 on 17 August. Huggins investigated the nova spectroscopically throughout this period, and showed that, just as in the case of Nova Cygni, the star's spectrum after its reappearance was different from its spectrum before it disappeared.

Nova Persei, a very bright nova discovered in 1901, followed a similar pattern. It went from magnitude +12.8 on 20 February to magnitude +0.1 on 23 February. It dimmed gradually, and by July it showed a spectrum characteristic of a cloud of gas being heated from within.

All novae, it seems, progress in more or less the same way. As the star brightens to its maximum, a process that can take anything from 1–15 days, it brightens by 11–13 magnitudes. At maximum light, the star 'bursts' and expels an expanding shell of gas. Over the next few days it expels more gas shells, and begins to dim. Often the decline is gradual and smooth. In a few cases, though, there are large fluctuations in magnitude that cause the star to disappear from view before reappearing. In either case, after a few months all we can detect is the spectrum of the gas shells being heated from within by the star. (I discuss the physical mechanism behind novae later in the chapter.)

We can now return to Shapley's argument.

In August 1885, the German astronomer Carl Ernst Albrecht Hartwig (1851–1923) discovered Nova S Andromedae in the bright central region of the spiral nebula M 31. Since astronomers had no way of knowing the distance to Nova S Andromedae it was possible that it just happened to lie in the same line of sight as M 31. However, the chances of this happening were very small. The nova was much more likely to be physically associated in some way with M 31. In 1911, the American astronomer Frank Washington Very (1852–1927) compared Nova S Andromedae with Nova Persei, whose distance had previously been estimated using the method of *expansion parallax* to be 460 pc*. Assuming that the maximum brightness of the two novae was the same (a plausible assumption, since, as we have seen, all novae seem to evolve in more or less the same way), Very concluded that M 31 was about 2.5 kpc distant. This was within the confines of the Galaxy. Shapley argued that if M 31 was an extragalactic object, then Nova S Andromedae would

---

*See the 'questions and problems' section at the end of the chapter.

have to have been incredibly luminous — millions of times more luminous than other novae.

Distribution of the spiral nebulae. Astronomers observed many more spiral nebulae away from the plane of the Galaxy than in the plane. If the spirals really were distant galaxies then their pattern of distribution on the sky should not refer in any way to our Galaxy. They should be found in equal numbers in all parts of the sky. The lack of spiral nebulae in the plane of the Galaxy indicated to Shapley that they were somehow associated with the Galaxy.

Surface brightness of the Galaxy. Frederick Hanley Seares (1873–1964), another of Shapley's colleagues at the Mount Wilson Observatory, compared the observed surface brightness of the spiral nebulae with his estimate of the surface brightness of the Galaxy. He showed that the Galaxy had to be much less bright than the spirals, if the spirals were at large distances. Since there was no reason to suppose that the Galaxy should be different in this way, he argued against a large distance for the spirals. (It is interesting that Shapley used this argument in the debate, because Seares' estimate of the surface brightness of the Galaxy used Kapteyn's estimate of the Galaxy's diameter. Shapley thought that Kapteyn's estimate of the Galaxy's diameter was far too small. Logically, he should have concluded that Seares' value for the surface brightness of the Galaxy also was too small.)

Speeds of the spiral nebulae. In the years following 1912, the American astronomer Vesto Melvin Slipher (1875–1969) studied the radial velocities of spirals. He found that many of them move away from the Galaxy at large speeds. Shapley thought this was difficult to understand if the spirals were large objects.

8.1.2 Curtis' argument

Curtis believed that Shapley's value for the size of the Galaxy was a gross overestimate. He preferred the Kapteyn universe, or something even smaller. To Curtis, therefore, the spiral nebulae were distant objects that lay outside our Galaxy — and were probably galaxies in their own right. He countered Shapley's arguments with some of his own.

Novae. In 1917, astronomers discovered four novae in M 31. They all had more or less the same apparent magnitude: between magnitude +15 and magnitude +18. This was very much dimmer than Nova S Andromedae, which had reached magnitude +7. Curtis argued that these four novae were typical of the novae seen in our Galaxy. Nova S Andromedae was atypical. It was some sort of 'super' nova. Assuming that the maximum brightness of the four novae was the same as the maximum brightness of Nova Persei, he concluded that M 31 was 150 kpc away. This put the Andromeda Nebula outside the Galaxy, even in Shapley's model.

The diameters of spiral nebulae. Spiral nebulae vary widely in apparent size. Curtis pointed out that the angular diameter of the largest known spiral was more than 1000 times greater than that of the smallest. If, as Shapley argued, the spirals were all at more or less the same distance from us, their intrinsic size had to vary by a factor of at least 1000. Curtis thought this unlikely, since all spirals had a very similar structure. He thought it more likely that the spirals had more or less the same intrinsic size, and that

Heber Doust Curtis

the variation in apparent size was due to their different distances from us. If this were the case the smallest spirals had to be about 1000 times farther away than the largest spirals. If the nearby spirals were 20 thousand light years distant, as Shapley argued, then the small spirals had to be 20 *million* light years distant. This would put them far out into the depths of space.

The distribution of spiral nebulae.   Curtis conceded that the distribution of spiral nebulae was a strong point in favour of Shapley's argument. But what if the plane of the Galaxy contained a lot of dust? The dust would create a 'zone of avoidance' in which spiral nebulae would be hidden from view.

Speeds of nebulae.   Curtis saw no reason why other galaxies should not have large speeds relative to our Galaxy. Slipher's results said nothing about the nature of the spiral nebulae.

Curtis began his professional life as a Professor of Greek and Latin, and was thus a more articulate and experienced public speaker than Shapley. He prepared a detailed technical presentation, whereas Shapley, who was angling for the job of director of the Harvard College Observatory, chose to pitch his talk at an elementary level. Curtis wrote to his family on 15 May 1920 that 'the debate went off fine in Washington, and I have been assured that I came out considerably in front'. As the months went by, though, it became clear that the result of the debate was inconclusive. Both men seemed to have strong arguments. The Great Debate stimulated discussion on the scale of the universe, and it brought the evidence out into the open. But it settled nothing.

8.2   HUBBLE AND GALAXIES

It was Edwin Powell Hubble (1889–1953), arguably the greatest astronomer of the twentieth century, who settled the questions raised in the Great Debate.

Edwin Powell Hubble

Despite his many accomplishments, Hubble seems to have been an insecure man. He embroidered many events from his early life, when the simple truth would have been impressive enough for most people. For instance, as a student he was a good all-round athlete and a fine boxer. In later life, though, he claimed to have fought Georges Carpentier to a draw in an exhibition bout. It is highly unlikely that such a bout ever took place. Hubble was a good amateur boxer, but he was not in this league; Carpentier was already a title holder at several weights, and would eventually become the world light-heavyweight champion. Hubble studied law at Queen's College, Oxford, as a Rhodes Scholar. He later claimed that, upon his return to America, he practised law part-time for one year and never lost a case. This seems to have been complete fabrication: Hubble never practiced law. Instead, he taught Spanish at a high school for a year before embarking on a PhD in astronomy at Yerkes Observatory. In 1919, he obtained a position at Mount Wilson Observatory. He remained there for most of his career — except for frequent and lengthy trips to his beloved England. (Throughout his life, Hubble affected an aloof Oxonian manner and tried to bury all traces of his Mid-Western roots. This irritated many of his colleagues, prime among them Shapley with whom he had in common a rural Missouri upbringing.)

Hubble's achievements in astronomy needed no embellishment. From the start he was interested in the nature of the spiral nebulae, and between 1919 and 1924 he uséd the 100-inch telescope at Mount Wilson to make a series of photographs of M 31 and M 33. The 100-inch telescope was then the biggest in the world, and Hubble made best use of it. He could resolve stars that were too dim to have been seen before. In February 1924 he discovered the first Cepheid variable star in a spiral nebula (M 31), and before the end of the year he found another 33 Cepheids in the spiral arms of M 31 and M 33. (Predating these discoveries was his discovery of Cepheids in NGC 6822, a distant nebula lacking in spiral arms. See figure 8.1.)

FIGURE 8.1: The irregular Local Group galaxy NGC 6822: the first galaxy in which Cepheids were detected.

Hubble found the absolute magnitudes of the Cepheids from the period–luminosity law. This directly gave their distances. He calculated that M 31 was 275 kpc (900 000 ly) distant and M 33 was 260 kpc (850 000 ly) distant. The irregular NGC 6822 was closer, but still extremely distant. There could be no question that these objects were well outside our Galaxy. When Shapley read Hubble's note to him, explaining the discovery, he turned to a colleague and said 'this is the letter that has destroyed my universe'. From their apparent size, Hubble estimated the diameters of M 31 and M 33 to be 12 kpc and 5 kpc respectively. It was clear, therefore, that the spiral nebulae were stellar systems just like the Galaxy but at very large distances. They were *spiral galaxies*. The term 'spiral nebula' became obsolete.

Once astronomers knew for certain that other galaxies existed, they could begin to better understand the points raised by the Great Debate. First, Curtis was right about the distribution of spiral galaxies. They populate all parts of the sky but, as Trumpler later showed, a layer of dust in the plane of the Galaxy prevents our seeing them there. Second, Curtis was also right about Nova S Andromedae. It *was* atypical. It was indeed a *supernova*. Third, the claim that M 101 rotated with a relatively short period was spurious. (In 1935, Hubble showed that the apparent rotations observed by van Maanen did not exist. No one knows for sure quite what van Maanen did wrong. His plates and equipment were not at fault.)

Though Curtis was right about the nature of spiral galaxies, he was wrong about the size and shape of our own Galaxy. Shapley's model of the Galaxy, and particularly our place in it, proved to be essentially correct. So who won the Great Debate? Both men had parts of the answer, but neither had the whole picture. In retrospect, then, the outcome of the debate might best be seen as a draw. (Such debates often take place in science, albeit on a much smaller scale, and their outcomes are often as inconclusive. Textbooks tend to present the progress of science as the march of a well-drilled platoon. In reality, it more resembles the progress of a drunken man staggering home. It often heads down blind alleys before finding the correct path.)

Hubble's discovery marked the beginning of the study of galaxies, and he remained at the forefront of this study. In 1937, for instance, he recognized that local galaxies are not distributed evenly over the sky. Their mutual gravitational attraction causes them to clump together. This gravitationally bound group of galaxies is called the *Local Group*. The distance scale of the Local Group forms the basis of this rung of the distance ladder. Before we can understand the Local Group, though, we need to understand more of Hubble's work on galaxies.

As Hubble investigated galaxies in more detail he found that they are not all spiral in shape. There are at least three types of galaxy: *spiral*, *elliptical* and *irregular*.

Spiral galaxies.   When seen 'head on', a spiral galaxy has spiral arms, typified by the spectacular M 83. (See figure 8.2.) When seen 'edge on', a spiral galaxy is noticeably flattened. But not all spiral galaxies are the same. Hubble classified a galaxy like M 31, with its large central nucleus and tightly wound spiral arms, as an Sa galaxy — 'S' for 'spiral' and 'a' to denote the tightness of the spiral arms. Spiral galaxies with a smaller nucleus and less tightly wound arms are Sb galaxies. Those with very open spiral arms are Sc galaxies.

Hubble found that about a third of all spiral galaxies are *barred* spirals. In a barred spiral the arms begin from a bar-shaped region rather than from a central spherical nucleus. Hubble classified the barred spirals as SBa, SBb and SBc, again depending upon the tightness of the spiral arms.

More recently, astronomers introduced the class of *lenticular* (S0) galaxies. These have flattened discs, like the spirals, but they lack the spiral arms. When viewed edge-on, they often appear to have the shape of a convex lens — a property that gives them their name.

FIGURE 8.2: The spiral galaxy M 83. Our own Galaxy probably looks much like this one. (Note that this is *not* a Local Group galaxy.)

Elliptical galaxies.   The elliptical galaxies are spheroidal or ellipsoidal in shape, and seem to differ from one another only in their degree of flattening. A spherical galaxy is an E0; a slightly flattened elliptical is an E1; and so on down to the most flattened elliptical, which is an E7. A really large elliptical galaxy can contain up to 100 times as many stars as our Galaxy.

The cD galaxies are luminous giant ellipticals surrounded by a large diffuse halo of faint starlight. They are found at the centres of rich clusters of galaxies. (Hubble missed cD galaxies from his classification scheme for the very good reason that he never saw one.)

In recent years, astronomers have found *dwarf ellipticals*, dE, and *dwarf spheroidals*, dSph. These are similar in structure to the ellipticals, but are often little bigger than a large globular cluster. In 1938, Shapley discovered the first two dwarf spheroidals, in Sculptor and Fornax, without realising what they were. Their low surface brightness was unlike any known galaxy, so he assumed that they were globular clusters. It was not until several years later that the German–American astronomer Wilhelm Heinrich Walter Baade (1893–1960) proved that Sculptor and Fornax are galaxies. We now know that such dwarf galaxies are the most common type of galaxy in the Local Group.

Irregular galaxies.   Hubble classified those galaxies in which he could see no obvious symmetry as *irregular* galaxies. A typical irregular galaxy is NGC 6822, the first galaxy in which Hubble discovered Cepheids. (See figure 8.1) When the true size of the Galaxy became clear, it followed that the Magellanic Clouds were themselves galaxies. Since the Clouds lack the definite structure of the spirals, or the symmetry of the ellipticals, Hubble classified them as irregulars. The irregular galaxies are often grouped into two subclasses: IrrI and IrrII. IrrI galaxies, like the Large and Small Magellanic Clouds, have an irregular appearance, but exhibit a mass distribution that is much more regular. In some ways they resemble spiral galaxies, and indeed some astronomers class the Large Magellanic Cloud as an SBc galaxy. IrrII galaxies are genuine irregulars. They probably owe their shape to gravitational interactions with nearby galaxies.

A few galaxies are so bizarre that they are not even classed as irregular. These *peculiar* galaxies are possibly the results of collisions.

Hubble classified the various types of galaxy then known in his famous *tuning fork* diagram* — see figure 8.3.

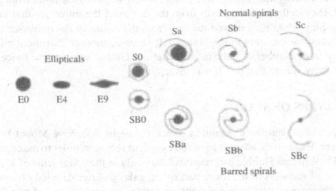

FIGURE 8.3: Hubble's 'tuning fork' classification of galaxies.

There was a problem with all this work, though. Hubble had Cepheid-based estimates of the distance to several nearby galaxies. He noticed that the brightest non-variable stars in these galaxies were always about 20 times brighter than the average Cepheid. As Shapley had done before him, he guessed that this was a general rule, which gave him a standard candle. In this way he soon estimated the distance to 40 galaxies. The farthest of these was, according to Hubble, at a distance of 1800 kpc (6 Mly). The problem was that, although galaxies clearly had a wide range of appearances, they all had one thing in common: when he calculated their size (which he could do, since he could measure their angular size and he believed he knew their distance) he found that *all* of them were much smaller than our Galaxy. Table 8.1 shows the mean diameter of different types of galaxy as given in an early paper by Hubble. Whether Shapley or Kapteyn were correct about the

---

*When he introduced this diagram he suggested that it might represent an evolutionary sequence for the galaxies, with a galaxy starting as an elliptical and evolving into a spiral. We now know that this suggestion is wrong — though our knowledge of the evolution of galaxies is still far from complete.

size of the Galaxy, it was clear from these figures that our Galaxy was gigantic compared to others. Only M 31 could begin to rival our own Galaxy in size.

TABLE 8.1: The mean diameter of different types of galaxy as determined by Hubble (1926b).

Type	Mean diameter (pc)	Type	Mean diameter (pc)
E0	360	Sa	1450
E1	430	Sb	1900
E2	500	Sc	2500
E3	590	SBa	1280
E4	700	SBb	1320
E5	810	SBc	2250
E6	960	Irr	1500

Astronomers throughout history had progressively removed mankind from privileged status. First, they dethroned the Earth from the centre of the universe: that honour belonged to the Sun. Then they moved the Sun from the centre of the universe: it was just one of a hundred billion stars in the Galaxy. It seemed, though, that there might be an unimaginably large number of galaxies and that our Galaxy was at least twice as big as any of them. Why should that be? It was disturbing.

8.3 TWO TYPES OF STAR

During 1942, with a wartime blackout in effect, the night sky above Mount Wilson was unusually clear. Baade took advantage of the excellent seeing in order to make an in-depth study of M 31. Whereas Hubble had resolved stars only in the spiral arms of M 31, Baade resolved some of the stars in the inner part of the galaxy. After detailed observations he decided that there were two completely different types, or populations, of star. The bluish stars in the spiral arms and the outskirts of M 31 were *Population I* stars. The reddish stars near the centre were *Population II* stars.

We now know that Population I stars are relatively young stars with high metallicity, which form in the dust clouds of the spiral arms. The Sun is a Population I star. Population II stars are old stars with low metallicity, which formed in elliptical galaxies, in globular clusters and in the dust-free central regions of spiral galaxies. The subdwarfs that Reid studied (see page 158) are Population II stars.

After the War, Baade continued his observations of M 31 with the new 200-inch Hale telescope at Palomar. He found more than 300 Cepheids in that galaxy, and discovered that they occurred in both populations. The two types of Cepheid have similar properties, but they are not identical. Their light curves show subtle differences, as do their spectra. By far the most important difference is that the period–luminosity relation for Population I Cepheids is not the same as the period–luminosity relation for Population II Cepheids. On average, Population I Cepheids are about four times more luminous (about 1.5 mag brighter) than Population II Cepheids. (Population I Cepheids are often called classical Cepheids, while Population II Cepheids are sometimes called W Virginis stars.)

Wilhelm Heinrich Walter Baade

Baade's discovery had a profound impact on the extragalactic distance scale. Until then, astronomers had used Shapley's calibration of the period–luminosity relation to calculate distance. Shapley calibrated the relation on the basis of the statistical parallax of 11 classical Cepheids, but due mainly to his neglect of interstellar absorption and galactic rotation, and to systematic errors in proper motions, he underestimated their luminosity by about 1.5 mag. Through sheer coincidence, his incorrect calibration of classical Cepheids turned out to be more or less correct for W Virginis stars. And since he detected W Virginis stars in globular clusters, his estimate of their distances was not seriously in error. This helps explain why the error was not picked up sooner: when he used the distance to globulars to calibrate the absolute magnitude of RR Lyrae stars, he accidentally obtained a reasonable answer. When astronomers used independent methods to determine the absolute magnitude of RR Lyrae stars they confirmed Shapley's answer, and thus seemed to confirm the Cepheid period–luminosity relation. Hubble found classical Cepheids in M 31 and other galaxies, but used in his analyses Shapley's period–luminosity relation. Since the zero point of this relation was too dim by about 1.5 mag, Hubble underestimated the luminosity of his Cepheids by a factor of four. This in turn meant that he underestimated the distances to the host galaxies by a factor of two. When Baade pointed out the error in 1952 the Galaxy remained the same diameter, but the universe doubled in size.

If the galaxies were twice as far away as originally thought, they had to be bigger than originally thought. After Baade's revision it remained the case that most nearby galaxies are smaller than the Galaxy. But M 31 is bigger. It contains perhaps 50% more stars than the Galaxy. Once again, mankind lost its special status in the universe.

8.3.1   New observational techniques

It took immense skill and patience for Hubble and Baade to resolve individual stars in other galaxies. Even with the biggest telescope in the world, the work was hard. Yet without accurate knowledge of the distance to nearby galaxies — which could be acquired only by detecting variable stars in those galaxies — it was impossible to calibrate other techniques that might work over longer distances. In recent years the situation has improved, and astronomers have a variety of new telescopes and observational techniques at their disposal. The most well known of these is the *Hubble Space Telescope*, of course, but there is also a host of new ground-based telescopes. The world's biggest telescope is now the twin Keck 10-m telescope, on Mauna Kea in Hawaii, but there are many other superb instruments. Some of them have adaptive optics installed, which helps combat the blurring effect of the atmosphere. Several telescopes look through different windows in the electromagnetic spectrum, thus complementing the view from optical telescopes. Best of all, telescopes now make use of *charge coupled device* technology — the most significant advance in astronomical imaging techniques since the invention of photography.

The Canadian physicist Willard Sterling Boyle (1924–) and the American physicist George Elwood Smith (1930–), working at Bell Laboratories, invented the charge coupled device (CCD) in 1970. A CCD is simply an array of semiconductors etched onto a doped silicon wafer. Underpinning the device is the photoelectric effect, a quantum process that enables the silicon wafer to convert light into electrons. The electrons then form an 'electronic image', analogous to the chemical image that forms on a standard photographic plate.

Imagine such a device placed at the end of a telescope that is pointing towards a galaxy. The silicon wafer converts incoming photons from the galaxy into electrons. Due to the way the wafer is made, the electrons cannot move within the silicon, and a coating of silicon dioxide prevents them from leaving the wafer. An electric charge therefore builds up in each cell of the semiconductor array. The size of the charge in a cell is directly proportional to the number of photons — in other words the brightness — falling upon that cell. At the end of an exposure the charges are measured, and the information fed directly to a computer. Because the information is in digital form the images can readily be processed and enhanced by computer software. The computer reconstructs an electronic image of the galaxy on a screen, with each pixel of the image corresponding to one of the semiconductor cells. Early CCD technology was limited to 800×800 arrays. Nowadays 2048×2048 arrays are not uncommon.

It was six years after Boyle and Smith's invention that astronomers used a CCD to obtain an astronomical image. In 1976, a team of American astronomers led by Bradford Adelbert Smith (1931–) took images of Jupiter, Saturn and Uranus using a CCD in conjunction with the 61-inch telescope at Mount Biglow, Arizona. Upon studying the images of Uranus they found the first evidence for particles high above the planet's methane layer. This discovery proved the usefulness of the technology in astronomy. The CCD has now replaced the conventional camera for much astronomical work. A modern CCD counts 80% of the photons falling upon it; in the red region of the spectrum the efficiency is close to 100%. Compare this with the fastest photographic emulsions available to Baade just after the War, which could count only 0.3% of incident photons. Even modern high-speed photographic emulsions count only 2–3% of incident photons. The remarkable optical efficiency of a CCD, working in tandem with the new generation of telescopes, means we can resolve stars even in quite remote galaxies. And this means we can obtain good estimates of their distances.

8.4 MEASURING THE DISTANCES TO NEARBY GALAXIES

Classical Cepheids are the best and most important distance indicator we have for nearby galaxies. For Cepheids to provide accurate distances, though, we must know their intrinsic brightness. It also helps if we have one or more independent checks on the Cepheid scale: we should keep in mind how astronomers were fooled until Baade discovered stellar populations. I discuss both these points in the sections below.

We should also note here that there are several complications in deriving the distance to a galaxy from the apparent magnitude of a star. For example, we must include a correction for obscuration by dust in our Galaxy. We must also make allowance for obscuration by dust *within* the galaxy being studied. This allowance is negligible for elliptical galaxies, since they are almost dust-free; for spirals, though, it is an important correction. I ignore the details here, and simply assume that all the appropriate corrections can be included.

8.4.1 Calibrating the Cepheid period–luminosity relation

In order to use Cepheids to derive the distances to galaxies, we must first calibrate the Cepheid period–luminosity relation*. Recall that Leavitt, from her original studies of Cepheids in the Magellanic Clouds, noticed that the relation between mean absolute magnitude $\langle M_V \rangle$ and period P (in days) was of the form

$$\langle M_V \rangle = a + b \log_{10} P. \tag{8.1}$$

We need to find values for a and b.

To determine the *slope* of this relation we just need a set of Cepheids with a large range in period and whose *relative* magnitudes are known. So the best place to establish the slope of the relation remains the LMC — even if we do not know the distance to the Cloud. (The LMC is better for this purpose than the SMC because it is less extended along the line of sight.)

To establish the *zero point* of the relation, and thus calibrate it absolutely, we must study Cepheids at a known distance. This means we must study Cepheids in our Galaxy (or else find an accurate and independent distance to the LMC). Cepheids are tantalisingly out of reach of trigonometrical parallax — except possibly, as we shall see later, for *Hipparcos* — so astronomers have resorted to other methods to obtain their absolute magnitude. Four of the commonest methods are statistical parallax, Cepheid-containing binary systems, the Baade–Wesselink method and main sequence fitting. The reading list at the end of the chapter includes references for each of these methods.

The statistical parallax method is the oldest method for determining the distance to a group of Cepheids: Hertzsprung and Shapley both used statistical parallax, and the method is still used. Although it produces relatively large errors, it provides a valuable check on other methods.

---

*The period–luminosity (PL) relation has some intrinsic scatter associated with it, since the instability strip on the HR diagram has a definite width. This led astronomers to develop the period–luminosity–colour (PLC) relation for Cepheids. The presence of a third parameter means that the PLC relation is much tighter than the PL relation alone. For our purposes, though, we need not delve into the details of the PLC relation.

In a few cases Cepheids have been detected in binary systems. We saw earlier how it is sometimes possible to estimate the distance to binary systems, and this enables us to deduce the luminosity of the Cepheid. The method has not been widely used, and the accuracy is poor, but again it serves as a check on other methods.

The Baade–Wesselink method is potentially an excellent way of deducing the distance to a Cepheid (and to some other types of celestial object). Baade initially proposed the method in 1926, and the Dutch astronomer Adriaan Jan Wesselink (1909–1995) improved the method in 1947. It works in the following way.

Consider a star of radius R at a distance d from us; suppose that the radiation flux at the star's surface is F_λ and that the radiation flux we observe here on Earth is f_λ. If for the moment we ignore the absorption of light from the star, then it can be shown that

$$f_\lambda = F_\lambda \left(\frac{R}{d}\right)^2. \tag{8.2}$$

In the case of a Cepheid, both the radius and the radiation flux are a function of time. Two observations at times t_1 and t_2 produce

$$\frac{f_{\lambda,1}}{f_{\lambda,2}} = \frac{F_{\lambda,1}}{F_{\lambda,2}} \left(\frac{R_1}{R_2}\right)^2$$

or

$$\frac{R_2}{R_1} = \sqrt{\frac{F_{\lambda,1} f_{\lambda,2}}{F_{\lambda,2} f_{\lambda,1}}}.$$

We can estimate $F_{\lambda,1}$ and $F_{\lambda,2}$ from the star's light curve (by assuming, for instance, that the star is a blackbody), and we can measure $f_{\lambda,1}$ and $f_{\lambda,2}$ directly, so that we know R_2/R_1: the ratio of the radii of the star at times t_1 and t_2.

If we record the spectrum of the star throughout its pulsation period we can determine $v_r(t)$: the radial velocity as a function of time. Simple integration gives us $R_2 - R_1$:

$$R_2 - R_1 = \int_{t_1}^{t_2} v_r(t)\,dt.$$

We know both R_2/R_1 and $R_2 - R_1$; two equations in two unknowns means we can solve for the radii. We determine the distance of the star from (8.2). A drawback with the method is that it needs very precise photometric and spectral data to be taken at the same time.

Perhaps the commonest way of obtaining the absolute magnitude of Cepheids is to detect them in clusters, and use main sequence fitting to determine the distance to the cluster. There are drawbacks with this method: we saw earlier how our understanding of the main sequence of the calibrating clusters is less than perfect; and in any case it is often difficult to be certain that a given Cepheid is a member of a particular cluster. Nevertheless, modern calibrations of the period–luminosity relation often use this method.

More than eight decades after the pioneering efforts of Hertzsprung and Shapley, obtaining the absolute calibration of the Cepheid period–luminosity relation remains a topic of research activity. During that time many workers have contributed to our understanding of Cepheids. It is impossible to credit all those workers, but we should mention Sandage and the American astronomer Robert Paul Kraft (1927–) who, in the late 1950s and early 1960s, helped establish the existence of the period–luminosity–colour relation for Cepheids. During the 1960s and 1970s, Sandage and his Swiss colleague Gustav Andreas Tammann (1932–) published a series of important papers on the Cepheid calibration. The Canadian-born husband-and-wife team of Barry Francis Madore (1948–) and Wendy Laurel Freedman (1957–) published an influential calibration of the period–luminosity relation in 1991, which has recently been superseded by a calibration by the English astronomer Nial Rahil Tanvir (1965–).

TABLE 8.2: Some influential calibrations of the Cepheid period–luminosity relation. $\langle M_V \rangle$ is the mean absolute magnitude; P is the period in days. (The period is the *fundamental* period, not the period of an overtone.)

Form of the PL relation	$\langle M_V \rangle$ at 10 days	Authors and year
$\langle M_V \rangle = -0.60 - 2.10 \log_{10} P$	-2.70	Hertzsprung (1913)
$\langle M_V \rangle = -0.72 - 2.10 \log_{10} P$	-2.82	Shapley (1918)
$\langle M_V \rangle = -1.67 - 2.54 \log_{10} P$	-4.21	Kraft (1961)
$\langle M_V \rangle = -1.43 - 2.80 \log_{10} P$	-4.23	Sandage and Tammann (1968)
$\langle M_V \rangle = -1.35 - 2.78 \log_{10} P$	-4.13	Feast and Walker (1987)
$\langle M_V \rangle = -1.40 - 2.76 \log_{10} P$	-4.16	Madore and Freedman (1991)
$\langle M_V \rangle = -1.38 - 2.77 \log_{10} P$	-4.15	Tanvir (1997)
$\langle M_V \rangle = -1.43 - 2.81 \log_{10} P$	-4.24	Feast and Catchpole (1997)

It is interesting to see how the commonly accepted form of the period–luminosity relation has changed over the years; table 8.2 shows some of the most influential estimates. It seems unlikely that we will see another change in the period–luminosity relation of the size introduced by Baade. The debate is at the level of 0.15 mag not 1.5 mag. But even small changes to the calibration can have a large impact on distances. For instance, the last entry in the above table refers to the recent work of Feast and his colleague Robin Michael Catchpole (1943–). Feast and Catchpole used data from *Hipparcos* to study the trigonometrical parallaxes of 223 Cepheids. Individually, the parallaxes are too small to derive a useful distance. The *closest* Cepheid in the sample was Polaris, which has a parallax of $\pi = 7.56 \pm 0.48$ mas. (Interestingly, Polaris turns out to be an overtone pulsator, so the period–luminosity relation is slightly more complicated in its case.) The Cepheid with the smallest parallax in the sample was U Sgr, which has a parallax of $\pi = 0.27 \pm 0.92$ mas. But even though these data are of little value in calculating the distances to each individual star (except perhaps for Polaris), Feast and Catchpole were able to analyse the parallax of the group as a whole and derive the Cepheid zero point. By combining their value for the zero point (i.e. -1.43 ± 0.10) with an existing value for the slope (i.e. -2.81 ± 0.06 based on 88 LMC Cepheids) they arrived at the period–luminosity relation

$$\langle M_V \rangle = -1.43 - 2.81 \log_{10} P. \tag{8.3}$$

This produces a 4% increase in the LMC distance compared to the Madore and Freedman calibration, but there is also another effect at work. If we use the Feast–Catchpole Cepheid scale to calculate the LMC distance, we also have to use their reddening scale. Taking everything into account, Feast and Catchpole estimate the LMC distance to be 54.9 kpc: a 10% increase on the Madore–Freedman estimate. Their distance to other galaxies depends on the specific corrections applicable to those galaxies. The M 31 distance, for instance, changes from the Madore–Freedman estimate of 767 kpc (about 2 500 000 ly) to the Feast–Catchpole estimate of 900 kpc (about 2 934 000 ly) — a 17% increase!

Example 8.2 The Cepheid period–luminosity relation

The table lists the period and absolute magnitude of 28 Cepheid variables in clusters and associations (absolute magnitudes are derived from estimates of the distance to the cluster; data from Gieren and Fouqué (1992)). Estimate a form for the period–luminosity relation.

Cepheid	Cluster or association	Period (days)	$\langle M_V \rangle$
EV Sct	NGC 6664	3.091	−2.86
SZ Tau	NGC 1647	3.149	−3.09
QZ Nor	NGC 6067	3.786	−3.16
CEb Cas	NGC 7790	4.479	−3.61
CF Cas	NGC 7790	4.875	−3.38
CEa Cas	NGC 7790	5.142	−3.70
VY Per	King 4	5.365	−3.34
CV Mon	Anon	5.379	−3.23
V Cen	NGC 5662	5.494	−3.22
CS Vel	Ruprecht 79	5.905	−3.58
V367 Sct	NGC 6649	6.294	−3.89
BB Sgr	Collinder 394	6.637	−3.05
U Sgr	M 25	6.745	−3.57
DL Cas	NGC 129	8.000	−3.88
S Nor	NGC 6087	9.754	−3.99
TW Nor	Lyunga 6	10.784	−4.14
V340 Nor	NGC 6067	11.288	−3.80
VY Car	Ass Car OB2	18.910	−4.76
RU Sct	Trumpler 35	19.702	−5.26
RZ Vel	Ass Vel OB1	20.399	−5.25
WZ Sgr	C1814-191a	21.847	−4.80
SW Vel	Ass Vel OB5	23.442	−5.02
T Mon	Ass Vel OB2	27.021	−5.59
KQ Sco	Ass Sco	28.695	−5.52
U Car	Ass Car OB2	38.806	−6.07
SV Vul	Ass Vul OB1	44.999	−6.46
GY Sge	Ass OB anon	51.062	−6.29
S Vul	Ass Vul OB2	67.593	−7.05

Solution. A data set like this needs a sophisticated statistical analysis, but we can obtain a rough estimate simply by fitting to the straight line

$$\langle M_V \rangle = a + b \log_{10} P.$$

If we do this we find that

$$\langle M_V \rangle = -1.3 - 2.9 \log_{10} P.$$

It is clear that slight revisions to the Cepheid period–luminosity relation can have significant effects on our distance estimates to nearby galaxies (although we should emphasize that the Feast–Catchpole calibration needs further verification before it can be accepted without question). It seems, though, that our calculations of Cepheid-based distances to some nearby galaxies are uncertain at the 10–15% level. Any uncertainty here will propagate through all later steps of the distance ladder, so it is important for us to look for other methods of estimating the distance to nearby galaxies. At the very least, such methods can act as a check on our Cepheid-based answers.

Example 8.3 A Cepheid distance

A careful study of a nearby galaxy finds ten Cepheids. The period, P, and mean apparent magnitude, $\langle m_V \rangle$, of each of the Cepheids is given in the table below. Using the Feast–Catchpole calibration of the period–luminosity relation, determine the distance to the galaxy. Try the same calculation with some of the other calibrations given in table 8.2.

P (days)	$\langle m_V \rangle$
4.3	21.80
6.7	21.22
7.2	21.15
9.8	20.81
10.1	20.74
12.2	20.53
12.3	20.53
21.7	19.79
32.4	19.34
41.9	19.00

Solution. Using the Feast–Catchpole calibration, we can take the period of each Cepheid and calculate its absolute magnitude. In the order given in the table above, we obtain

$$\langle M_V \rangle = -3.21, \ -3.75, \ -3.84, \ -4.22, \ -4.25,$$
$$-4.48, \ -4.49, \ -5.19, \ -5.67, \ -5.99.$$

We can then use (4.8) to calculate the distance of each Cepheid. Averaging these results we find that the galaxy is 1 Mpc distant. (In practice we would find much more scatter in the data than is included in this example!)

8.4.2   Other stellar standard candles

Although it is vital to have a check on Cepheid distances, so that we may have confidence in our estimates of the scale of the Local Group, we face a problem. The problem is to find suitable standard candles: few types of star shine brightly enough for us to detect them over extragalactic distances.

Novae.   We have already met one type of star that *does* have sufficient luminosity for us to detect them over the required distances: novae. Indeed, we saw earlier how novae were among the first stars to be used to estimate extragalactic distances. In 1917, Curtis compared novae in M 31 with those in the Galaxy to conclude that M 31 was very distant. He reached this conclusion by assuming that all novae reach the same maximum brightness. Unfortunately, he was wrong in this: the peak magnitude ranges from −5 to −10. In other words, at maximum brightness, the brightest nova is over 100 times more luminous than the dimmest nova. We cannot use the maximum brightness of novae as a standard candle, but there *is* a property of novae that we can use to determine distance.

Astronomers now know much more about the origin of novae. In 1964, Kraft showed that novae always occur in binary systems in which one of the stars is a hot white dwarf and the other is a main sequence star. The gravitational field of the dwarf captures any gas emitted from its companion. This gas spirals round the dwarf before falling onto its surface, where it is compressed and heated to a high temperature. When the temperature is high enough, nuclear reactions begin and a hydrogen-burning shell forms on the surface. As more material falls onto it, this burning shell of hydrogen becomes unstable. Eventually, the instability causes an explosion — a nova.

Physicists have used computer programs to model the energy output of novae, and these programmes suggest that the greater the mass of the white dwarf the greater the peak luminosity. Also, the more massive the dwarf, the less material needs to be captured before the shell of burning hydrogen develops the instability. Since a low-mass shell is more easily ejected into space than a high-mass shell, the decline in brightness at visible wavelengths is more rapid for low-mass shells. There is thus a relationship between the rate at which a nova declines in brightness and its peak absolute magnitude. So by measuring its light curve we can deduce the peak absolute magnitude of a nova, and this in turn gives us its distance. In 1978, the French–American astronomer Gérard Henri de Vaucouleurs (1918–1996) used this method to estimate various extragalactic distances. He calculated the distance of the LMC to be 49 kpc, the SMC to be 52 kpc and M 31 to be 617 kpc. His distance estimates of the Magellanic Clouds are in good agreement with the Cepheid method. His estimate of the distance to M 31 does not agree so well, but it is at least of the right order of magnitude.

Our understanding of the physics of nova explosions is less well developed than our understanding of the Cepheid pulsational mechanism, and in any case novae are not as common as Cepheids, so nova-based distances are unlikely to supplant Cepheid-based distances. Nevertheless, since the physics of the two processes are so different it is encouraging that there is this general level of agreement. We seem to be on the right track.

RR Lyrae stars.   From the discussion in the previous chapter one might suppose that RR Lyrae stars would make good extragalactic distance indicators. Unfortunately, although they are about 50 times more luminous than the Sun they are about 50 times less luminous than a typical Cepheid. They are thus much too dim to be seen easily over large distances. For many years it was impossible to resolve them in any galaxy beyond the Magellanic Clouds. Baade, for instance, expended much time and effort with the 200-inch Hale telescope in an unsuccessful search for RR Lyrae stars in M 31. It was not until 1987 that astronomers found RR Lyrae stars in M 31. The Dutch–Canadian astronomer Sidney van den Bergh (1929–) and the Canadian astronomer Christopher John Pritchet (1950–) found them using a CCD attached to the Canada–France–Hawaii telescope on

TABLE 8.3: The distance to four Local Group galaxies using Cepheids, RR Lyrae stars and the TRGB (adapted from Lee *et al.* (1993)). There are large errors associated with these measurements so, within errors, the distances are the same for each method. Note that the Cepheid distances given here are based on the Madore–Freedman calibration of the period–luminosity law. (Superscripts indicate references.)

Galaxy	Distance (kpc)		
	Cepheids	RR Lyrae	TRGB
LMC	50[a]	45[b]	48[c]
IC 1613	766[a]	714[d]	714[e]
M 31	733[a]	740[f]	773[g]
M 33	843[a]	875[h]	871[g]

[a] Madore and Freedman (1991).
[b] Walker (1988). [c] Reid *et al.* (1987).
[d] Saha *et al.* (1992). [e] Freedman (1988).
[f] Pritchet and van den Bergh (1987).
[g] Mould and Kristian (1988). [h] Pritchet (1988).

Mauna Kea. On the basis of these observations they estimated a distance to M 31 of 740 ± 50 kpc. They later resolved RR Lyrae stars in M 33, and estimated the galaxy's distance to be about 870 kpc. More recently, using the advances in CCD technology and the better telescopes mentioned earlier, astronomers have found RR Lyrae stars in several other nearby galaxies.

Table 8.3 shows RR Lyrae-based distance estimates of four Local Group galaxies: LMC, IC 1613, M 31 and M 33. It also compares these estimates with those derived from Cepheids (and also with the TRGB method, which we will discuss shortly). There is reasonable agreement between RR Lyrae and Cepheid distances, with perhaps some indication that RR Lyrae stars produce smaller estimates than Cepheids. The measurements have uncertainties associated with them, of course, and it may be that the true distances of these galaxies lie somewhere in the overlapping error bars.

Leaving aside the slight uncertainty in their apparent absolute magnitude, RR Lyrae stars are very useful distance indicators within the Local Group — if for no other reason than that several Local Group galaxies do not contain Cepheids. If we use RR Lyrae stars as a distance indicator in these cases, though, we should remember that they belong to a different stellar population than the Cepheids. RR Lyrae stars are Population II stars; classical Cepheids belong to Population I. When we continue extending our distance ladder we will actually have *two* ladders: a Population I ladder and a Population II ladder. We must be careful to tie the two structures together, or at least distinguish between them, or we risk being fooled.

What about those galaxies for which we can detect Cepheids, but not RR Lyrae stars? How can we check the Cepheid distance in these cases? One possible method is to use the *tip of the red giant branch* (TRGB).

The tip of the red giant branch. This method relies on identifying some of the brightest stars in a galaxy. The idea of using bright stars to measure a galaxy's distance is, of course, an old one. As we saw on page 173, Hubble used the idea when he assumed that the brightest stars are 20 times brighter than the average Cepheid. And Baade, in 1944, suggested that bright red stars might be standard candles. The present method, developed by Madore, Freedman and several of their colleagues, took these suggestions and put them on a firmer theoretical footing.

The reason we can use a population of red giant stars as a standard candle is as follows. A low-mass star becomes a red giant after it exhausts its supply of core hydrogen and starts to burn hydrogen in a thin shell surrounding a helium core. As time goes by the shell becomes bigger and the luminosity of the star increases. The star therefore moves up the red giant branch on the HR diagram. While this is happening, helium ash from the hydrogen-burning shell rains down on the core and adds to the helium already there. When the helium core reaches a critical mass, which is about $0.4 M_\odot$, the helium starts to burn furiously. The event is called the *helium core flash*. This large energy output lasts for only a short while; the outer parts of the star quickly contract and the star becomes less luminous. It moves back down the red giant branch. Eventually the star begins a new stage of its life: it moves over to a horizontal branch on the HR diagram, where it stays until it exhausts its supply of helium. A plot of red giant stars on a HR diagram therefore shows a discontinuity at the tip of the red giant branch. The discontinuity marks the helium core flash and shows the maximum luminosity of red giant stars. The TRGB has a luminosity that depends on the helium core mass. Since the core mass is almost constant (there is a dependence on metallicity, but it is only a slight dependence) the TRGB has a constant and uniform luminosity: if we can determine this luminosity we can use the TRGB as a standard candle.

Example 8.4 A TRGB distance

A discontinuity in the red giant branch of a galaxy is detected at an apparent visual magnitude of 21 mag. If the absolute visual magnitude of the TRGB is -3.5 mag, how far away is the galaxy? What uncertainties are there in this distance estimate?

Solution. The distance modulus of the galaxy is given by $(m - M)_0 = 21 + 3.5 = 24.5$ mag. This corresponds to a distance of $10^{(24.5+5)/5}$ pc $= 794$ kpc.

There are several possible sources of error in the measurements, as well as possible systematic errors in the analysis. For instance, uncertainty in the position of the tip may cause an error of ± 0.1 mag; uncertainty in the amount of absorbtion will probably be less than this, but it may not be negligible. We have already mentioned that the uncertainty in the RR Lyrae distance scale is one source of systematic error at the level of tenths of a magnitude. Uncertainty in the metallicity dependence of the TRGB magnitude is a much smaller effect. Measurements are usually made in the I-band. You can probably think of other effects that should be accounted for.

Unfortunately, the zero point of the TRGB magnitude depends upon the calibration of RR Lyrae stars which, as we have seen, is still uncertain. It seems likely, though, that when observed through an infrared filter with a central wavelength of $0.9\,\mu$m the TRGB has an absolute magnitude of about -4. Not only is the discontinuity luminous, it is easily observed. Unlike the hunt for Cepheids, which can take hours of telescope time over a period of months, the TRGB method needs just a single observation (though the single exposure is typically much longer than each Cepheid exposure). Finally, we can apply the method widely: all Local Group galaxies contain red giant stars. This makes the TRGB potentially a good standard candle for work with nearby galaxies.

In the early 1990s, Freedman, Madore and co-workers argued that the TRGB method compares well in accuracy with the Cepheid method. (See table 8.3.) Distances derived from the two methods seemed to agree to within 5%. This is important because the two methods are based on different observations, different physics, different calibrations and different types of stars. Since independent methods agree with each other we can have some confidence in both methods. Like the RR Lyrae method, the TRGB method is a Population II indicator, because it works with old stars.

Red clump stars.   A recent method of determining distances, developed by the Polish-born astronomer Bohdan Paczyński (1940–) and his colleagues at Princeton, is to use *red clump* stars as a standard candle. Red clump stars have undergone the helium flash and descended back down the red giant branch. They are thus helium core burning stars. They remain in a particular part of the HR diagram for a long time before evolving away; so they tend to accumulate — or clump — in that area of the HR diagram.

Hipparcos measured accurate trigonometrical parallaxes of about 2000 nearby red clump stars, of which about 600 had accurate I-band photometry. (As with the TRGB method, it is best to use I-band observations when using red clump stars as a standard candle.) The local red clump luminosity function is well described by a Gaussian with a peak at $M_I = -0.23 \pm 0.03$ and a dispersion of about 0.15 mag. The *Hubble Space Telescope* measured the apparent magnitude of about 6300 red clump stars in M 31. If the absolute magnitude of the clump depends only weakly on age and chemical composition, and if the stellar populations observed in M 31 are similar to the red clump population observed in the solar neighbourhood, then we can compare the *Hipparcos* and *Hubble Space Telescope* measurements and calculate the distance to M 31 in a single step. Astronomers using this technique recently obtained a distance to M 31 of 784 ± 30 kpc. (The same method gives the distance to the LMC of 41.02 ± 2.33 kpc. This distance is much smaller than other distance indicators.)

Red clump stars are only beginning to be used as distance indicators, and much work remains to be done on understanding and verifying the underlying assumptions. The method has great potential, though, because red clump stars can be found in great numbers and so statistical errors are small. And it is yet another way of checking our ideas of the distance scale in the Local Group.

8.5   THE LOCAL GROUP

By detecting Cepheids in nearby galaxies (or, where this is impossible, by detecting RR Lyrae stars, novae or the tip of the red giant branch) we can deduce galaxian* distances — and hence start to draw a map of the Local Group. It turns out that the Local Group contains about 30 galaxies. There are no lenticular or cD galaxies among these objects.

---

*So that they can use the word 'galactic' to refer adjectivally to our Galaxy, astronomers have invented the word 'galaxian' to refer adjectivally to other galaxies.

In fact, most of our Local Group neighbours are small irregular and dwarf spheroidal galaxies. Slightly more impressive are the irregular Magellanic Clouds and the elliptical galaxy M 32. But three galaxies dominate: M 31, M 33 and our own Galaxy. These three spirals emit more than 90% of the light from the Local Group.

By far the biggest galaxy in the Local Group is the majestic spiral M 31. If the Feast–Catchpole calibration is correct then M 31 lies at a distance of 900 kpc (2 930 000 ly). In any case, it the most distant object that can be seen with the naked eye, since under good conditions it can just be distinguished as a fuzzy patch of light. This light set out on its journey almost three million years ago.

Table 8.4 shows estimated distances to all the Local Group galaxies. Until recently, textbooks agreed that our nearest neighbours in the Local Group were the Magellanic Clouds. In 1994, three Cambridge-based astronomers discovered a galaxy that is much closer. Gerard Francis Gilmore (1951–), Michael John Irwin (1952–) and Rodrigo Alec Ibata (1967–) discovered SagdEg — the Sagittarius dwarf elliptical galaxy. SagdEg is only about 25 kpc (82 000 ly) from Earth, and about 15 kpc (50 000 ly) from the centre of our Galaxy. It is so close that the gravitational effects of our Galaxy will one day tear it apart.

SagdEg remained hidden all these years for two reasons. First, it is so spread out over the sky that its maximum surface brightness is less than that of the night sky. It does not grab one's attention on a photographic plate! Only recently have we been able to detect objects of such low surface brightness. Second, it lies in the zone of avoidance, the 5–10° band along the plane of the Galaxy where dust obscures visibility. There may be several other nearby dwarf galaxies hiding in that zone.

The latest[*] galaxy to be admitted to the ranks of the Local Group is the Antlia dwarf galaxy. This had been seen several times by astronomers as a dim, diffuse glow in the southern constellation of Antlia (the Air Pump). In March 1997, Irwin and two of his research students used the Cerro Tololo Inter-American Observatory in Chile to obtain CCD images of this glow. They succeeded in resolving red giant stars in the object. From the apparent brightness of these stars, Irwin deduced that the Antlia dwarf lies at a distance of about 1 Mpc (3.3 Mly). The Antlia and Tucana dwarf galaxies are the only dwarf spheroidals in the Local Group that do not dance attendance upon the big spirals.

Table 8.4 also shows the distance of the various Local Group galaxies from the centre of the cluster. As is clear from the table, nearly all of the galaxies are less than 900 kpc from the centre of the Local Group. A working definition of the size of the Local Group, therefore, is that it is a sphere of radius 1 Mpc. This definition would exclude only three galaxies: SagdIg (not to be confused with SagdEg), Antlia and Tucana. It is difficult to be more precise than this. A galaxy belongs to the Local Group if the gravitational effects of the rest of the Local Group determine its motion through space. Galaxies farther than

---

[*]In fact, it has recently been confirmed that three dwarf spheroidal galaxies discovered in 1998 are companions to M 31, and are thus members of the Local Group. The Russian astronomers Igor D. Karachentsev and Valentina E. Karachentseva found two dwarf galaxies that they called Pegasus dwarf and Cassiopeia dwarf. The American astronomers George H. Jacoby, Taft E. Armandroff and James E. Davies found two dwarf galaxies in the same region, and named them And V and And VI. It turns out that And VI is the same galaxy as Pegasus dwarf. So the three most recent additions to the Local Group are And V, And VI (Pegasus dwarf) and And VII (Cassiopeia dwarf).

TABLE 8.4: The distance to Local Group galaxies, adapted from van den Bergh (1994). The table also shows the distance of the galaxies from the 'centre' of the Local Group, assuming that the centre lies at a distance of 400 kpc in the direction of M 31. (The Galaxy and M 31 are the largest objects in the Local Group, so we can take the 'centre of mass' of the cluster to lie somewhere on a line connecting them. Since M 31 is more massive than the Galaxy, the centre is closer to M 31 than to the Galaxy.)

The list of galaxies in the table is complete at the time of writing, although it is extremely likely that more Local Group galaxies will soon be discovered. (Astronomers have discovered no less than seven Local Group galaxies in the 1990s: Tucana and Sextans in 1990, SagdEg in 1994, Antlia in 1997 and And V, And VI and And VII in 1998.) In addition, there are about 20 existing candidates for Local Group inclusion. At present these candidates are deemed to be too distant, or to be moving too fast, to belong to the Local Group. However, as more information about these galaxies becomes available, some of them may be admitted to the group. Note that the distances do not include the Feast and Catchpole correction. If their re-calibration of the Cepheid period–luminosity relation turns out to be correct, then many of the distances in this table may need to be increased by about 10%. Note also that some of the distances may be inconsistent with other values given in this book. This reflects the fact that distances on this rung of the ladder are uncertain! Different workers obtain different values.

Name	Type	Distance (kpc)	Distance from centre of Local Group (kpc)
Galaxy	Sb/Sc	—	400
SagdEg	dSph	25	425
LMC	IrrIII–IV	49	370
SMC	IrrIV/IV	58	430
Ursa Minor	dSph	63	380
Draco	dSph	63	370
Sculptor	dSph	78	390
Sextans	dSph	79	450
Carina	dSph	87	450
Fornax	dSph	131	400
Leo II	dSph	215	520
Leo I	dSph	273	580
Phoenix	dIrr/dSph	390	540
NGC 6822	IrrIV–V	540	660
NGC 185	dSph/dE3	620	230
NGC 147	dSph/dE5	660	270
NGC 205	S0/E5	725	320
M 31	SbI–II	725	320
M 32	E2	725	320
And I	dSph	725	320
And II	dSph	725	320
And III	dSph	725	320
And V	dSph	725	320
And VI	dSph	725	320
And VII	dSph	725	320
LGS 3	dIrr	760	410
IC 1613	IrrV	765	520
M 33	ScII–III	795	420
Aquarius	dIrr	800	800
Tucana	dSph	870	1060
WLM	IrrIV–V	940	800
Antlia	dSph	1000	1030
SagdIg	dIrr	1100	1180
IC 10	dIrr	1250	880

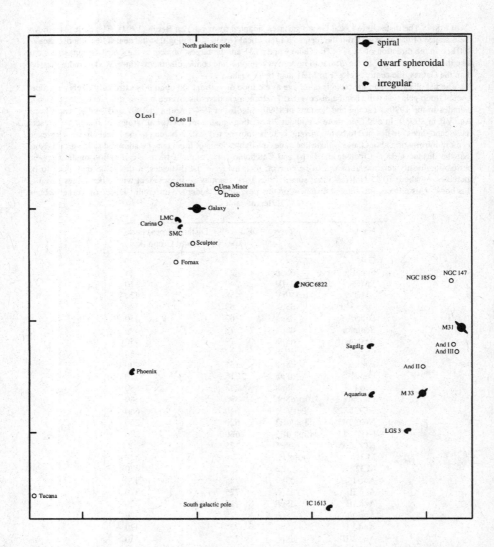

FIGURE 8.4: A side-on view of the Local Group. The zero of galactic longitude points out of the page towards the reader. The scale of the diagram is such that successive ticks on the axes are separated by 250 kpc. The area covered by the diagram is thus about 1 × 1 Mpc. At this scale, six galaxies cannot be shown. SagdEg, the closest galaxy to our own, is too close to be visible on this diagram. Similarly, at this scale the elliptical galaxies M 32 and NGC 205 are hidden from view by M 31. On the other hand, the dwarf spheroidal galaxy Antlia, and the irregular galaxies WLM and IC 10, are too remote to show at this scale. Antlia lies to the far left of this diagram, IC 10 to the far right, and WLM to the bottom right. The newly discovered And V, And VI and And VII are not shown. The clustering of small galaxies around the two dominant members of the Local Group, M 31 and our Galaxy, is clearly evident.

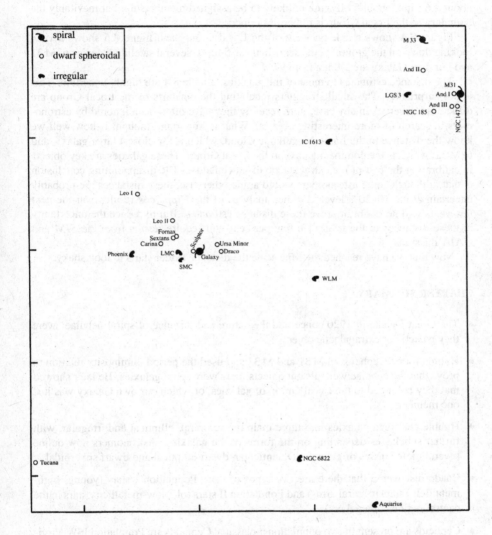

FIGURE 8.5: Looking down on the Local Group. The scale of the diagram is such that successive ticks on the axes are separated by 250 kpc. The area covered by the diagram is thus about 1×1 Mpc. At this scale, six galaxies cannot be shown. SagdEg, the closest galaxy to our own, is too close to be visible on this diagram. Similarly, at this scale the elliptical galaxies M 32 and NGC 205 are hidden from view by M 31. On the other hand, the dwarf spheroidal galaxy Antlia, and the irregular galaxies SagdIg and IC 10, are too remote to show at this scale. Antlia lies to the far left of this diagram, IC 10 to the top right, and SagdIg to the bottom right. The newly discovered And V, And VI and And VII are not shown. The clustering of small galaxies around the two dominant members of the Local Group, M 31 and our Galaxy, is clearly evident.

about 1.6 Mpc (about 5 Mly) are unlikely to be assigned membership, but inevitably the boundary of the Local Group is rather ill-defined.

Figure 8.4 shows a side-on view of the Local Group, and figure 8.5 shows a view looking down on the group. At the scale in these figures, several satellite galaxies of M 31 and our own Galaxy are hidden from view.

The distance estimates to most of the galaxies in table 8.4 are highly uncertain. This is not surprising. The small stragglers inhabiting the outskirts of the Local Group are difficult to observe. In any case, until recently they have often been ignored by astronomers in search of more interesting objects. What is important, though, is how well we know the distance to the Large Magellanic Cloud, which is the closest large galaxy, and to M 31, which is the dominant galaxy in the Local Group. These galaxies are key objects in calibrating the rest of the extragalactic distance ladder. The disappointing conclusion is that, although some astronomers would argue otherwise, these distances are probably uncertain at the 10–20% level. Further analyses of the *Hipparcos* results over the next few years will no doubt improve these distance estimates. But to reduce the uncertainty in these distances to the level of a few per cent will require results from the *SIM* and *GAIA* missions.

Now that we have reached the Mpc scale the distance ladder starts to look shaky.

CHAPTER SUMMARY

- The Great Debate in 1920 concerned the nature and distance of spiral nebulae: were they galactic or extragalactic objects?

- Hubble found Cepheids in M 31 and M 33 and used the period–luminosity relation to prove that the nebulae were distant objects: they were spiral galaxies. He later showed that they belonged to the Local Group of galaxies, of which our own Galaxy was just one member.

- Hubble classified galaxies into three main types: spiral, elliptical and irregular, with further subclasses depending on the forms of the galaxies. Astronomers now define several more classes of galaxy: cD, lenticular, dwarf elliptical and dwarf spheroidal.

- Baade discovered that there are two types of star: Population I stars (young, high-metallicity stars in spiral arms) and Population II stars (old, low-metallicity stars in the central regions of a galaxy).

- Cepheids are present in two populations: classical Cepheids are Population I; W Virginis stars are Population II. Baade's work showed that the zero-point of the commonly used period–luminosity relation for classical Cepheids was too dim by about 1.5 mag. Therefore the derived distance to galaxies doubled.

- The period–luminosity relation for Cepheids is of the form

$$\langle M_V \rangle = a + b \log_{10} P.$$

This relation exhibits some scatter. A period–luminosity–colour relation has much less scatter.

- The slope of the period–luminosity relation (b in the above equation) is best determined from LMC Cepheids.

- The zero point of the period–luminosity relation (a in the above equation) has been estimated using several methods: statistical parallax, the Baade–Wesselink method, Cepheid-containing binary systems and main sequence fitting. More recently, Feast and Catchpole used *Hipparcos* data on the trigonometrical parallaxes of 26 Cepheids to deduce a value for the zero point.

- The Feast–Catchpole calibration of the period–luminosity law is

$$\langle M_V \rangle = -1.43 - 2.81 \log_{10} P.$$

This relation needs further verification, since the *Hipparcos* data were at the limit of usefulness. If it is correct, the Local Group distance scale increases by about 10% over previously accepted values.

- Detecting Cepheids is the best way of determining the distance to Local Group galaxies, but not all galaxies contain Cepheids. For these galaxies we can use RR Lyrae stars and the TRGB as distance indicators. Novae have also provided a cross-check on the Cepheid distance scale. A new technique is to use red clump stars.

- The Local Group contains about 30 galaxies. The three largest members are the spirals M 31, M 33 and the Galaxy. Most Local Group members are dwarf galaxies.

- The closest galaxy to our own is the recently discovered dwarf spheroidal galaxy SagdEg, which is 25 kpc distant. There may well be other undiscovered Local Group galaxies hiding in the zone of avoidance.

- The Local Group galaxy farthest from our own is IC 10, which is about 1.25 Mpc distant.

- A working definition of the size of the Local Group is that it is a sphere of radius 1 Mpc, centred on the Local Group centre of mass.

QUESTIONS AND PROBLEMS

8.1 Compare and contrast the arguments given in the Great Debate with those presented during the Diamond Jubilee debate on the distance scale of γ-ray bursters.

8.2 Outline a programme of observation for detecting Cepheids in an external galaxy, and using them to determine the galaxy's distance. For instance, how long would the programme of observation take? How many observations would be required? Exactly what factors would you need to take into account?

8.3 Calculate the distance modulus of each of the galaxies in the Local Group.

8.4 Consider the Local Group galaxies in turn. In each case find out which distance indicator produces the most accurate distance determination.

8.5 The LMC is one of the key objects in establishing the extragalactic distance ladder. Write an account of how distance estimates of the LMC have changed during the twentieth century.

8.6 If the Sun were in the LMC what would be its apparent magnitude? Repeat your calculation for each of the galaxies in the Local Group.

8.7 When calculating the distance to a galaxy using Cepheids (or some other stellar tool) we allow for dimming caused by dust in our Galaxy and dust in the host galaxy. Do you think we need to allow for absorbtion *between* galaxies? Explain your answer.

8.8 Nova Persei exploded in 1901 and emitted an expanding shell of gas. Precise astrometric and spectral observations since then have shown that the shell expands by $0.5''$ per year and that the radial velocity of the shell is $1100 \, \mathrm{km \, s^{-1}}$. Use this information to calculate the distance of Nova Persei. If the peak apparent magnitude of the nova was $+0.1$, what was the peak absolute magnitude?   460pc; $M = -8.21$

8.9 If Nova Persei had occurred in M 31, what would have been its peak apparent magnitude? (Take the distance to M 31 to be 725 kpc.) Nova S Andromedae reached a peak apparent magnitude of $+7$. How much more luminous was Nova S Andromedae than Nova Persei? $m = +16.09$; 4325 times more luminous

8.10 You study a galaxy and find a Cepheid of period 10 days. It has an apparent magnitude of $+17$. How far away is the galaxy (neglecting absorption)? What are the sources of uncertainty in your answer? ≈ 177 kpc

FURTHER READING

GENERAL

Christianson G E (1997) *Edwin Hubble: Mariner of the Nebulae* (Institute of Physics: Bristol)
— A recent biography of the twentieth century's greatest astronomer.

Fernie J D (1969) The period–luminosity relation: a historical review. *Pub. Astron. Soc. Pacific* **81** 707–731
— A fascinating account of the development of the period–luminosity relation. It is particularly clear on the source of Shapley's miscalibration of the Cepheid period–luminosity relation, a topic that is frequently reported incorrectly in textbooks.

Hetherington N S (1996) *Hubble's Cosmology: A Guided Study of Selected Texts* (Pachart: Tucson, AZ)
— An annotated guide to several of Hubble's key papers. Particularly useful if you do not have access to the original papers (some of which are referenced below).

Lake G (1992) Cosmology of the Local Group. *Sky & Telescope* **84 (12)** 613–619
— A well-illustrated non-mathematical account of the Local Group.

Smith R W (1982) *The Expanding Universe* (Cambridge University Press: Cambridge)
— An in-depth account of the Great Debate, and how Hubble helped to settle the questions posed in it.

CLASSIC PAPERS

Hertzsprung E (1913) Über die räumliche Verteilung der Verändlichen vom Delta Cephei-Typus. *Astron. Nachr.* **196** 201
— The first calibration of the Cepheid period–luminosity relation.

Hubble E P (1925) NGC 6822, a remote stellar system. *Astrophys. J.* **62** 409–433
— The first Cepheids found outside our Galaxy.

Hubble E P (1926a) A spiral nebula as a stellar system, Messier 33. *Astrophys. J.* **63** 236–274
— Proof that the spiral nebulae are galaxies in their own right.

Hubble E P (1926b) Extragalactic nebulae. *Astrophys. J.* **64** 321–369
— Hubble initiates the systematic study of galaxies.

Shapley H (1918) Studies based on the colors and magnitudes in stellar clusters. Sixth paper: on the determination of the distances of globular clusters. *Astrophys. J.* **48** 89–124
— Shapley's first attempt at calibrating the Cepheid PL relation, which contained the infamous 1.5 mag error.

Shapley H (1938) Two stellar systems of a new kind. *Nature* **142** 715–716
— The discovery of the first dwarf spheroidal galaxies, Sculptor and Fornax. We now know that such galaxies are the most common type of galaxy in the Local Group.

CEPHEIDS

Evans N R (1991) Classical Cepheid luminosities from binary companions. *Astrophys. J.* **372** 597–609
— The author obtains luminosities for the Cepheids η Aql, W Sgr and SU Cas from *International Ultraviolet Explorer* spectra of their binary companions.

Feast M W and Catchpole R M (1997) The Cepheid PL zero-point from *Hipparcos* trigonometrical parallaxes. *MNRAS* **286** L1–L5
— This recalibration of the Cepheid period–luminosity relation, if correct, has important consequences for the rest of the distance ladder.

Feast M W and Walker A R (1987) Cepheids as distance indicators. *Ann. Rev. Astron. Astrophys.* **25** 345–375
— An in-depth pre-*Hipparcos* review of the Cepheid yardstick.

Kraft R P (1961) Color excesses for supergiants and classical Cepheids V. The period–color and period–luminosity relations: a revision. *Astrophys. J.* **134** 616–632
— An early and influential discussion of the period–luminosity–colour relations.

Laney C D and Stobie R S (1994) Cepheid period–luminosity relations in K, H, J and V. *MNRAS* **266** 441–454
— The authors derive a period–luminosity relation based on 26 galactic Cepheids and 115 Cepheids in the Magellanic Clouds.

Madore B F and Freedman W L (1991) The Cepheid distance scale. *Pub. Astron. Soc. Pacific* **103** 933–957
— A clear account, by two leading practitioners, of the practicalities of using Cepheids as extragalactic distance indicators.

Madore B F and Freedman W L (1998) *Hipparcos* parallaxes and the Cepheid distance scale. *Astrophys. J.* **492** 110–115
— The authors discuss the implication of the *Hipparcos* results for the LMC distance modulus and the Cepheid-based extragalactic distance scale. They conclude that the *Hipparcos* results need not necessarily lead to a drastic alteration in their preferred distance scale.

Oudmaijer R D, Groenewegen A T and Schrijver H (1998) The Lutz–Kelker bias in trigonometric parallaxes. *MNRAS* **294** L41–L46
— A discussion of the Lutz–Kelker bias in *Hipparcos* parallaxes; the authors argue that the Feast–Catchpole calibration is biased, and that when the bias is corrected for the period–luminosity relation is closer to the usually accepted value.

Sandage A R and Tammann G A (1968) A composite period–luminosity relation for Cepheids at mean and maximum light. *Astrophys. J.* **151** 531–545
— An influential paper. Sandage and Tammann have contributed more than most astronomers in elucidating the Cepheid distance scale.

Tanvir N R (1997) Cepheids as distance indicators. In: *The Extragalactic Distance Scale* ed. M. Livio, M. Donahue and N. Panagia. (Cambridge University Press: Cambridge) pp 91–112
— Tanvir derives a new calibration of the Cepheid PL relation. This volume contains 24 chapters, covering all aspects of determining extragalactic distances. For the graduate student.

http://ddo.astro.utoronto.ca/
— A database of over 500 galactic Cepheids, compiled by J D Fernie, B Beattie, N R Evans and S Seager.

DISTANCES TO NEARBY GALAXIES

Armandroff T E, Davies J E and Jacoby G H (1998) A survey for low surface brightness galaxies around M 31. I. The newly discoverd dwarf Andromeda V. *Astron. J.* **116** 2287–2296
— One of the three dSph companions to M 31 discovered in 1998.

Freedman W L (1988) New Cepheid distances to nearby galaxies based on *BVRI* photometry. I. IC 1613. *Astrophys. J.* **326** 691–709
— Based on observations of 11 Cepheids the author obtains $\mu_0(\text{IC}1613) = 24.3 \pm 0.1$.

Freedman W L (1988) Stellar content of nearby galaxies. I. *BVRI* photometry for IC 1613. *Astron. J.* **96** 1248–1306
— An early application of the TRGB method, which produces μ_0(IC1613) $= 24.2 \pm 0.2$.

Ibata R A, Gilmore G F and Irwin M J (1994) A dwarf satellite galaxy in Sagittarius. *Nature* **370** 194–196
— The announcement of the discovery of the closest galaxy to our own.

Lee M G, Freedman W L and Madore B F (1993) The tip of the red giant branch as a distance indicator for resolved galaxies. *Astrophys. J.* **417** 553–559
— The authors argue that the TRGB method is comparable in accuracy to the Cepheid method.

Mateo M (1998) Dwarf galaxies of the Local Group. *Ann. Rev. Astron. Astrophys.* **36** 435–506
— A census of the Local Group dwarfs, based on recent distance and radial velocity determinations.

Mould J R and Kristian J (1986) The stellar population in the halos of M 31 and M 33. *Astrophys. J.* **305** 591–599
— The authors use the TRGB method to deduce μ_0(M31) $= 24.4 \pm 0.25$ and μ_0(M33) $= 24.8 \pm 0.3$.

Pritchet C J (1988) RR Lyrae stars in nearby galaxies. In: *The Extragalactic Distance Scale* ed. S. van den Bergh and C. J. Pritchet. (Astronomical Society of the Pacific: San Francisco) pp 59–68
— Observations of RR Lyrae stars in M 31, M 33, NGC 147 and NGC 185. (This book is a useful review of the status of many different distance indicators as of 1988. It is interesting to see how much progress has been made in one decade.)

Pritchet C J and van den Bergh S (1987) Observations of RR Lyrae stars in the halo of M 31. *Astrophys. J.* **316** 517–529
— The first observation of RR Lyrae stars in a galaxy beyond the Magellanic Clouds.

Reid N, Mould J R and Thompson I (1987) The stellar populations of Shapley constellation III. *Astrophys. J.* **323** 433–450
— The authors derive μ_0(LMC) $= 18.4 \pm 0.15$ on the basis of TRGB method.

Saha A, Freedman W L, Hoessel J G and Mossman A E (1992) RR Lyrae stars in Local Group galaxies. IV. IC 1613. *Astron. J.* **104** 1072–1085
— The authors used a CCD attached to the Hale 200-inch telescope to find 15 RR Lyrae stars in IC 1613. They found $\mu_0 = 24.10 \pm 0.27$.

Stanek K Z and Garnavich P M (1998) Distance to M 31 with the *HST* and *Hipparcos* red clump stars. *Astrophys. J.* **503** L131–L134
— By comparing nearby red clump stars observed by *Hipparcos* with M 31 red clump stars observed by *HST* the authors deduce a distance to M 31 of $784 \pm 13 \pm 17$ kpc.

Stanek K Z, Zaritsky D and Harris J (1998) A 'short' distance to the LMC with the *Hipparcos* red clump stars. *Astrophys. J.* **500** L141–L144
— The red clump method produces a distance to the LMC of only 41.02 ± 2.33 kpc.

van den Bergh S (1994) The outer fringes of the Local Group. *Astron. J.* **107** 1328–1332
— The author lists the distance to Local Group galaxies.

Walker A R (1988) CCD photometry of the Magellanic Cloud RR Lyrae variables. In: *The Extragalactic Distance Scale* ed. S. van den Bergh and C. J. Pritchet. (Astronomical Society of the Pacific: San Francisco) pp 69–70
— Observations of RR Lyrae stars in Magellanic Cloud clusters are used to determine the distance to the LMC.

9

Seventh step: more distant galaxies

Once astronomers realised that spiral nebulae were galaxies, it was clear that there were many galaxies to study. Before the turn of the century, the American astronomer James Edward Keeler (1857–1900) of the Lick Observatory had estimated that about 120 000 nebulae were within photographic range of his telescope. Charles Dillon Perrine (1868–1951), continuing Keeler's work, later estimated that there were 500 000 nebulae within photographic range. A survey begun in 1998 hopes to acquire images of 500 000 000 galaxies. Estimates suggest that there may be 10^{11} galaxies in the universe. The Local Group contains only 30 galaxies. There is still a long way to go on the distance ladder.

We can use the techniques described in the previous chapter to probe space just beyond the Local Group. In this way astronomers have discovered that the Local Group interacts gravitationally with similar small groups that are a few millions of parsecs distant. The Maffei 1 group is closest; it contains galaxies that may once have been members of the Local Group. Not much farther away are the Sculptor (South Polar) group, the M 81 group and the M 83 group. All of these, including our own Local Group, inhabit the outskirts of the mighty *Virgo cluster*. The Virgo cluster contains thousands of galaxies.

We can see Cepheids as far away as 24 Mpc (78 Mly), which takes us deep into the heart of the Virgo cluster. Unfortunately, not all galaxies contain Cepheids, and those that do require extremely hard and time-consuming study if their Cepheids are to be resolved. The Virgo cluster is far beyond the reach of the RR Lyrae and TRGB methods, so these methods do not help us much on this rung of the ladder. And what about the myriad of galaxies beyond the Virgo cluster, in which we can no longer resolve Cepheids? We could use novae: they are very bright, so we can see them over large distances. It is not easy to calibrate the method, though, and the task of searching for bright novae is logistically difficult. There must be better distance indicators than novae.

The last two decades have seen the development of several new distance indicators, which have improved our knowledge of this step of the distance ladder. I describe nine of them here, in varying degrees of detail. They range from 'local' indicators that complement a Cepheid-based study of our galactic neighbourhood, to 'long range' indicators that take us deep into the universe. I leave a comparison of the accuracy of the various methods to the final chapter.

The farther out we go, the more difficult it becomes to measure the distance to a galaxy. Indeed, on this step of the ladder we have to abandon hope of measuring the distance to every single galaxy that we can see. There are too many of them, and they are too far away. Some of the indicators on this step of the ladder therefore concentrate on giving the distance to whole *clusters* of galaxies. We begin, though, with a method that gives the distance to a particular type of star.

9.1 PLANETARY NEBULA LUMINOSITY FUNCTION

A *planetary nebula** occurs at the end of the red giant phase of a star. As a giant star dies, its outer layers slowly expand into space and expose the dense central core. During this process the surface temperature of the core increases, and becomes hot enough to emit ultraviolet radiation. These ultraviolet photons are energetic enough to ionise the shell gases, and cause the shell to glow.

The characteristic spectra of planetary nebulae, particularly in the optical region, enable astronomers to identify a planetary nebula even when it is too distant for photometric identification. The spectrum of a planetary nebula consists of a few intense emission lines, and many more weaker emission lines, superimposed upon a weak continuous background. The two brightest emission lines in nearly all planetary nebulae have a wavelength of 5006.8Å (500.68 nm) and 4958.9Å (495.89 nm); these are green lines, and give rise to the greenish appearance of many of these objects. At first it was thought that the λ5007 and λ4959 lines came from an unknown element (which Huggins named 'nebulium'). Russell argued that it was more likely that the lines originated from known elements radiating under extreme conditions — so-called *forbidden lines*, which are downward transitions from metastable states. In 1927, the American astronomer Ira Sprague Bowen (1898–1973) showed that the lines corresponded to transitions in doubly ionised oxygen, O^{++}. Nowadays, astronomers often refer to the lines as λ5007 [O III] and λ4959 [O III]; the square brackets indicate that the line corresponds to a forbidden transition. The spectra of many planetary nebulae also contain lines from ionised hydrogen and nitrogen, which give rise to a reddish hue. Because of these bright emission lines, planetary nebulae are among the most beautiful objects in the sky. Figures 9.1, 9.2 and 9.3 show examples of typical planetary nebulae, and spectacular colour images of these and similar objects can be found on many Web sites.

Any particular planetary nebula is visible for only 25 000 years or so, which is an eye-blink in cosmic terms. Although the shell expands quite slowly, with a typical velocity of just 25 km s^{-1}, after a few thousand years the material of the nebula dissipates and becomes part of the general interstellar medium. The nebula disappears from view. Nevertheless, planetary nebulae are quite common. One estimate puts the number of planetaries in our Galaxy as high as 50 000, although only about 2000 have so far been discovered and less than 100 are close enough to have had their distances measured directly.

There are several ways we can estimate the distance to a nearby planetary nebula, but in many cases the distance estimates are rather crude. Trigonometric parallax works for a few planetaries. For instance, astronomers at the US Naval Observatory have measured

---

*Herschel called these objects *planetary* nebulae because some of them, when viewed through the telescopes of his time, had a greenish disk-like appearance that resembled the distant planet Uranus. Planetary nebulae have no direct connection with planets. Interestingly, though, recent work suggests that we can use the properties of certain nebulae to learn something about the possible existence of planets near distant stars!

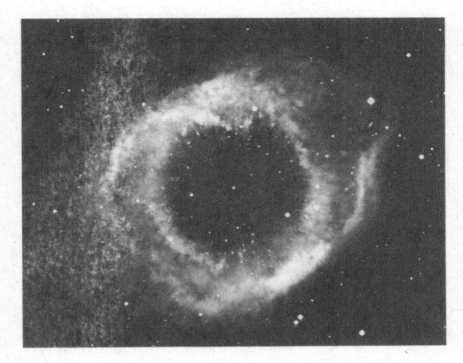

FIGURE 9.1: The Helix Nebula, NGC 7293, is the nearest planetary nebula to Earth. It is about 400 ly away, and its angular diameter is about the same as that of the Moon.

the trigonometric parallax of the central stars of 11 planetary nebulae; this work has determined the distances to four central stars with an accuracy of better than 20%, and to a further five central stars with an accuracy of better than 50%. In some other cases the planetary nebula forms part of a binary system, and we can use spectroscopic parallax to determine the distance to the secondary star. Perhaps the most widely used method is expansion parallax: this works if we can measure the rate of angular expansion of the nebula, $\dot{\theta}$. (This is small: typically only 0.2″ per century for a nebula that is 1 kpc away.) If the expansion is spherically symmetric, a spectroscopic determination of the radial velocity of expansion, v, gives us enough information to calculate the distance through the relation

$$d(\text{pc}) = \frac{211v\,\left(\text{km s}^{-1}\right)}{\dot{\theta}\,\left(\text{mas yr}^{-1}\right)}. \tag{9.1}$$

The method is the same as that used to estimate the distance to Nova Persei; see page 166.

One result of all this work is that we know that planetary nebulae are bright. In the

FIGURE 9.2: The Egg Nebula, CRL 2688, is 3000 ly from Earth.

late 1970s, the American astronomers Holland Cole Ford (1940–) and David Charles Jenner (1943–) suggested that planetary nebulae might all have a similar maximum intrinsic brightness; the upper limit of absolute magnitude seemed to be −4.5. This would make them potential standard candles for determining galaxian distances. In 1978, Ford and Jenner published measurements on eight planetaries in M 81. They noted that these nebulae appeared to be about 20 times fainter than those in M 31. Since they knew the distance to M 31, and assuming that the planetaries were of the same intrinsic brightness, they could derive the distance to M 81. Their estimate of about 3 Mpc agreed quite well with other estimates.

An estimate based on just eight objects could be subject to large errors, of course. The emergence of CCD technology, combined with better telescopes, enabled Ford and the American astronomers George Howard Jacoby (1950–) and Robin Bruce Ciardullo (1954–) to develop a better way of using planetaries as a distance indicator.

Jacoby and Ciardullo use a narrowband filter to photograph the sky at a wavelength of 5007 Å — the [O III] emission line. Since planetary nebulae shine particularly brightly at this wavelength, the contrast between planetaries and the background sky is increased. By subtracting a similar image taken just away from the λ5007 line they can isolate and distinguish planetaries from stars even in quite remote galaxies. The technique works so well that Jacoby and Ciardullo quickly found several hundred extragalactic planetaries. There are now more extragalactic than galactic planetaries in the catalogues.

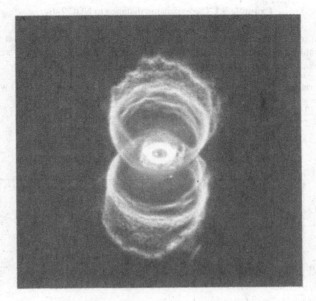

FIGURE 9.3: The Hourglass Nebula, MyCn18, is 8000 ly from Earth.

Jacoby and Ciardullo noted that when they studied a galaxy in this way they detected many faint planetaries but only few bright planetaries. By counting the number of planetaries of each brightness they defined the *planetary nebula luminosity function* (PNLF). Although it was clearly impossible to use an individual planetary nebula as a standard candle, the shape of the luminosity function seemed to be the same for every galaxy. The number of planetaries decreased smoothly with increasing luminosity, and there was a limiting luminosity beyond which they could detect no planetaries. The cut-off was sharp. Figure 9.4 shows the PNLF for M 31.

Jacoby and his colleagues proposed that the PNLF was of the form

$$N(M) \propto e^{0.307M}\left(1 - e^{3(M^*-M)}\right) \tag{9.2}$$

where $N(M)$ is the number of planetaries with absolute magnitude M, and M^* is the absolute magnitude of the brightest planetary. From studies of M 31 it seemed that $M^* = -4.48$. Figure 9.5 shows a graph of (9.2); notice the similarity to the curve in figure 9.4.

By detecting enough planetaries in the 'flat' part of the distribution we can use (9.2) to accurately determine the position of the cut-off. The characteristic cut-off in the PNLF thus potentially provides an excellent standard candle: a comparison of the PNLF in a remote galaxy with that in M 31 gives the distance to the galaxy (in terms of the M 31 distance). Although M 31 remains the main calibrator for the method, there are now seven galaxies with a Cepheid distance in which the PNLF method also works.

Example 9.1 The expansion distance to NGC 246

A long-term study of the planetary nebula NGC 246 has shown that it expands at the rate of 1.4″ per century. Doppler studies indicate that the radial velocity of expansion is $38\,\mathrm{km\,s^{-1}}$. If the expansion is spherically symmetrical, how far away is the nebula?

Solution. A rate of 1.4″ per century is equivalent to 14 mas per year. From (9.1) we obtain

$$d = (211 \times 38)/14 = 573\,\mathrm{pc}.$$

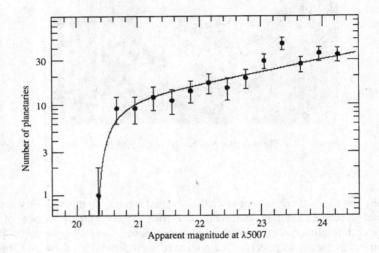

FIGURE 9.4: The planetary nebula luminosity function for M 31. The sharp cut-off at the bright end of the function is characteristic of the PNLF in all galaxies.

Why should the luminosity function be so sharply truncated? More importantly, why should this cut-off be at a constant luminosity? Since planetary nebulae are seen with many different shapes, are the end result of many different types of star, and have very different luminosities (the luminosity function spans about 8 magnitudes), at first glance it seems unlikely that the cut-off should be so sharp and so constant. And unless we have confidence in the constancy of the cut-off, we cannot have confidence in PNLF distances. Several factors may be at work, but perhaps the key is that all planetary nebulae have a similar central star. In particular, the central star has a much smaller range of mass than the progenitor star, so all planetaries may contain about the same maximum number of ultraviolet photons to ionise the gases of the nebula.

The PNLF method works out to about 20 Mpc (65 Mly). Beyond that, the dimmer planetary nebulae become too faint to see. This technique therefore does not extend the

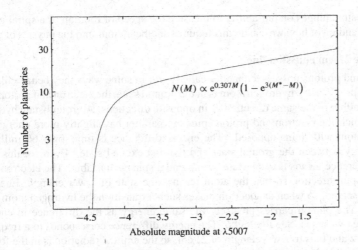

$$N(M) \propto e^{0.307M} \left(1 - e^{3(M^* - M)}\right)$$

FIGURE 9.5: A graph of (9.2) showing the relative number of planetaries against absolute magnitude.

Cepheid range, but it is a useful complement to Cepheids. And it is good enough to reach to the Virgo and Fornax clusters of galaxies. In 1990, for instance, Jacoby and his collaborators used the PNLF method to determine the distance to six Virgo galaxies (M 49, M 60, M 84, M 85, M 86 and M 87). They found the average distance to these galaxies to be 14.7 ± 1.0 Mpc (about 47 Mly). After further analysis, they revised the distance of the Virgo core to 15.1 ± 0.9 Mpc. The PNLF distance to the Fornax cluster is 17.7 ± 0.5 Mpc. In total, the PNLF has so far produced the distance to 34 galaxies.

The technique works particularly well in elliptical galaxies, where the nebulae are easy to distinguish and reddening due to dust is rare. This is just as well, since ellipticals do not contain Cepheids. Planetaries thus provide our best distance estimates to nearby ellipticals. Indeed, until recently, planetaries were used almost exclusively as a Population II indicator. The situation has changed, though, and the spirals M 101, M 51 and M 96 now have PNLF distances. The planetary nebula luminosity function may help to tie together the Population I and Population II strands of the distance ladder.

The PNLF is a secondary distance indicator: we first have to calibrate it in a galaxy of known distance. As already mentioned, M 31 is the main calibrating galaxy. Any change in the distance to M 31 — through a recalibration of the Cepheid period–luminosity relation for instance — feeds through to distances derived using the PNLF.

9.2   THE TULLY–FISHER METHOD

A key question in the Great Debate was whether or not van Maanen had seen the rotation of the spiral galaxy M101. Hubble eventually showed that van Maanen had been mistaken. Nevertheless, spiral galaxies *do* rotate (though their rotation rates are impossible to measure using van Maanen's techniques). One of the most widely used galaxian distance indicators hinges on our ability to measure the speed of rotation of a spiral galaxy. An understanding of how we can do this requires another detour into the physics of atoms.

9.2.1   The 21-cm emission line

Electrons and protons possess a quantity called *spin*: in some ways they behave like tiny spinning tops*. When a hydrogen atom is in the ground state, the electron and proton spins may point either in the same direction or in opposite directions. A ground-state hydrogen atom in which the electron and proton spins are parallel has slightly more energy than the same atom with spins opposed. The energy difference is tiny: only 60 millionths of the energy between the ground state and the first excited state. Even so, this small energy difference means that the state with parallel spins is unstable. The electron tends to flip its spin direction so that the atom reaches the state of lower energy. Energy is always conserved, so when an electron makes such a transition the hydrogen atom emits a photon. The photon carries precisely the same energy as the difference in energies between the two states. See figure 9.6. The energy difference corresponds to a frequency of 1420 Mhz and thus to a wavelength of 21 cm, so the emitted radiation is in the form of radio waves. (Physicists can calculate these quantities with extraordinary precision. To be exact, when a hydrogen atom flips between the parallel-spin state to the opposite-spin state, it emits radiation with a frequency of 1420.406 MHz.)

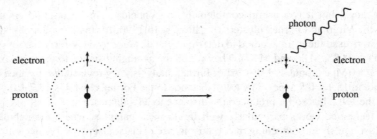

FIGURE 9.6: The origin of the hydrogen 21-cm line. When the electron flips its spin, the atom emits a photon with a wavelength of 21 cm.

This spin-flip process is rare. In 1944, the Dutch astronomer Hendrick Christoffell van de Hulst (1918–) calculated that, on average, a given atom of hydrogen would take about 11 million years to make the transition. He also showed that the slow rate of this spontaneous transition speeds up considerably if the hydrogen atoms interact with each other. By making a reasonable estimate for the interaction rate in interstellar clouds it turns out that a given hydrogen atom there should emit a 21-cm photon about once every 400 years. (It might be wondered why all the hydrogen atoms have not long-since dropped down to the state with anti-parallel spins. The answer is that cosmic rays and energetic electromagnetic radiation continuously ionise the hydrogen atoms. When the electrons

---

*Electrons and protons are *not* miniature spinning tops! They inhabit the quantum world, and the behaviour of quantum particles is bizarre — unlike anything in our everyday experience. Nevertheless, the analogy between the spin of an electron and the spin of a top is good enough for our present purposes.

and protons recombine, 75% of hydrogen atoms end up in the state with parallel spins.)

The distribution of hydrogen atoms in interstellar space is quite uneven. The average density of hydrogen is about one atom per cubic centimetre, which is better than any man-made vacuum, but there are regions with essentially no hydrogen and there are regions where the density is as high as 10–100 atoms per cubic centimetre. The high-density regions of neutral hydrogen are called H I regions. Although any given hydrogen atom in an H I region emits a photon only rarely, there are so *many* atoms in such a region that at any given instant lots of them are emitting photons. So, according to the van de Hulst calculations, these H I regions should emit radiation strongly at a wavelength of 21 cm. A radio telescope tuned to 21 cm should 'see' an H I region as a very bright object.

Immediately after the Second World War, several groups raced to be the first to detect the 21-cm line. In 1951, two American physicists, Edward Mills Purcell (1912–1997) and Harold Irving Ewen (1922–), won the race. After their discovery, radio astronomy using the 21-cm line quickly became a special branch of astronomy. Since spiral galaxies are packed with neutral hydrogen, astronomers could use the 21-cm line to map the distribution of hydrogen in galaxies where the more usual observation of stars was impossible.

Using large radio telescopes, astronomers can detect the 21-cm line even in very distant spiral galaxies. In most cases when they do this they find that the line is not sharp. A spiral galaxy rotates so that, even if it is just slightly edge-on to our line of sight, some of its H I regions move away from us while others move towards us. It is rather like looking at a carousel: the rotation of the carousel carries some of its passengers away from you, while passengers on the opposite side move towards you. The Doppler effect thus broadens the 21-cm line into a double peak. It redshifts the radiation from regions that move away from us, and it blueshifts the radiation from regions that move towards us. The faster the rotation of the galaxy, the greater the line broadening. We denote the width of the line (usually measured at 20% of the peak) by W. Using the Doppler equations it is fairly easy to show that

$$W = \frac{2\Delta\lambda}{\lambda} \approx \frac{2V_{\max}\sin i}{c} \tag{9.3}$$

where $V_{\max}$ is the maximum rotational velocity and i is the angle of inclination of the galaxy to our line of sight (for a galaxy seen edge-on, $i = 90°$). Radio observations of the 21-cm linewidth can thus determine the value of the rotational velocity of a spiral galaxy.

9.2.2 The Tully–Fisher relation

In 1977, the Canadian-born astronomer Richard Brent Tully (1943–) and his colleague the American astronomer James Richard Fisher (1943–) discovered an empirical relation — the Tully–Fisher relation — between the luminosity of a spiral galaxy and its rotational velocity. Such a relation could form the basis of a distance indicator: measure the H I 21-cm linewidth to find the rotational velocity, use this to calculate the luminosity, and compare with the apparent magnitude of the galaxy to deduce its distance.

Why, though, should there be a relation between luminosity and rotation? We can obtain a rough idea as follows.

When astronomers measure the *rotation curve* of a spiral galaxy — a curve of rotational velocity against distance from the nucleus — they always find that beyond the central region the curve is nearly flat. A study of rotation curves tells us much about the mass distribution in a system. Simply from Newton's law of gravity we know that a flat rotation curve implies that most of the mass in a galaxy is spherically distributed with a density that is proportional to r^{-2}, where r is the distance from the nucleus. (Note that this in turn implies the existence of a great deal of non-luminous material in the outer reaches of spiral galaxies. The nature of this non-luminous material is a mystery.) To be precise, if the galaxy is of mass M and of radius R, then we can show that

$$M = \frac{V_{\max}^2 R}{G}$$

where $V_{\max}$ is the maximum rotation speed as given by the flat part of the rotation curve. Now, suppose that the mass-to-light ratio, M/L, is the same for all spirals, i.e. $M/L = k_{ML}$. Then

$$L = \frac{1}{k_{\mathrm{ML}}} \times \frac{V_{\max}^2 R}{G}.$$

If we also suppose that all spirals have the same surface brightness, i.e. $L/R^2 = k_{\mathrm{SB}}$, then

$$L = \frac{1}{k_{\mathrm{ML}}^2 k_{\mathrm{SB}} G^2} \times V_{\max}^4.$$

If we combine the constants into one constant term k, we obtain

$$L = k V_{\max}^4. \tag{9.4}$$

In other words, the luminosity should be proportional to the fourth power of rotational velocity. This is the essence of the Tully–Fisher relation. (This derivation perhaps raises more questions than it answers. The reason why there should be a tight correlation between luminosity and rotation, for spirals of all types and luminosities, remains unknown. We are unlikely to fully understand the Tully–Fisher relation until we understand the details of galaxy formation.)

We can find the absolute magnitude, M, of a spiral galaxy by combining (9.4) with Pogson's equation; we obtain

$$M = -10 \log_{10} V_{\max} + b \tag{9.5}$$

$$= -10 \log_{10} \left(\frac{W}{2 \sin i} \right) + b \tag{9.6}$$

where b is a constant. Since the factor of -10 comes from a consideration of an idealised case, it is safer to write the Tully–Fisher relation as

$$M = a \log_{10} \left(\frac{W}{2 \sin i} \right) + b \tag{9.7}$$

where a and b are constants to be determined experimentally. Work on calibrating the relation in (9.7) has been continuing since Tully and Fisher first proposed it.

9.2.3   Calibration of the Tully–Fisher method

Tully and Fisher first had to check that their proposed relation between linewidth and luminosity worked. They did this by observing spiral galaxies in nearby clusters. By comparing galaxies in the same cluster they tried to ensure that all the galaxies they studied — bright or dim, large or small — were essentially at the same distance. They began by measuring the magnitudes and rotational velocities of spiral galaxies in the Virgo and Ursa Major clusters. If their relation was valid, a plot of magnitude against rotational velocity would show a correlation. If it was not valid, such a diagram would show a scattered set of points. When they drew the plot they found a clear correlation: their method worked. Finally, by using their method on galaxies at a known distance (in other words, galaxies for which one or more of the other distance indicators had been used) Tully and Fisher were able to calibrate their relation.

The typical uncertainty in the method seemed to be about 20–25% for an individual galaxy. This may seem to be quite poor for a potential standard candle. After all, you would be unhappy if you bought a light bulb rated at 100 W only to find that it was really 75 W. Apply the method to groups of galaxies, though, where several spirals are at more or less the same distance from us, and the uncertainties begin to average out. For groups of galaxies, the Tully–Fisher method has a typical uncertainty of just 10%.

As we noted in previous chapters, spiral galaxies seen edge-on often have a band of dust running across them. The dust band looks like a black line that cuts the galaxy in two. Dust reddens and dims starlight, and it can even blot out starlight completely. In the early 1980s, the American astronomer Marc A. Aaronson* (1950–1987) suggested that it made more sense to apply the Tully–Fisher method using the infrared luminosity of a galaxy rather than its optical luminosity. He argued that, since dust is transparent to infrared radiation, a plot of infrared luminosity against linewidth should show much less scatter. He went on to show that the infrared Tully–Fisher method is indeed more accurate than the same method using optical measurements. Today, when astronomers apply the Tully–Fisher method, they are more likely to use infrared than optical measurements.

Example 9.2 A Tully–Fisher distance

A typical formulation of the infrared Tully–Fisher relation is

$$M = -9.5 \log_{10}\left(\frac{W}{2 \sin i}\right) + 2. \tag{*}$$

The quantity $W/(2 \sin i)$ is measured to have a value of $250\,\mathrm{km\,s^{-1}}$ for a particular galaxy. In the infrared the galaxy has an apparent magnitude of $+10$. How distant is the galaxy? (Ignore absorbtion effects.)

Solution. From (*) we determine the absolute magnitude to be -20.78. Thus the distance to the galaxy is $d = 10^{(10+20.78+5)/5)} = 14.3\,\mathrm{Mpc}$.

---

*Aaronson, like Goodricke, died at a tragically early age whilst studying Cepheids. Aaronson died in a freak accident while preparing to make observations of Cepheids in M 101.

More than 2000 spiral galaxies have had their distance measured using the Tully–Fisher method. Until recently, most of these distances relied on a calibration of the relation in just a few nearby spirals. The infrared Tully–Fisher calibration, for instance, rested on the Cepheid distances to M31, M33, NGC3109, M81, NGC2403 and NGC300. One of the goals of a *Hubble Space Telescope* Key Project has been to obtain Cepheid distances to 18 inclined spirals in order to improve the infrared Tully–Fisher calibration.

Radio astronomers can detect the 21-cm line over large distances, so the Tully–Fisher method works as far out as 100 Mpc (325 Mly). It is a Population I method, since it works only on galaxies with spiral arms — which are home to Population I stars like the Sun.

9.3 THE FUNDAMENTAL PLANE

The Tully–Fisher method works because spiral galaxies rotate, and they contain clouds of neutral hydrogen gas that let us measure the rotation. What about elliptical galaxies? Most large elliptical galaxies show little or no rotation. Even if they did rotate, they contain little or no gas and so we could not pick out the 21-cm line broadening of neutral hydrogen. The Tully–Fisher method works *only* for spirals. To estimate the distance to nearby ellipticals we can use the planetary nebula luminosity function, but for more remote elliptical galaxies we need a new method. In 1976, the American astronomers Sandra Moore Faber (1944–) and Robert Earl Jackson (1949–) proposed a new distance indicator for ellipticals.

Faber and Jackson pointed out that stars in the central parts of a large elliptical galaxy will move about at random, like balls in a tombola. Some stars will move fast and some will move slow, and we can say little about the velocity of any individual star. We can, however, define an average stellar velocity in the central region. The velocity dispersion — denoted by σ — is a measure of how much the stellar velocities in the centre of an elliptical galaxy depart from the average velocity. If σ were zero, for instance, the stars would all have the same velocity. If σ were large, the stars would have a large spread of velocities. We can obtain an expression for σ by using the *virial theorem*, which states that in any gravitationally bound system the total energy of the system is exactly half of the time-averaged potential energy[*]. Physicists have shown that this is a completely general result; it applies to an ideal gas, to a star in equilibrium, and to a cluster of galaxies. Applying the virial theorem to an elliptical galaxy, assuming that it is a uniform distribution of mass M in a sphere of radius R, we can show that

$$\sigma^2 = \frac{GM}{5R}.$$

If we make exactly the same assumptions that Tully and Fisher made, namely that the mass-to-light ratio is the same for all ellipticals (i.e. $M/L = k_{ML}$) and that all ellipticals have the same average surface brightness (i.e. $L/R^2 = k_{SB}$), then we obtain

$$\sigma^4 = \tfrac{1}{25} k_{ML}^2 k_{SB} G L.$$

---

[*]The virial theorem was established in 1870 by the German physicist Rudolf Clausius (1822–1888). The word 'virial' comes from a Latin word meaning 'forces', since Clausius arrived at the theorem by considering the forces acting on a system of particles.

Combining all the constants into a single constant k gives us the *Faber–Jackson relation* for elliptical galaxies:

$$L = k\sigma^4. \tag{9.8}$$

Note the similarity between the Faber–Jackson and Tully–Fisher relations. Both relate the luminosity of a galaxy to the fourth power of a velocity. The physical mechanism behind the Faber–Jackson relation is even more mysterious than that behind the Tully–Fisher relation. The latter presumably arises from unknown details of galaxy formation; the former has to do with unknown details of the dynamics of a galaxy in equilibrium.

Since the stars in the central few kiloparsecs of an elliptical galaxy move randomly, their radial velocities are also random. Some move directly toward us, some directly away from us, and most have a component of velocity along our line of sight. We can obtain the velocity dispersion from the Doppler broadening of spectral lines. Typically, the dispersion is $100\text{–}125\,\mathrm{km\,s^{-1}}$. From the Faber–Jackson relation we then determine the luminosity. We have a potential distance indicator for ellipticals.

Astronomers checked the validity of (9.8) by measuring L and σ for elliptical galaxies in the same cluster, which were presumably at more or less the same distance from us. They studied several different clusters, and in each case they found that, although there was a power-law relation between L and σ, there was considerable scatter in the relation. The scatter was not due to measurement error; it was *intrinsic*, and to do with galaxian properties. In other words, elliptical galaxies did not follow the Faber–Jackson relation particularly closely. The relation could not safely be used as a distance indicator. (The uncertainty in distance to a single galaxy using the Faber–Jackson relation is over 30%.)

If there is scatter in a relationship between two quantities it is natural to search for a third parameter to account for the scatter. In 1987, the American astronomers Marc Davis (1947–) and Stanislav George Djorgovski (1956–) replaced the luminosity of an elliptical galaxy with two new parameters: an effective radius of the galaxy, R_e, and the average surface brightness within that radius, I_e. They found a relationship between the three parameters of the form

$$R_e = k\sigma^{1.36} I_e^{-0.85} \tag{9.9}$$

where k is a constant.

Two parameters establish a line; three parameters establish a plane. Davis and Djorgovski called their relationship between effective radius, average surface brightness and velocity dispersion the *fundamental plane* for ellipticals.

The fundamental plane gives the distance to a galaxy in a different way from the Faber–Jackson method. With the Faber–Jackson method, σ is measured and the luminosity deduced; ellipticals are thus used as a standard candle. With the fundamental plane, σ and I_e are measured in order to deduce an effective radius: ellipticals are thus used as a standard rod. (The reason we use an 'effective radius' — actually, the semi-major axis of an ellipse — rather than the physical radius of a galaxy is that a galaxy has no well-defined edge. On photographic plates a typical elliptical galaxy looks like a fuzzy blob.)

At the same time that Davis and Djorgovski developed the fundamental plane, a group of seven American and British astronomers produced a different modification of the

Faber–Jackson method. The Seven Samurai (as the astronomers became known) found that there is a good correlation between σ and a quantity they called D_n. The quantity D_n is the diameter of a central circular region of an elliptical galaxy within which the total average surface brightness is some particular value. The surface brightness they chose was 20.75 magnitudes per square second of arc. This quantity can be measured quite easily with modern CCD-based techniques. The D_n term incorporates both the luminosity and the surface brightness of a galaxy, so we might expect the relation between D_n and σ to be much tighter than the original Faber–Jackson relation; and it is.

The Seven Samurai studied 97 elliptical galaxies and showed that the D_n–σ method gave relative distances with a precision of about 25% for a single galaxy and about 10% for clusters of galaxies.

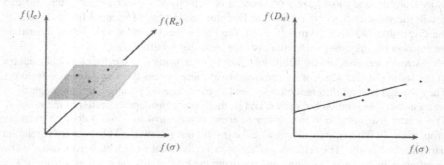

FIGURE 9.7: The relationship between velocity dispersion, effective radius and average surface brightness within the effective radius defines a plane. The relationship between D_n and velocity dispersion defines a line. Elliptical galaxies can lie at quite a distance from the line, but most seem to lie on the fundamental plane.

The D_n–σ relation is really an approximation to the fundamental plane. It gives us an 'edge on' view of the plane. Unfortunately, the view is not *directly* edge on, so there is some extra scatter in the relation. (Take a sheet of paper and look at it from the side: you see a very thin line. Now look at it from a few degrees above the plane of the paper: the line thickens considerably. The same thing happens with the D_n–σ relation. Because it is a few degrees away from the fundamental plane, galaxies can lie at quite a distance from the line. See figure 9.7.) The fundamental plane is therefore a slightly more reliable distance indicator than the D_n–σ relation — and very much better than the Faber–Jackson relation.

The fundamental plane is a Population II method. In terms of precision, it compares reasonably well with the Tully–Fisher method, particularly for clusters of galaxies. Fortunately, ellipticals are often strongly clustered. Since ellipticals are very bright objects, the fundamental plane and D_n–σ methods have a slightly greater range than the Tully–Fisher method. Unfortunately, there are no close bright ellipticals for us to calibrate the relations. For this reason the fundamental plane and D_n–σ methods are best suited for determining relative distances of clusters of elliptical galaxies.

The Seven Samurai have been active in using these methods, and have built up a large store of data. These data contain many interesting features. For instance, by combining

D_n–σ distances and information on galaxian velocities, the Seven Samurai showed that there is a large-scale flow of galaxies — including our own — to a point in the Hydra–Centaurus Supercluster. One of the Samurai, the American astronomer Alan Michael Dressler (1948–), dubbed this point 'The Great Attractor'. When we take the final step on the distance ladder, it is important to know whether such large-scale flows exist.

9.4   THE SURFACE BRIGHTNESS FLUCTUATION METHOD

Many of the methods for measuring galaxian distances depend upon resolving individual stars in a galaxy. In 1988, the American astronomers John Landis Tonry (1953–) and Donald P. Schneider (1955–) described a method to measure galaxian distances when individual stars cannot be resolved. The method, called the *surface brightness fluctuation* (SBF) method, has since been developed extensively by Tonry.

The idea behind the method is that, although a telescope may not resolve a population of bright stars in a galaxy, the discrete nature of the stars causes fluctuations in the surface brightness of the galaxy. The galaxy appears 'blotchy'. By measuring the degree of brightness fluctuations, we can estimate the galaxy's distance.

The amount of blotchiness in a galaxy's image may seem to be a rather loose concept, but it is possible to assign to it a precise and quantitative definition. The technicalities are rather involved, but we can obtain a rough idea of the principles involved as follows.

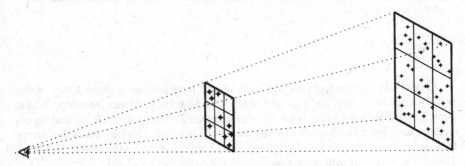

FIGURE 9.8: The symbol ✦ represents an unresolved giant star in a galaxy. When a CCD images a distant galaxy each pixel will contain many unresolved giant stars and so the Poisson fluctuations are small: the galaxy appears smooth. The fluctuations are larger in a nearby galaxy: the galaxy appears blotchy.

Suppose we image a galaxy with a CCD. We can measure the mean flux per pixel and the root mean square (RMS) variation in flux from pixel to pixel. The brightness fluctuations are caused mainly by the brightest red giants in the galaxy. Suppose that on average there are N of these unresolved stars per pixel. If the number of bright stars per pixel obeys Poisson statistics, then the RMS variations will be $\sqrt{N}$. If the mean flux per star is $\bar{f}$, the surface brightness per pixel is $N\bar{f}$ and the RMS variation of surface brightness is $\sqrt{N}\bar{f}$. The surface brightness does not depend upon distance, d, because $N \propto d^2$ and $\bar{f} \propto d^{-2}$. The brightness fluctuations, on the other hand, scale with distance as d^{-1} because $\sqrt{N} \propto d$. This forms the basis of a distance indicator. Imagine two identical galaxies, one twice as far as the other. Both will have the same surface brightness, but the more distant galaxy will appear twice as smooth. See figure 9.8. (It is rather like the picture on a television, which is smooth from a distance but which close up appears to be grainy. Just as we can deduce the distance to a television set from the smoothness of the picture, so we can deduce a galaxy's distance from the fluctuations in surface brightness.)

Once astronomers have made the observations, they can quickly calculate the distances involved. The difficulty with the method is in making the observations. Measuring surface brightness fluctuations takes more time and effort than the Tully–Fisher and fundamental plane methods (though a lot less effort than detecting Cepheids). On the other hand, it shows much less dispersion than the Tully–Fisher and fundamental plane methods, so its results should be more reliable.

Example 9.3 Surface brightness fluctuations

Two galaxies, A and B, are imaged with a CCD. For galaxy A the RMS fluctuations from pixel to pixel vary by 9%; for galaxy B the variation is 3%. What are the relative distances of A and B?

Solution. Surface brightness fluctuations scale as d^{-1}. Thus A is three times closer than B. In practice, when using this method we have to take care that the variations are due to populations of stars in the galaxies being studied. For example, we must account for fluctuations due to foreground stars.

In principle, we can apply the SBF method over large distances: about 40 Mpc using ground-based telescopes, and beyond 125 Mpc using the *Hubble Space Telescope*. So far, though, it has been used mainly on galaxies less than 25 Mpc distant. Tonry and his co-workers have recently completed an SBF survey of about 400 galaxies. From this survey they deduced a distance to the core of the Virgo cluster of 15.8 Mpc (about 51.5 Mly). This agrees very well with the distance derived from the PNLF method.

We can apply the SBF method to the central bulges of spiral galaxies, but it works best on elliptical galaxies, S0 galaxies and globular clusters. It is thus a Population II method. Unfortunately, there are very few suitable nearby galaxies that can be used to calibrate the method. Apart from this caveat, it seems that the surface brightness fluctuation method may be a particularly useful secondary distance indicator.

9.5   GLOBULAR CLUSTER LUMINOSITY FUNCTION

Globular clusters played an important role in elucidating the scale of the Galaxy. Perhaps we can use them to measure distances to other galaxies.

Globular clusters are found in the halos of all large galaxies. We know of over 160 globulars in our Galaxy, and more than 250 have been found in our neighbour, M 31. Globulars are particularly prevalent in giant elliptical galaxies: a typical giant elliptical may contain over 1000 globular clusters. The average absolute magnitude of globular clusters seems to be about $M = -7.5$, which is as bright as the brightest supergiant stars. The brightest individual clusters reach an absolute magnitude of -11, which is brighter than any star ever appears (except for supernovae). So in principle we can see them in galaxies over very large distances. But how can we use them to measure distances?

In 1955, the American astronomer William Alvin Baum (1924–) made the first attempt to use globular clusters to measure a galaxian distance. He simply compared the brightest globulars in M 87 with those in M 31, assumed that the luminosities of these objects were identical, and determined a distance to M 87 from the inverse-square law. We saw earlier that Ford and Jenner did exactly the same thing with planetary nebulae to estimate a distance to M 81. But just as individual planetary nebulae are poor standard candles, so too are individual globular clusters. We need something better.

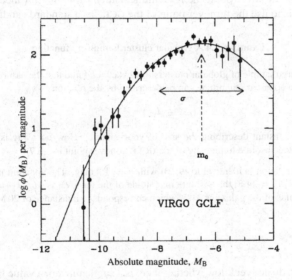

FIGURE 9.9: The globular cluster luminosity function in the Virgo cluster. To construct this function, about 2000 bright clusters were studied in four giant elliptical galaxies. The apparent magnitude of the clusters was converted to an absolute magnitude (in the B band) by adopting a value of 17 Mpc for the distance to the Virgo cluster. The curve is a Gaussian with $\sigma = 1.45$ mag.

In the early 1970s, the Canadian astronomer René Racine (1939–) suggested that the *globular cluster luminosity function* (GCLF) would be a better distance indicator than the luminosity of a single cluster. This is the same idea behind the PNLF method, of course, but the shapes of the GCLF and PNLF are very different. When we study globulars as a function of magnitude we see few bright objects, lots of medium-bright objects, and few dim objects. See figure 9.9.

Astronomers can fit a Gaussian curve to these observations. The relative number of globular clusters as a function of magnitude, $\phi(m)$, is given by

$$\phi(m) = Ae^{-(m-m_0)^2/2\sigma^2} \tag{9.10}$$

where m_0 is the turnover point — the magnitude at which most globulars are found — and σ is a measure of the width of the distribution; the normalisation constant A represents the total number of globulars in the galaxy.

The relation (9.10), with a dispersion of $\sigma \approx 1.4\,\text{mag}$, is found to hold in all the globular cluster luminosity functions so far studied. It is important to remember, though, that this is an *empirical* relation; there is no known physical reason why the relation should hold. But suppose it does hold for all galaxies. If we can sample enough clusters in a galaxy then we can determine the turnover point quite precisely — to about $\pm 0.2\,\text{mag}$ for a single galaxy. The situation improves for rich clusters. If we know the absolute magnitude of the turnover point then we immediately know the distance to the galaxy. The idea, then, is to use the turnover point of the GCLF as a standard candle.

Example 9.4 A globular cluster luminosity function

You study a large sample of globular clusters in a galaxy and find that the number of clusters with any given apparent magnitude can be described by the relation

$$\phi(m) = Ae^{-(m-24.8)^2/3.92}$$

where A is a constant describing the size of your sample. How far away is the galaxy? (Assume that the absolute luminosity of the GCLF turnover point is -6.7.)

Solution. The relation is identical to (9.10) with $\sigma = 1.4\,\text{mag}$. The apparent magnitude of the turnover is $m_0 = 24.8$; the absolute magnitude of the turnover is $M_0 = -6.7$. Thus the distance modulus of the galaxy is 31.5, which corresponds to a distance of 20 Mpc.

Astronomers do not yet know whether there is a single universal value for the absolute magnitude of the turnover point; probably there is not. But for all the luminosity functions so far studied, the difference is small. For our Galaxy, $M_0 = -6.8 \pm 0.17$ (this value depends on knowing the distance to galactic clusters through RR Lyrae stars; as we have seen, the magnitude of RR Lyrae stars is uncertain). The turnover for M 31 is about 0.2 magnitudes brighter (depending on the distance used for Andromeda). If we take PNLF and SBF values for the distances to the Leo, Fornax and Virgo galaxy clusters, then we can calibrate the absolute magnitude of the turnover in giant ellipticals in the cluster. The average for the Galaxy, M 31 and seven giant ellipticals in Leo, Fornax and Virgo is -6.6 ± 0.26. So with luck we will not be too far wrong if we take $M_0 = -6.7$; as more work takes place on this method we may eventually discover whether the turnover depends upon galaxy type and, if it does, the form of this dependence.

The GCLF is a Population II indicator since it works best in giant elliptical galaxies, which are the home to old stars. The method has been used less often than other galaxian distance indicators, but it may be important in the future. We can measure the turnover out to about 50 Mpc using ground-based telescopes, and even farther out with the *Hubble Space Telescope*. If we can learn how to use the bright end of the luminosity function then we may be able to probe out to 200 Mpc, which is even farther than the Tully–Fisher and fundamental plane methods.

9.6 MEGAMASERS

The methods described so far are all secondary distance indicators: we need to calibrate them, usually by finding a Cepheid-based distance to a galaxy (which in turn requires that we calibrate the period–luminosity relation, which in turn requires that we carefully chart the local stellar neighbourhood). Each step of the distance ladder introduces further uncertainty. Would it not be better to use primary indicators to calculate galaxy distances, and thus remove the need for the treacherous distance ladder? The problem, of course, is in finding a good primary indicator that we can use over distances of megaparsecs. The rest of this chapter reviews several possible primary distance indicators. The first, a recent technique that has great potential, is the study of megamasers — masers that are powerful enough for us to see over cosmic distances.

Water masers occur near the core of our Galaxy. (As noted on page 153, radio observations of these masers have provided us with an estimate of the size of the Galaxy.) They occur in the central regions of other galaxies, too. The most famous examples of extragalactic masers are those originating in the central regions of NGC 4258.

Early radio observations of NGC 4258 suggested that the maser emission came from a disk of material rotating about the galaxian centre. To learn more, astronomers required better resolution. On 26 April 1994, a team of Japanese and American astronomers observed the maser emission for 14 hours using the Very Long Baseline Array (VLBA). The VLBA is a dedicated group of ten 25-m radio telescopes positioned all over the United States, which produces an effective diameter of 8000 km. The VLBA provides superb angular resolution.

The team observed spikes of maser emission, with each spike corresponding to a lump of masing material. There were spikes from masers with a radial velocity equal to the average velocity of the galaxy, but also spikes from masers with higher and lower radial velocities. From the details of the observations it was clear that the simplest explanation was that a disk of material orbits a central mass, just as the planets of the solar system orbit our Sun. The high- and low-velocity spikes originated from masers at the edge of the disk as we see it. The observations of the masers were precise enough to deduce that the disk rotates with a rotational velocity of $1080 \, \text{km s}^{-1}$ and, from the way that certain features drifted, that their centripetal acceleration due to gravity is $9.5 \pm 1.1 \, \text{km s}^{-1} \, \text{yr}^{-1}$. To be able to whip the disk around like this the central region must be very massive: at least $3.6 \times 10^7 \, M_\odot$. On the other hand, the volume of the central region is small: all this mass is inside a region less than 0.13 pc in radius. The density at the centre of NGC 4258 is therefore very large: thousands of times more dense than the densest globular cluster. The central material would certainly collapse under its mutual gravitational attraction. There can be only one conclusion: there is a black hole in the centre of NGC 4258.

The discovery of a black hole candidate made the news, of course, but the VLBA observations also gave the distance to NGC 4258 very precisely. The astronomers found that the disk has a radius of 0.13 pc, and from its angular radius (4.1 mas) they could deduce its distance: 6.4 ± 0.9 Mpc (21.2 ± 2.9 Mly). This was a clear improvement on the precision of previous distance estimates for the galaxy, which ranged from 3.3 Mpc to 7 Mpc.

Example 9.5 A galaxian distance from a circumnuclear maser

The circumnuclear disk of NGC 4258 rotates at 1080 km s^{-1} under a centripetal acceleration due to gravity of 9.5 km s^{-1} yr^{-1}. The angular radius of the disk is 4.1 mas. Calculate the distance to the galaxy.

Solution. Since the rotation is simple Keplerian motion we have $a = v^2/R$, where R is the radius of the disk:

$$R = \frac{v^2}{a} = \frac{\left(1080 \, \text{km s}^{-1}\right)^2}{9.5 \, \text{km s}^{-1} \, \text{yr}^{-1}} = 3.87 \times 10^{12} \, \text{km} = 0.13 \, \text{pc}.$$

To calculate the distance d:

$$d = \frac{R}{\theta} = \frac{0.13 \, \text{pc}}{4.1 \, \text{mas}} = \frac{0.13 \, \text{pc}}{1.99 \times 10^{-8} \, \text{rad}} = 6.4 \, \text{Mpc}.$$

This is a geometric method. It requires no assumptions or calibrations — just a measurement of angles and velocities. Since the hearts of most galaxies are thought to contain a massive black hole, conditions in NGC 4258 may not be unusual. So astronomers may well find masers in the central regions of other galaxies. If this happens, the technique will prove to be very useful. At the time of writing, though, it has yielded only one galaxian distance.

9.7   THE SUNYAEV–ZEL'DOVICH EFFECT

In 1972, the Russian cosmologists Rashid Aliyevich Sunyaev (1943–) and Yakov Borisovich Zel'dovich (1914–1987) described a method for estimating the distances to remote clusters of galaxies. The method works *only* for clusters, and at present it is far less precise than the other methods so far described. On the other hand, it has the advantage that it is a primary distance indicator — and we can use it to measure distances of a billion light years or more.

The Sunyaev–Zel'dovich (SZ) effect involves the interaction of two elements: the cosmic microwave background radiation and the hot intergalactic gas found in rich galaxy clusters.

9.7.1   Cosmic microwave background radiation

One of the major astronomical discoveries of recent decades occurred in 1965, when radio engineers found that a cold sea of blackbody radiation bathes the entire universe.

Astronomers have since studied the spectrum of the cosmic background radiation in great detail. We now know that it comes from an almost perfect blackbody, and that the temperature of the radiation is 2.728 ± 0.04 K. From (5.4) we see that the peak wavelength at this temperature is in the microwave region; astronomers therefore refer to the *cosmic microwave background*. An important point to note is that the radiation is exceptionally isotropic. To better than one part in 10^4 the radiation is the same throughout space, in whichever direction we look.

The origin of this radiation lies in the origin of the universe itself; I explain this in more detail in the last chapter. To understand the SZ effect, though, it is enough to know that a cosmic microwave background exists and that radio astronomers can measure its temperature with good precision.

centre: Yakov Borisovich Zel'dovich

9.7.2   Intergalactic gas in galaxy clusters

Clusters of galaxies contain hundreds or thousands of galaxies. The two nearest clusters are the mighty Virgo cluster, which contains over 2000 galaxies, and the more compact Fornax cluster. As we venture farther out into the universe we encounter even richer clusters. A large fraction of all galaxies belong to clusters.

In recent years, astronomers have studied rich galaxy clusters by using X-ray detectors* as well as more conventional optical telescopes. The *Uhuru* satellite, launched in 1970, was the first mission dedicated solely to X-ray astronomy. One of its first discoveries was that the space between the galaxies in a cluster is not empty: a very hot gas permeates the galaxies of a rich cluster. Later satellites, particularly *OSO 8*, *Ariel 5* and *Einstein*, confirmed the discovery. A hot dilute gas, filling the space between galaxies, seems to be an intrinsic feature of all rich clusters. (The gas probably forms within galaxies, perhaps through the action of planetary nebulae and supernovae. If a galaxy has a high speed relative to the intergalactic medium — and many of them do — the resistance of the medium strips away the gas. The gas disperses, and itself becomes part of the intergalactic medium. The process is called *ram-pressure stripping*.)

This intracluster gas is *extremely* hot. The X-rays detected by *Uhuru* and other satellites emanate from collisions in a gas that is at a temperature of 10^8 K! At such high temperatures the gas is completely ionised; it is a plasma of fast-moving electrons and positive ions. (If the electrons and ions moved any faster — or, which is saying the same thing, if the gas was any hotter — they would leave the cluster. As it is, the gravitational pull of all the cluster galaxies is just enough to confine the gas to the cluster.)

9.7.3   The Sunyaev–Zel'dovich effect

Within the confines of a rich cluster we thus find both microwave photons and energetic electrons. Sunyaev and Zel'dovich pointed out that these photons and electrons can interact, with observable consequences.

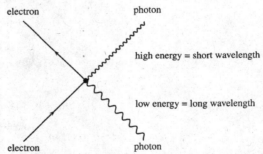

FIGURE 9.10: The inverse Compton process. A low-energy photon scatters off a high-energy electron. In the process the photon changes direction. It also gains energy, and shifts to a higher frequency (shorter wavelength).

That photons can interact with electrons has long been known. The American physicist Arthur Holly Compton (1892–1962) first investigated the effect in 1923. In his original experiment, photons collided with free electrons in graphite. In such collisions the electron and photon exchanged momentum and moved off in different directions. They

---

*The Earth's atmosphere is opaque to X-rays, which is just as well for us, of course. But it means that, in order to study celestial sources of X-rays, astronomers must place their instruments above most of the atmosphere. Early instruments were borne by high-altitude balloons and sounding rockets. Much better results come from orbiting X-ray observatories.

also exchanged energy: the photon lost energy and the electron gained energy. This is now called the *Compton effect*. The *inverse* Compton effect can also occur. It happens when a low-energy photon scatters off a high-energy electron. The photon gains energy and the electron loses energy. See figure 9.10.

From time to time, the energetic electrons in the intracluster gas will collide with photons from the cosmic microwave background. Consider a photon heading in our direction. If it interacts with an electron it will never make it to our detectors: the inverse Compton collision will knock it away from our line of sight. The result is that there will be a deficit of microwave photons in the line of sight towards a rich galaxy cluster. The isotropy of the microwave background radiation will break down in the direction of the cluster. One can also describe this by saying that the patch of sky near a cluster will be slightly cooler (by about 0.5 mK) than the surrounding regions of sky. This is the Sunyaev–Zel'dovich effect. See figure 9.11.

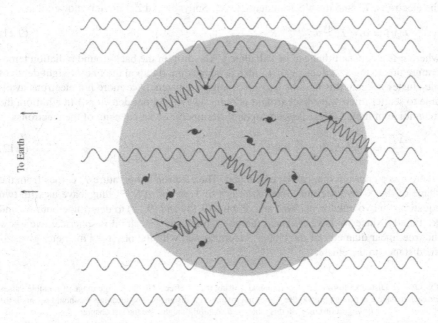

FIGURE 9.11: The Sunyaev–Zel'dovich effect. Electrons from the hot intracluster gas interact with microwave background photons, scattering them away from our line of sight.

It could also be argued that just as many photons should scatter *into* our line of sight as are scattered *from* our line of sight, and this would be correct. However, the photons scattered into our line of sight will have more energy, and therefore a shorter wavelength, than unscattered microwave photons. So the SZ effect actually predicts a *deficit* of background radiation at wavelengths greater than 1.38 mm and an *excess* of background radiation at wavelengths shorter than 1.38 mm. Most work focuses on the decrement in the background.

Workers tried to detect the effect soon after Sunyaev and Zel'dovich pointed out its significance. Unfortunately, it is not easy to make the necessary measurements. It was not until 1984 that astronomers carried out the first convincing detection of the SZ effect. A team led by the British astronomer Mark Birkinshaw (1954–), using the 40-m radio telescope at Owens Valley Radio Observatory in California, saw the SZ effect in the galaxy clusters A 665, A 2218 and 0016+16. Following Birkinshaw's lead, other teams have now measured the effect. The effect is tiny. The 'bite' taken out of the microwave background is typically just one part in 10^4.

How can we use the SZ effect to determine the distance to a cluster of galaxies? Suppose that the cluster is spherical, with radius R, and that the gas cloud has a constant density. The X-ray brightness of the cluster at a particular frequency, $I(v)$, depends directly on the line-of-sight depth of the cluster. (This is to be expected. Bigger clusters have more hot electrons available to create X-rays.) It also depends on the temperature of the electrons, T_e, and the electron density, N_e. Sunyaev and Zel'dovich showed that

$$I(v) = aN_e^2 T_e^{-1/2} e^{-hv/kT_e} R \tag{9.11}$$

where a is a constant that can be calculated. The drop in the background radiation temperature due to the SZ effect, $\Delta T_r/T_r$, also depends directly upon the line-of-sight depth of the cluster. (Again this is to be expected. Bigger clusters have more hot electrons available to scatter microwave background photons away from our detectors.) In addition, the magnitude of the SZ effect depends upon the temperature and density of the electrons:

$$\frac{\Delta T_r}{T_r} = -bT_e N_e R \tag{9.12}$$

where b is a constant that can be calculated. The electron temperature T_e comes from the shape of the X-ray spectrum; measurements of $I(v)$ and $\Delta T_r/T_r$ thus leave us with two equations in two unknowns. We can combine (9.11) and (9.12) to determine both N_e and R. Finally, since we have assumed that the intracluster gas cloud is spherical, we know the true linear diameter of the cloud. A comparison with the observed diameter gives us the distance to the cluster*.

TABLE 9.1: Distances to four clusters of galaxies using the SZ effect. Note the large range of possible values for each distance; at present the SZ effect does not give precise distances. The meaning we should assign to the largest distances in this table needs careful thought; see the last chapter.

Cluster	Distance (Mpc)	Distance (Bly)
A 1656	70–150	0.23–0.49
A 2163	45–1200	0.15–3.91
A 2218	570–1280	1.86–4.17
A 665	790–1650	2.58–5.38

---

*If the cluster is cigar-shaped rather than spherical our answer will be wrong. The size of the error will depend on how much the cluster departs from sphericity and which axis points towards us. Another source of error occurs if the gas forms dense clumps; such clumps produce more X-rays than our simple model assumes.

At the time of writing, astronomers have measured the distance to nine clusters using this method. Table 9.1 gives the distance to four such rich clusters. As the table makes clear, the SZ effect does not give precise distances; but this will change. The joint UK–US *Planck* satellite, which has a provisional launch date of 2004, will make sensitive measurements of the SZ effect.

9.8   GRAVITATIONAL LENSES

The idea that the gravitational field of a massive object might bend light-rays is not new. In 1704, Newton wrote:

> 'Do not Bodies act upon Light at a distance, and by their action bend its Rays; and is not this action strongest at the least distance?'

Exactly 100 years later, the German astronomer Johann Georg von Soldner (1776–1833) used Newton's ideas on optics and mechanics to calculate that a light-ray grazing the Sun's surface would be deflected through an angle of 0.875″.

By 1915, Einstein had developed a new theory of gravitation: the *general theory of relativity*. One of the first problems to which he applied his theory was the deflection of light by the Sun. He calculated that a light-ray grazing the Sun's surface would be deflected by an angle $4GM_\odot/c^2R_\odot = 1.75''$ — exactly twice the angle predicted by Soldner. The deflection of starlight thus seemed to offer a way of testing between Newton's and Einstein's theories of gravity.

Eddington believed that he could detect a deflection of 1.75″. The problem, though, was that the experiment required a total solar eclipse: he had to record the positions of stars close to the Sun's disc, and those stars would only become visible at the moment of totality. See figure 9.12. The first solar eclipse he could use for the experiment occurred in May 1919. Eddington organised simultaneous expeditions to the African island of Principe and the Brazilian town of Sobral in order to observe from the extreme points of the eclipse. Six months later, he and his collaborators took photographs of the same stellar field. By comparing the positions of the stars on the plates he showed that starlight passing close to the Sun was indeed deflected* by 1.75″. The expedition not only seemed to confirm Einstein's general theory of relativity; it proved that gravity could bend the path of a light-ray. (It was also an excellent example of international scientific cooperation. Here was an Englishman organising a long and difficult trip in the aftermath of World War I in the hope of confirming the theory of a German physicist. The expedition caught the imagination of the public, and Einstein became a world-famous figure.)

Eddington realised that the Sun had acted as a *gravitational lens*. Just as glass bends light, so does mass. The deflection of light by the mass of the Sun is tiny. However, the magnitude of the effect depends upon the amount of mass causing the deflection, and there are objects with very much more mass than the Sun: galaxies, for instance. In 1937, the Swiss–American astronomer Fritz Zwicky (1898–1974) suggested that a galaxy could

---

*We now know that Eddington's observations were not as clear-cut as originally thought; his experiment had only 30% accuracy. But modern experiments using the same ideas show that the deflection angle is indeed 1.75″ to better than 1%.

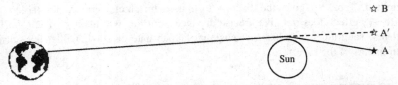

FIGURE 9.12: The bending of starlight by the gravitational field of the Sun. The Sun's gravity bends light from
star A so that it appears in position A′. The small angles involved can be measured relative to the reference
star B. (Not to scale.)

focus the light from a distant background object, making it possible to observe objects that
are otherwise too faint to see. A galaxy could act as a natural cosmic telescope.

Zwicky argued that a good method to look for gravitational lenses is to search for two
or more images of the same object. Figure 9.13 illustrates the idea. It shows a distant
quasar. (Quasars are described in the final chapter of the book. The nature of quasars is
unimportant in the context of the present discussion. It is enough to know that they are
highly luminous objects, and that in some cases they can alter in brightness over a period
of a few weeks.) Between us and the quasar lies a galaxy, with a typical mass of $10^8 M_\odot$.
The gravitational field of the galaxy bends light from the quasar, just as the mass of the
Sun bends starlight. Due to the lensing effect of the intervening galaxy we see several
images of the same quasar.

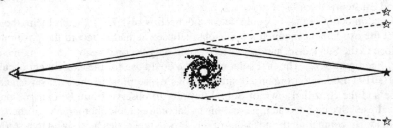

FIGURE 9.13: The gravitational lensing effect of a galaxy on a distant quasar. We do not see the quasar itself;
instead, we see several different images of the quasar.

More than 40 years after Zwicky's suggestion, astronomers discovered a lens of the
type shown in figure 9.13. In 1979, the English astronomer Dennis Walsh (1933–), the
New Zealand-born astronomer Robert Francis Carswell (1940–) and the American astronomer Raymond John Weymann (1934–), found two quasars separated by just 6.1″.
The quasars were on opposite sides of the brightest galaxy within a cluster, and they had
identical spectra. Such a coincidence was highly unlikely. The explanation had to be that
the two quasars, Q0957+561A and Q0957+561B, were the double image of the single
quasar Q0957+561. Many other gravitational lenses are now known to exist; astronomers
have found about 40 multiply imaged quasars. Some gravitational lenses produce several
images of an object; the quasar Q2237+031, for instance, has four visible images and

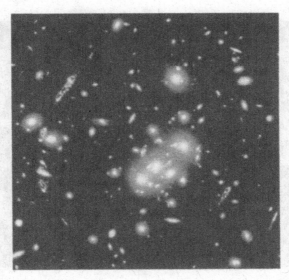

FIGURE 9.14: The galaxy cluster 0024+1654 acts as a gravitational lens: the five loop-shaped objects are five images of the *same* galaxy.

is called the *Einstein cross*. Others smear out the light into rings or arcs; figures 9.14 and 9.15 illustrate this effect, and show how a galaxy cluster can act as a gravitational lens. Figure 9.15 shows an image of a galaxy that has been smeared into an arc by the cluster CL1358+62. Until recently, this galaxy was the most distant known object in the universe.

The light from Q0957+561 reaches us by two different paths. The path that forms image A may be quite different in length from the path that forms image B. Furthermore, the two paths will in general traverse through different gravitational field strengths*. Suppose then that Q0957+561 flickers in brightness. Because of the difference in path lengths, and because of the phenomenon of gravitational time dilation predicted by Einstein's theory of gravity, we might see Q0957+561A flicker *before* Q0957+561B flickers. This is exactly what happens. The southern image of the quasar varies in exactly the same way as its northern twin, but more than a year later.

In 1964, 15 years before the first gravitational lens was discovered, the Norwegian astronomer Sjur Refsdal (1935–) studied such a system theoretically as part of his PhD thesis. He showed that, if assumptions are made about the mass distribution of the lensing galaxy (which is like making assumptions about the shape of a traditional glass lens), then by measuring the time delay between the two paths the distance to the source and to the lensing galaxies can be calculated. Suppose the distance to the lens is d_L, the distance to

---

*In addition to an effect due to the difference in path lengths, there is an effect due to the gravitational field of the lens. The contribution of the gravitational time delay must be accounted for when calculating distances.

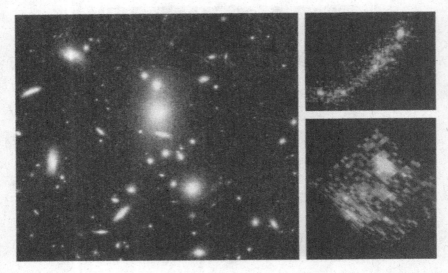

FIGURE 9.15: Cluster CL1358+62 acts as a gravitational lens, smearing the light from the distant galaxy into an arc. The tiny smudge briefly held the record for the most distant known object in the universe. The picture was taken by the *HST*; spectroscopic confirmation of the redshift was obtained by the Keck 10-m telescope.

the source is d_S, and the time delay is Δt. Refsdal showed that

$$\Delta t \propto \left(\frac{d_L d_S}{d_S - d_L} \right).$$

The constant of proportionality depends upon the angular separation of the images and the mass distribution of the lens, both of which can be measured or modelled. If we know the ratio of the distances to the source and lens (the following chapter describes how we can find this) a measurement of Δt gives us d_L and d_S.

In 1991, the Indian astronomer Ramesh Narayan (1950–) was one of the first to use Refsdal's idea. He calculated the distance to the galaxy responsible for lensing Q0957+561, and determined a value of 900–1700 Mpc. (When such large distances are involved, we should also cite the cosmological model we are using. This complication is left to the final chapter.)

Narayan's value was uncertain to almost a factor of two. Part of the uncertainty was in modelling the mass distribution of the lensing system: the bright central galaxy is part of a cluster, and the rest of the cluster also acts as a lens. The time delay was also uncertain. Some workers said that the delay was a little over 400 days, others that it was about 540 days, and still others that the data were insufficient to measure the delay at all. This may seem strange. Surely it is easy to measure the time delay between two events. The problem lies in the way in which modern observatories operate. They tend to assign an observer a block of nights to work on a programme. What this particular observing programme needs is just 30 minutes of telescope time during the night — but it needs the same equipment night after night for months at a stretch. The established observatories cannot accommodate this sort of working practice. However, in 1994 a team led by the American astronomer Edwin Lewis Turner (1949–) had access to a new observatory in New Mexico. This observatory encouraged exactly the sort of observing programme required in a study of gravitational lenses. Turner and his team watched Q0957+561 every night, and they eventually realised that they could predict what image B would do by carefully watching image A. They decided that the time delay was 415 days. In 1995, they published a prediction in the *Astrophysical Journal* of how image B would fluctuate in 1996. They were right. It turned out that the time delay between the two images was 417 ± 3 days. This meant that they could calculate the distance to the lensing galaxy with a little more precision.

Turner and his team calculated a distance to the lensing galaxy of 1400–2100 Mpc (with the same warning as given earlier about the danger of taking such large distances at face value). Light from the quasar set off on its journey to us about the time that the solar system was forming. Billions of years later we can see the light today — or in 417 days' time if the light takes the longer route.

There are other systems in which the mass distribution of the lens seems to be simpler than in the case of Q0571+561. It should be easier to apply Refsdal's idea in these systems. In the future, then, as more gravitational lenses are discovered, this method of distance determination will increase in importance.

The gravitational lens method has the advantage that it can measure huge distances — much larger than all the secondary distance indicators discussed so far, and larger even than the Sunyaev–Zel'dovich effect. It has the added advantage that, like the Sunyaev–Zel'dovich effect, it is a primary indicator and quite independent of the rest of the distance ladder.

9.9 SUPERNOVAE

Supernovae are exploding stars. They are among the most violent explosions in the universe. For a short while a supernova can rival its parent galaxy in brightness. These destructive events represent the death throes of a star, but they also make possible our own existence. Of the elements that make up our world, only hydrogen and helium were present in the early universe. Carbon, oxygen, iron and so on — the heavier elements upon which life depends — were 'cooked' inside stars. Billions of years ago, when those stars exploded, they seeded our region of the Galaxy with heavy elements. And when the solar system formed, it did so from material that contained those heavy elements. It is a fascinating thought: our bodies are made of stuff from a star that died billions of years ago.

Supernovae are rare. A supernova should occur in the Galaxy about once every 50 years on average but, because interstellar dust hides most of the Galaxy from view, galactic supernovae mostly shine unseen. In the past 1000 years, observers have recorded only five galactic supernovae. These were visible in 1006, 1054, 1181, 1572 (recorded in detail by Tycho) and 1604 (recorded in detail by Kepler). Infuriatingly, no nearby supernovae have occurred since the invention of the telescope. The closest was discovered on

23 February 1987, when light from the explosion of the star SK−69°202 reached Earth after a journey time of some 180 000 years. SN1987A, as the supernova was officially named, occurred in the Large Magellanic Cloud. Although not in our Galaxy, this was the best observed supernova since Kepler's. See figure 9.16.

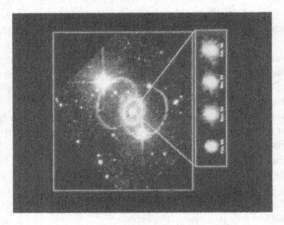

FIGURE 9.16: The aftermath of SN1987A, a Type II supernova that occurred in the LMC.

Supernovae could be the best of all distance indicators. We can see them over a huge expanse of space, and they provide us with the distance to a particular galaxy rather than a rough distance estimate to a cluster. First, though, we must be certain that they are a good standard candle. (A more appropriate choice of phrase in this case might be 'standard bomb'.) If supernovae were unique events, each arising from a different set of circumstances and developing in different ways, then they would be of little use as yardsticks. It turns out that there are only a few distinct types of supernovae, a fact first realised in 1940 by the American astronomer Rudolph Leo Minkowski (1895–1976) while working under Zwicky at Mount Wilson.

Zwicky and Minkowski classified supernovae as Type I and Type II. We now divide Type I supernovae into three subclasses: Type Ia, Type Ib and Type Ic. Although they are all called Type I, they originate from very different physical processes. Fortunately, we can distinguish between the three subclasses by spectroscopic methods, so no confusion arises. For our purposes, we need consider only Type Ia supernovae, which are the most luminous of all supernovae. Type II supernovae are classed as either Type IIL or Type IIP, depending upon whether their light curves fade linearly or whether there is a plateau in brightness 2–3 months after the explosion. The distinction need not concern us; we can consider Type II supernovae to be a single class.

There are four main differences between Type Ia and Type II supernovae. First, as already stated, Type Ia explosions are more luminous than Type II. Second, after peak brightness has been reached Type Ia supernovae decline in luminosity in a regular way; Type II supernovae do not. Third, the two types have different spectra; hydrogen lines are absent in the spectra of Type Ia supernovae, whereas Type II supernovae show an

abundance of hydrogen. Fourth, we find Type Ia supernovae in all locations in all types
of galaxy; Type II supernovae almost always occur in the arms of spiral galaxies.

These observations — particularly the last one — tell astronomers a great deal about
the physical mechanisms of the two types of explosion. It turns out that Type Ia supernovae originate from old low-mass stars in binary systems, while Type II supernovae
represent the evolutionary endpoint of young single high-mass stars.

9.9.1 Type II supernovae

Type II supernovae originate from stars that have more than about eight times the mass of
the Sun. Such high-mass stars spend only a short time on the main sequence. Once they
exhaust their supply of hydrogen they shine through a series of nuclear reactions involving
progressively heavier elements: carbon and oxygen nuclei fuse into neon and magnesium
nuclei; neon and magnesium nuclei fuse into silicon and sulphur nuclei; silicon and sulphur nuclei fuse into iron nuclei. Towards the end of its life the interior of a high-mass
star resembles the layered structure of an onion, with the chemical composition of each
layer depending upon the maximum temperature it reached. See figure 9.17.

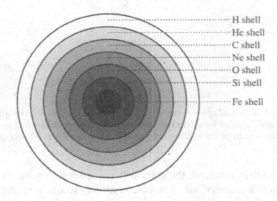

FIGURE 9.17: At the end of its life the interior of a high-mass star resembles an onion in its layered structure.
(The shell widths are not drawn to scale.)

A high-mass star cannot long sustain its profligate lifestyle by burning heavy elements: it burns its supply of oxygen in less than a year, and silicon burning lasts for only
a couple of days. In fact, the star is doomed when it begins to develop an iron core. Any
nuclear reaction involving iron *absorbs* energy, so the fuel supply switches off. And if
the core no longer produces energy, there is nothing to support the star against the relent-
less force of its own gravity. When the central core consists almost entirely of iron, a
catastrophe occurs. The iron core begins to collapse under its own weight. The collapse
creates high-energy photons, which rip apart the atomic nuclei in the core in a process

called *photodisintegration*, which in turn causes the following reactions to take place:

$$^{56}\text{Fe} + \gamma \rightarrow 13\,^{4}\text{He} + 4\text{n}$$

$$^{4}\text{He} + \gamma \rightarrow 2\text{p} + 2\text{n}.$$

The iron nuclei, which took hundreds of millions of years to form, are dismantled within an instant into their subatomic components — protons and neutrons. The collapse continues and the densities become tremendous: the core density becomes 10^{14} times more dense than water. Protons and electrons fuse at such densities, forming neutrons and neutrinos. The neutrinos, moving close to the speed of light, escape into space. See figure 9.18.

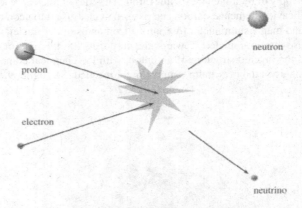

FIGURE 9.18: As the iron core collapses, a soup of subatomic particles forms. Gravity continues to press down on this particle soup, and it quickly forces electrons and protons to fuse. Each fusion of an electron and a proton creates a neutron and an elusive neutrino. The neutrino shoots off into space.

In about one tenth of a second, the inner core collapses into what is essentially a large neutron — we have a neutron star. The core contracts so quickly, in fact, that it 'overshoots' and briefly becomes even more dense than an atomic nucleus. At this point the strong nuclear force stiffens the core and, as long as the core is not too massive, prevents further collapse. The core bounces back, in the same way that a rubber ball springs back if you squash it tightly in your hands. What of the outer core? The material of the outer core implodes, rushing supersonically toward the inner core at speeds approaching $70\,000\,\text{km s}^{-1}$. The rebounding inner core smashes into the infalling outer layers, creating a shock wave that travels through the star. In some stars the shock triggers an explosion in less than 0.01 s; this is called a *prompt explosion*. In more massive stars the shock wave may stall, but the stalled shock wave can receive an extra boost from the flood of neutrinos fleeing the core. (Neutrinos interact only rarely with matter. In this case, though, there are so *many* neutrinos that enough of them interact with the stellar material to add energy to the shock.) In this case the explosion takes slightly longer to trigger. This is a *delayed explosion*. In either case, the shock wave moves through the star and the explosion becomes so violent that all manner of nuclear fusion reactions take place in the outer layers.

These fusion reactions create heavy nuclei. The explosion then propels between a third and a half of the star's mass out into space. This is what we call a Type II supernova.

Type II supernovae leave behind remnants. Where once there was an ordinary star, after the explosion there is a neutron star (or a black hole if the core mass is large enough). Such neutron stars spin rapidly and, as discussed on page 109, we can detect the radio pulses they emit. A more visible record of a Type II supernova is the shell of expanding gas it ejects. The famous Crab Nebula, the first object on Messier's list, is the gaseous remnant of the 1054 supernova.

Type II supernovae are luminous so we can see them over vast distances. Unfortunately, they do not all reach the same peak luminosity. There are several possible reasons for this. The core and the envelope may contain asymmetries; the initial star may have a mass in the range $10-100 M_\odot$; different stars rotate at different rates, have different chemical compositions, and so on. Type II events are messy. They are certainly not 'standard bombs', which at first sight makes them poor distance indicators.

The American astronomers Robert Paul Kirshner (1949–) and John Ying-Kuen Kwan (1947–) suggested another way of using Type II supernovae to measure the distance to remote galaxies. The American astronomer Robert Vernon Wagoner (1938–) later improved upon the method. These astronomers argued that precise measurements of the expansion rate of a supernova's photosphere would provide us with its distance.

We can easily measure the radial speed of the ejecta from the Doppler shift of its spectral lines. (For instance, redshifted spectral lines indicate that the outer layers of SN1987A that happen to be moving directly towards us do so at about $30\,000\,\mathrm{km\,s^{-1}}$, which is 10% of light speed). If a supernova explosion is symmetrical, so that it ejects material at the same speed in all directions, we know the expansion speed at right angles to our line of sight: it is the same as the radial speed. We can also measure the rate at which the shell expands in terms of seconds of arc per year. In principle, this is an easy measurement: simply observe the supernova carefully over several years and record how its apparent diameter increases. If we then combine the rate of expansion (in seconds of arc per year) with the speed of expansion (in kilometres per second) we obtain the distance to the supernova. Only at one particular distance will a particular speed of expansion produce a particular yearly change in the angular size of the shell. (This is the same method that gave the distance to Nova Persei — see page 166. More recently, in 1998 four American astronomers used this method to calculate the distance to the novae QU Vulpeculae and V351 Puppis. The former erupted in 1984 and ejected a shell at over $3200\,\mathrm{km\,s^{-1}}$; the latter erupted in 1991 and ejected a shell at over $5100\,\mathrm{km\,s^{-1}}$. From measurements of the angular expansion rate using the *Hubble Space Telescope* it turns out that QU Vulpeculae is 5.6 kpc (18 300 ly) distant and V351 Puppis is 4.5 kpc (14 800 ly) distant.)

Unfortunately, there is a major flaw in this method. As the worked example shows, 'the angles involved are tiny. Modern ground-based telescopes struggle to measure angles as small as $0.05''$ because of the blurring effect of the atmosphere. Even the *Hubble Space Telescope* cannot measure the expansion rate of a supernova outside the Local Group. In principle, the idea of directly measuring the rate of expansion is a good one. In practice, it does not work for distant supernovae.

Kirshner and his colleagues were well aware of the difficulty, of course. To obviate it, they suggested a modification of the Baade–Wesselink method. From (8.2) we can write

Example 9.6 Expansion parallax of a supernova

The angular diameter of a supernova remnant is observed to increase by $0.1''$ in five years. Spectroscopic measurements indicate that the ejecta moves at $30\,000\,\mathrm{km\,s^{-1}}$. How far away is the supernova? Repeat the calculation for an angular expansion rate of $0.05''$ in five years.

Solution. The distance is given by $d = v/\dot{\theta}$, where θ is measured in radians. Using the numbers given in the question, a $0.1''$ increase in diameter in five years corresponds to a distance of $10^6\,\mathrm{ly}$. An expansion rate of $0.05''$ in five years corresponds to a distance of $2 \times 10^6\,\mathrm{ly}$.

θ, the angular size of the supernova photosphere, as

$$\theta = \frac{R}{d} = \sqrt{\frac{f_\lambda}{F_\lambda}}$$

where f_λ is the observed flux and F_λ is the actual flux, measured at wavelength λ. The Baade–Wesselink method when applied to stars assumes that the star radiates as a blackbody. This assumption does not work for Type II supernovae: particle scattering causes the supernovae to radiate with less flux than a blackbody at the same temperature. We must therefore write $F_\lambda = \zeta_\lambda B_\lambda(T)$, where $B_\lambda(T)$ is the Planck blackbody function evaluated at temperature T and ζ_λ is a correction factor that takes into account the reduced flux. We can thus write

$$\theta = \frac{R}{d} = \sqrt{\frac{f_\lambda}{\zeta_\lambda B_\lambda(T)}}. \tag{9.13}$$

The advent of supercomputers has enabled theorists to produce sophisticated models of Type II supernovae, so that ζ_λ can be evaluated for a variety of cases. We can therefore calculate θ.

Since the interstellar medium exerts very little pressure on the ejecta, the shell expands freely at velocity v. At any time t after the explosion at t_0 the radius of the photosphere is represented simply by

$$R = v(t - t_0) + R_0 \tag{9.14}$$

where R_0 is the initial radius. We can safely ignore R_0, since it is negligible except when $t \approx t_0$. Combining (9.13) and (9.14) produces

$$d = \left(\frac{v}{\theta}\right)(t - t_0). \tag{9.15}$$

If we make at least two measurements at different times of v (from spectra) and θ (from fluxes and computer models) we can solve for both t_0 and d. In other words we can calculate the date of the explosion and the distance to the supernova. This is the *expanding photosphere method*. If we make measurements at several different times we can check the consistency of the method: θ, R and v will all change, but the distance d must remain the same.

The expanding photosphere method has been applied most successfully by Kirshner and his colleagues. They have used it to measure distances up to 180 Mpc (nearly 600 Mly). The method can be used as a primary distance indicator. It is better thought of as a secondary indicator, though, since the results are usually checked against observations of Type II supernovae in galaxies of known distance. The calibration is still subject to some uncertainty, since there have been few nearby Type II supernovae to check the method. The recent SN1987A is the best check. We know the distance to the LMC from Cepheids, and there is an independent estimate of the distance to SN1987A based on observations of ultraviolet emission from material surrounding the star (see the 'Questions and Problems' section). Two other galaxies with Cepheid-based distances also serve as calibrators: M 81 (which hosted SN1993J) and M 101(which hosted SN1970G).

Eleven galaxies have had their distances measured by both the Tully–Fisher and expanding photosphere methods. The distances are in good, though not perfect, agreement.

At the time of writing, the expanding photosphere method has produced distances to 18 galaxies. In years to come the method is likely to become one of our best yardsticks for distant galaxies. Since Type II supernovae occur primarily in spiral arms, the expanding photosphere method is a Population I indicator.

9.9.2   Type Ia supernovae

We can distinguish between Type Ia and Type II supernovae spectroscopically: the spectrum of a Type Ia supernova lacks hydrogen lines. This lack of hydrogen, and the observation that Type Ia supernovae occur in all parts of a galaxy and in all types of galaxy, suggests that they do not result from the catastrophic final phase of high-mass stars. Nor do they result from single lower-mass stars, which eject a planetary nebula and then gently cool. It is most likely that a Type Ia supernova is the explosion of a carbon–oxygen white dwarf star.

A white dwarf by itself is a stable object, but many white dwarfs exist in binary systems. The white dwarf Sirius B, for instance, happens to orbit the brightest star in our sky. In such binary systems the white dwarf can feed off its companion. The intense gravitational field of the dwarf sucks gaseous material away from the companion star, and this material eventually settles on the surface of the dwarf. The accretion rate is small: typically, each year the mass of material accumulating on the dwarf will be much less than $10^{-6} M_\odot$. Over time, though, the material piles up. As more and more gas rains down upon the dwarf, the material at the surface becomes compressed to a high density. Quantum mechanical effects cause the nuclei to develop high speeds, so the effective temperature rises. The accreted material begins to burn violently. This is a nova. The explosion stops the infall of material, at least temporarily, and it may be thousands of years before the nova explodes again.

Suppose, though, that the white dwarf is close to the Chandrasekhar limit. In this case, the intense gravity of the white dwarf can keep the burning shell close to the surface, and material continues to fall from the companion star. Hydrogen fuses into helium, and then, as the temperature rises still further, helium fuses into carbon. If the original white dwarf was rich in carbon then at this point it consists almost entirely of carbon nuclei. As even more material falls from the companion star onto the white dwarf, the mass of the white dwarf increases inexorably towards $1.4M_\odot$: the Chandrasekhar limit. The dwarf becomes denser and hotter. Eventually, the temperature reaches nearly a billion degrees and carbon nuclei can fuse together. An explosion occurs. The details of the explosion are not fully understood, but in one scenario the carbon burning starts in the core and moves towards the surface at subsonic speeds: this is a process called *deflagration*. Whatever the exact details may be, what is certain is that the explosion destroys the star. Unlike Type II supernovae, which leave behind remnants, Type Ia supernovae leave few traces. The white dwarf progenitor disintegrates completely.

The total energy release in a Type Ia supernova is no greater than in a Type II supernova. The difference is in the form the energy takes. A Type II event releases most of its energy in a burst of invisible neutrinos. Most of the energy in a Type Ia event goes into the dispersal of the star. The subsequent light curve is powered by radioactive decay of heavy elements, in particular the decay chain $^{56}Ni \rightarrow ^{56}Co \rightarrow ^{56}Fe$. Type Ia supernovae are thus much more luminous than Type II supernovae. Such explosions are so luminous that we can see them in galaxies too remote to have been given names.

Since Type Ia supernovae all arise from the same type of event there is a good chance that they all reach the same peak brightness. This seems even more likely when it is considered that their light curves are remarkably similar, as is their spectral evolution. (Figure 9.19 shows the light curve of a typical Type Ia supernova.) If they all shine with the same peak brightness then they would be the best possible type of standard candle. Fortunately, there is evidence that they *do* all have about the same peak brightness.

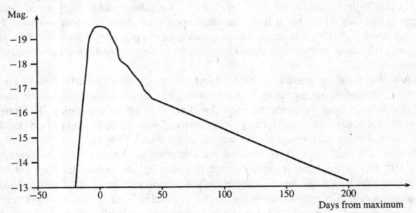

FIGURE 9.19: A typical light curve of a Type Ia supernova. The supernova rises to maximum brightness in about 20 days. The maximum absolute magnitude is about -19.5. A few days after maximum, the brightness starts to drop quickly. The decline in brightness slows and becomes linear 44 days after maximum. In the linear part of the light curve the supernova loses about 0.017 magnitudes per day, until it eventually fades from view. The light curve after carbon burning is powered by the radioactive decay of ^{56}Ni.

We know of three galaxies — NGC 1316, NGC 3913 and NGC 5253 — that have hosted two Type Ia supernovae. In each of these galaxies the brace of supernovae had similar apparent magnitudes. Since in each case the supernovae were at essentially the same distance they must have had similar absolute magnitudes. Further evidence that Type Ia supernovae all have a similar absolute magnitude is provided by supernovae in Virgo cluster galaxies. Although the Virgo cluster may have an appreciable line-of-sight depth, to obtain a rough estimate of supernova luminosity we can consider all the galaxies to be more or less at the same distance. Again, astronomers see that Type Ia supernovae in the cluster have similar apparent magnitudes, and thus similar absolute magnitudes. There is some evidence that a few Type Ia events can be peculiar. Some of them (e.g. SN1991bg) seem to be intrinsically dim and red, rather than simply being reddened by dust. A very few (e.g. SN1991T) seem to be intrinsically brighter than normal. But the majority of Type Ia supernovae hold no surprises, and we can probably identify those that are abnormal. It therefore seems likely that Type Ia supernovae are good standard candles, with a dispersion in the peak luminosity of about 0.2 magnitudes. To use them as a distance indicator all we need is a value for their peak absolute magnitude.

In 1992, the American astronomer David Reed Branch (1942–) considered all estimates for the peak absolute magnitude of Type Ia supernovae and concluded that the value lay in the region -19.6 ± 0.2. Some of the evidence Branch used came from theoretical models of the explosion. For instance, physicists can model the radioactive decay chain $^{56}Ni \rightarrow {}^{56}Co \rightarrow {}^{56}Fe$, which continues the energy release after the carbon burning, and they deduce that the peak magnitude must be about -19.4 ± 0.2. Other evidence came from observations of galactic supernovae. From eyewitness accounts, the Type Ia supernovae seen in the Galaxy (SN1006, SN1572 and SN1604) seem to have had a peak magnitude of about -19.7 ± 0.3*.

The value obtained by Branch agrees well with the peak brightness derived from supernovae in galaxies whose distance we know from Cepheid measurements. Between 25 April 1995 and 3 July 1995, a team of astronomers led by Sandage used the *Hubble Space Telescope* to find 20 Cepheids in the Virgo cluster galaxy NGC 4639. From the period–luminosity relation, Sandage calculated the absolute magnitude of the Cepheids and thus their distance. It turns out that the galaxy NGC 4639 is about 25 Mpc (82 Mly) distant. This is, to date, the farthest galaxy in which astronomers have detected a Cepheid. Five years before the Sandage team made these measurements, NGC 4639 was host to the Type Ia SN1990N. Sandage knew the peak apparent magnitude of SN1990N and also its distance. He could now calculate its peak absolute magnitude: -19.3 ± 0.23. The discovery of Cepheids in NGC 4639 made this the sixth galaxy in which astronomers had seen both Cepheids *and* a Type Ia supernova. (Two more galaxies have been added to the list since then; table 9.2 lists them all.) When Sandage averaged the peak magnitudes of the supernovae in these galaxies he obtained a value of -19.47 ± 0.07.

It seems that all the different methods tell the same story. First, Type Ia supernovae are one of the most reliable candles in astronomy. Second, their peak absolute magnitude is about -19.5. Their fantastic luminosity means we can see them 1000 times farther away

---

*It is difficult to be too confident in this value, even when one of the observers was Tycho. They used phrases like 'dazzling in brightness' to describe the magnitude, which lacks the precision of modern photometric techniques.

TABLE 9.2: There are eight galaxies that have hosted Type Ia supernovae and in which we have detected Cepheids. The galaxy NGC 5253 has been host to two visible supernovae.

Supernova	Host galaxy
SN1895B	NGC 5253
SN1937C	IC 4182
SN1960F	NGC 4496
SN1972E	NGC 5253
SN1974G	NGC 4414
SN1981B	NGC 4536
SN1989B	NGC 3627
SN1990N	NGC 4639
SN1998bu	M 96

than we can see Cepheids, so at present they are perhaps our best method of measuring distances deep into space. We still must be careful when using them as distance indicators; we need to take account of reddening in the host galaxy, for instance. But the main drawback with using Type Ia supernovae as distance indicators is that they are rare and that they do not last long at their peak brightness.

A new development called the Supernova Cosmology Project (SCP), which is led by the American astronomer Saul Perlmutter (1959–), is finding many new supernovae in galaxies at *huge* distances from us. The SCP is a collaborative venture between workers based in several countries. It uses a combination of advanced CCD technology, large telescopes, computing power and the internet that did not even exist five years ago. In five more years it will perhaps be routine. First, astronomers use an ultrasensitive CCD attached to the Inter-American Cerro Tololo 4-m telescope in Chile to image thousands of distant galaxies just after a new Moon. Three weeks later, they image the same galaxies once again. Computers subtract light from the old image away from the new image. Any remaining light must come from supernovae that occurred during those three weeks, supernovae that are probably still rising to peak brightness. This procedure guarantees that there are supernovae available for study during the best nights for observing: immediately before the new Moon. Brightness measurements of the supernovae are made with large telescopes around the world; if necessary, the *Hubble Space Telescope* is used. The Keck 10-m telescope in Hawaii carries out the final identification of the supernova spectra.

The method is effective: in one 48-hour period in November 1995, the SCP found 11 supernovae. In total, the team have so far discovered more than 80 supernovae. A similar project, called the High-z SN Search, has found 35 distant Type Ia supernovae using the same sort of techniques. If the peak brightness of these supernovae was -19.5, then some of them were 2200 Mpc (7 Bly) distant! (At the time of writing, the most distant supernova ever seen is SN1998eq: the SCP found it to have a redshift of 1.2. The conversion of redshift into distance is the subject of the final chapter.) This is the largest distance so far mentioned in this book, and again such a distance comes with a proviso: we have reached a rung on the distance ladder where we are probing the geometry of the universe as a whole. From now on we must be careful about what we mean by distance.

CHAPTER SUMMARY

- Astronomers use a variety of methods to estimate galaxian distances. Secondary indicators require calibration in a galaxy whose distance is already known, usually from Cepheids. Any error in the distance to the calibrating galaxy propagates through to distances derived from secondary indicators. Primary indicators, based on geometrical arguments or well-understood physics, do not require intermediate calibration.

- The planetary nebula luminosity function is very similar in all galaxies. Its use as a distance indicator is based on finding the apparent magnitude of the cut-off. The absolute magnitude of the cut-off is about -4.48. The method can be applied to both spirals and ellipticals.

- The Tully–Fisher relation for spiral galaxies relates the luminosity, L, with the maximum rotational velocity, v_{max}:

$$L \propto V_{max}^4.$$

The rotational velocity may be found from radio observations of the 21-cm linewidth. The relation is cleaner if the infrared luminosity is used.

- The Faber–Jackson relation for elliptical galaxies relates the luminosity, L, with the central velocity dispersion, σ:

$$L \propto \sigma^4.$$

With a third parameter the relation is cleaner; we then obtain a fundamental plane for elliptical galaxies. The D_n–σ relation, where D_n is the diameter of a central circular region within which the total average surface brightness is 20.75 magnitudes per square second of arc, is one projection of the fundamental plane.

- The surface brightness fluctuation method uses a CCD to image a galaxy. The fluctuation in the number of photons from pixel to pixel is a function of the galaxy's distance. The closer the galaxy, the greater the fluctuations in surface brightness. The method works for ellipticals.

- The globular cluster luminosity function is found empirically to be Gaussian. Its use as a distance indicator is based on finding the apparent magnitude of the turnover. The absolute magnitude of the turnover is about -6.8. The method works for ellipticals.

- Megamasers can sometimes be found in a circumnuclear disk. Careful measurements of the velocities of the masers enable us to deduce the radius of the disk; if the apparent size is known, the distance follows. It is a primary distance indicator.

- The Sunyaev–Zel'dovich effect is a distortion of the cosmic microwave background caused by the interaction of microwave photons with energetic electrons from the intracluster gas in a rich galaxy cluster. Combining the magnitude of the distortion with the X-ray brightness of the cluster determines the line-of-sight depth of the cluster. Taking this to be the diameter of the cluster yields the distance to the cluster. It is a primary distance indicator.

- By measuring the time delay between the two images of a gravitationally lensed quasar we can, if we know the mass distribution of the lensing galaxy, calculate the distance to the galaxy. It is a primary distance indicator.

- The distance to Type II supernovae can be deduced using the expanding photosphere method, which is essentially a modification of the Baade–Wesselink method. It is potentially a primary distance indicator but, since the physics of the explosions is not completely known, it is best used operationally as a secondary indicator.

- Type Ia supernovae are potentially excellent standard candles, with an absolute peak magnitude of -19.5.

QUESTIONS AND PROBLEMS

9.1 What qualities define a good extragalactic distance indicator?

9.2 Suppose that astronomers discover that M 31 is 20% farther away than previously believed. Consider the effect that such a discovery would have on each of the distance indicators mentioned in this chapter.

9.3 A torus of material surrounds SK$-69°202$, the star in the LMC that became SN1987A. The *IUE* satellite observed ultraviolet emission from the ring about 80 days after the supernova event, and they grew in strength for about 400 days. It is assumed that the emission was due to material from the shock wave hitting the torus. Further observations show that the torus is inclined at angle of $43°$. The angular diameter of the major axis of the torus is $1.66 \pm 0.03''$. From the time delay, estimate the size of the torus and thus the distance to the supernova. (Note: this is a difficult question!) Compare your answer with distances for the LMC derived by other distance indicators.

9.4 A Type Ia supernova is observed to have an apparent magnitude of $+15$ at peak brilliance. The extinction to the host galaxy is 0.5 magnitudes. How far away is the supernova?   63 Mpc

9.5 The galactic planetary nebula NGC 7009 expands at the rate of $0.7''$ per century. The radial velocity of expansion is $21\,\mathrm{km\,s^{-1}}$. If the expansion is spherically symmetric, how far away is the nebula? Could we use this technique to measure the nebula's distance if it were in another galaxy?   633 pc; no

9.6 You observe tens of planetary nebulae in a distant galaxy and find that $N(m)$, the number of planetaries with apparent magnitude m, is well described by the relation $N(m) \propto e^{0.307}\left(1 - e^{3(m^*-m)}\right)$. The cut-off m^* is equal to $+25$. How far away is the galaxy?   7.9 Mpc

9.7 Find, in your local library or from the Web, photographs of the spiral galaxies M 101 and NGC 4762. If the maximum rotational velocity of these two galaxies were similar, which would display the larger 21-cm linewidth? Explain your answer.   NGC 4762, which is edge-on

FURTHER READING

GENERAL

Jacoby G H *et al.* (1992) A critical review of selected techniques for measuring extragalactic distances. *Pub. Astron. Soc. Pacific* **104** 599–662

— A review of the indicators mentioned in this chapter (except for gravitational lenses, the SZ effect and masers) in more technical detail than provided here.

Heck A and Caputo F (1999) *Post-Hipparcos Cosmic Candles* (Kluwer: Dordrecht)

— The implications of the *Hipparcos* results for the distance ladder are still the subject of debate. This book contains several reviews by key workers in the field.

PLANETARY NEBULAE

Bottinelli L *et al.* (1991) A systematic effect in the use of planetary nebulae as standard candles. *Astron. Astrophys.* **252** 550–556
— Does the PNLF method lead to systematic errors in distance estimates for distant galaxies?

Ciardullo R *et al.* (1989) Planetary nebulae as standard candles. II. The calibration in M 31 and its companions. *Astrophys. J.* **339** 53–69
— In the second of a long series of papers, the authors show that the PNLF cut-off at λ5007 is about −4.5.

Feldmeier J J, Ciardullo R and Jacoby G H (1997) Planetary nebulae as standard candles. XI. Application to spiral galaxies. *Astrophys. J.* **479** 231–243
— The authors use the PNLF method to obtain the distances of three spiral galaxies: M 101 (7.7 ± 0.5 Mpc from 65 planetaries), M 51 (8.4±0.6 Mpc from 64 planetaries) and M 96 (9.6±0.6 Mpc from 74 planetaries). Prior to this, the PNLF method had been used mainly with elliptical galaxies.

Ford H C *et al.* (1996) The stellar halo of M 104. I. A survey for planetary nebulae and the PNLF distance. *Astrophys. J.* **458** 455–466
— The authors deduce a distance of 8.9 ± 0.6 Mpc for M 104 from 294 planetaries; from this they deduce a (large) value for H_0.

Gurzadyan, G. A. (1997) *The Physics and Dynamics of Planetary Nebulae* (Springer: Berlin)
— A useful reference for the postgraduate student.

Harris H C *et al.* (1997) Trigonometric parallaxes of planetary nebulae. In: *Planetary Nebulae* ed. H. J. Habing and H. J. G. L. M. Lamers (Kluwer: Dordrecht)
— The authors describe measurements of planetary nebulae made as part of the USNO parallax programme.

Jacoby G H (1989) Planetary nebulae as standard candles. I. Evolutionary models. *Astrophys. J.* **339** 39–52
— This paper introduced the PNLF method as a galaxian distance indicator.

Jacoby G H, Ciardullo R and Harris W E (1996) Planetary nebulae as standard candles. X. Tests in the Coma region. *Astrophys. J.* **462** 1–12
— Provides PNLF distances to three galaxies: NGC 4494 (12.8 Mpc), NGC 4565 (10.5 Mpc) and NGC 4278 (10.2 Mpc). The errors are about ±1 Mpc in each case.

McMillan R, Ciardullo R and Jacoby G H (1993) Planetary nebulae as standard candles. IX. The distance to the Fornax cluster. *Astrophys. J.* **416** 62–73
— From PNLF distances to three Fornax galaxies, the authors provide an estimate for H_0.

Soffner T *et al.* (1996) Planetary nebulae and H II regions in NGC 300. *Astron. Astrophys.* **306** 9–22
— From 34 planetaries in this late-type spiral the authors deduce its distance modulus to be 26.9 ± 0.4.

http://www.aao.gov.au
http://seds.lpl.arizona.edu/billa/bb
— These sites contain colour images of planetary nebulae and links to other sites devoted to planetaries.

SURFACE BRIGHTNESS FLUCTUATIONS

Lauer T *et al.* (1998) The far-field Hubble constant. *Astrophys. J.* **499** 577–588
— The authors use the surface brightness fluctuation technique to deduce $H_0 = 89 \pm 10 \, \mathrm{km \, s^{-1} \, Mpc^{-1}}$.

Thomsen B *et al.* (1997) The distance to the Coma cluster from surface brightness fluctuations. *Astrophys. J.* **483** L37–L40
— From the surface brightness fluctuations of NGC 4881 the authors obtain a distance to the Coma cluster of 102 ± 14 Mpc and a Hubble constant of $H_0 = 71 \pm 11 \, \mathrm{km \, s^{-1} \, Mpc^{-1}}$.

Tonry J L and Schneider D P (1988) A new technique for measuring extragalactic distances. *Astron. J.* **96** 807–815
— This paper introduced the SBF distance indicator.

TULLY–FISHER RELATION

Giovanelli R *et al.* (1997) The Tully–Fisher relation and H_0. *Astrophys. J.* **477** L1–L4
— The authors use data for 24 clusters to deduce that $H_0 = 69 \pm 5$.

Theureau G *et al.* (1997) Kinematics of the local universe. V. The value of H_0 from the Tully–Fisher B and log D_{25} relations for field galaxies. *Astron. Astrophys.* **322** 730–746
— A clear treatment of the problems of Malmquist bias in using the Tully–Fisher relation. The authors deduce $H_0 = 53.4 \pm 5.0\,\mathrm{km\,s^{-1}\,Mpc^{-1}}$.

Tully R B and Fisher J R (1977) A new method of determining distances to galaxies. *Astron. Astrophys.* **54** 661–673
— This paper introduced one of the most widely used methods for measuring galaxian distances.

THE FUNDAMENTAL PLANE

Djorgovski S and Davis M (1987) Fundamental properties of elliptical galaxies. *Astrophys. J.* **313** 59–68
— In this paper the authors introduced the idea of the fundamental plane of elliptical galaxies.

Dressler A *et al.* (1987) Spectroscopy and photometry of elliptical galaxies. I. A new distance estimator. *Astrophys. J.* **313** 43–58
— This paper by the Seven Samurai introduced the idea of the D_n–σ relation.

Faber S M and Jackson R E (1976) Velocity dispersions and mass-to-light ratios for elliptical galaxies. *Astrophys. J.* **204** 668–683
— This paper introduced the Faber–Jackson distance indicator.

Hjorth J and Tanvir N R (1997) Calibration of the fundamental plane zero point in the Leo I group and an estimate of the Hubble constant. *Astrophys. J.* **482** 68–74
— The authors find that there is a small scatter in the fundamental plane in the Leo I group; they use this to deduce a distance to the Coma cluster of $108 \pm 12\,\mathrm{Mpc}$, and derive $H_0 = 67 \pm 8\,\mathrm{km\,s^{-1}\,Mpc^{-1}}$.

GLOBULAR CLUSTER LUMINOSITY FUNCTION

Hanes D A and Whittaker D G (1987) Globulars as extragalactic distance indicators: maximum-likelihood methods. *Astron. J.* **94** 906–916
— An early application of the GCLF method to some elliptical galaxies.

Harris W E (1991) Globular cluster systems in galaxies beyond the Local Group. *Ann. Rev. Astron. Astrophys.* **29** 543–579
— A nice review of extragalactic globular clusters, including the application of the GCLF.

MEGAMASERS

Miyoshi M *et al.* (1995) Evidence for a black hole from high rotation velocities in a sub-parsec region of NGC 4258. *Nature* **373** L127–L129
— In 1995, this was the best direct evidence for the existence of a black hole; that aspect of the paper has been superseded. However, the authors remain the only workers to calculate a galaxian distance from observations of megamasers.

SUNYAEV–ZEL'DOVICH EFFECT

Birkinshaw M, Hughes J P and Arnaud K A (1991) A measurement of the Hubble constant from the X-ray properties and the Sunyaev–Zel'dovich effect of Abell 665. *Astrophys. J.* **379** 466–481
— The authors deduce $H_0 = 40 \pm 9\,\mathrm{km\,s^{-1}\,Mpc^{-1}}$.

Birkinshaw M and Hughes J P (1994) A measurement of the Hubble constant from the X-ray properties and the Sunyaev–Zel'dovich effect of Abell 2218. *Astrophys. J.* **420** 33–43
— The authors deduce $H_0 = 55 \pm 17\,\mathrm{km\,s^{-1}\,Mpc^{-1}}$.

Furuzawa *et al.* (1998) *ASCA* observation of the distant cluster of galaxies CL 0016+16 and implication for H_0. *Astrophys. J.* **504** 35–41
— At $z = 0.541$, this cluster is very distant indeed. The authors obtain $H_0 = 47 \pm 14\,\text{km s}^{-1}$.

Herbig T *et al.* (1995) A measurement of the Sunyaev–Zel'dovich effect in the Coma cluster of galaxies. **449** L5–L8
— The authors obtain $H_0 = 71^{+30}_{-25}\,\text{km s}^{-1}\,\text{Mpc}^{-1}$.

Holzapfel W L *et al.* (1997) Measurement of the Hubble constant from X-ray and 2.1 millimeter observations of Abell 2163. *Astrophys. J.* **480** 449–465
— The authors obtain $H_0 = 78^{60}_{-40}\,\text{km s}^{-1}\,\text{Mpc}^{-1}$.

McHardy I M *et al.* (1990) *GINGA* observations of Abell 2218: implications for H_0. *MNRAS* **242** 215–220
— The authors obtain $H_0 = 24^{+13}_{-10}\,\text{km s}^{-1}\,\text{Mpc}^{-1}$.

Myers S T *et al.* (1997) Measurements of the Sunyaev–Zel'dovich effect in the nearby clusters A478, A2142 and A2256. *Astrophys. J.* **458** 1–21
— The authors obtain $H_0 = 54 \pm 14\,\text{km s}^{-1}\,\text{Mpc}^{-1}$.

Sunyaev R A and Zel'dovich Y B (1972) The observation of relic radiation as a test of the nature of X-ray radiation from the clusters of galaxies. *Comments Astrophys. Space. Phys.* **4** 173–178
— The first description of the Sunyaev–Zel'dovich effect.

Tsuboi M *et al.* (1998) Measurement of the Sunyaev–Zel'dovich effect toward Abell 2218 at 3 GHz. *Pub. Astron. Soc. Japan* **50** 169–173
— The authors obtain $H_0 = 54^{+51}_{-21}\,\text{km s}^{-1}\,\text{Mpc}^{-1}$.

GRAVITATIONAL LENSES

Einstein A (1936) Lens-like action of a star by the deviation of light in the gravitational field. *Science* **84** 506–507
— Einstein described what is now known as an Einstein ring. He considered the case of a lensing star, not a galaxy, and stated that 'there is no great chance of observing this phenomenon'.

Keeton C R and Kochanek C S (1997) Determination of the Hubble constant from the gravitational lens PG 1115 + 080. *Astrophys. J.* **487** 42–54
— The authors obtain $H_0 = 51^{+14}_{-13}\,\text{km s}^{-1}\,\text{Mpc}^{-1}$.

Kundić T *et al.* (1995) An event in the light curve of 0957 + 561 A and a prediction of the 1996 image B light curve. *Astrophys. J.* **455** L5–L8
— From image A fluctuations in 1995 the authors predicted how image B would fluctuate in February 1996.

Kundić T *et al.* (1997) A robust determination of the time delay 0957 + 561 A, B and a measurement of the global value of Hubble's constant. *Astrophys. J.* **482** 75–82
— Their prediction was correct; the authors obtain $H_0 = 64\,\text{km s}^{-1}\,\text{Mpc}^{-1}$.

Refsdal S (1964) The gravitational lens effect. *MNRAS* **128** 295–306
— The first clear account of gravitational lenses in astronomy.

Refsdal S (1964) On the possibility of determining Hubble's parameter and the masses of galaxies from the gravitational lens effect. *MNRAS* **128** 307–310
— A companion to the above paper that describes the use of gravitational lenses as a distance indicator.

Refsdal S and Surdej J (1994) Gravitational lenses. *Rep. Prog. Phys.* **56** 117–185
— A very clear review of all aspects of gravitational lenses, including their use as a distance indicator.

Soldner J (1804) *Berliner Astron. Jahrb. 1804* 161
— The first calculation of the magnitude of a gravitational lens effect.

Turner E L (1988) Gravitational lenses. *Sci. American* **259** (1) 26–32
— A nice introduction to gravitational lenses.

Walsh D, Carswell R F and Weymann R J (1979) 0957 + 561 A, B: twin quasistellar objects or gravitational
lens? *Nature* **279** 381–384
— The first observation of a gravitationally lensed quasar.

SUPERNOVAE
Branch D (1998) Type Ia supernovae and the Hubble constant. *Ann. Rev. Astron. Astrophys.* **36** 15–55
— The author concludes that $M_V \approx -19.4$ or -19.5, and that $H_0 = 60 \pm 10\,\mathrm{km\,s^{-1}\,Mpc^{-1}}$. Compare with
the 1992 review by Branch and Tammann, to see how rapidly progress has been made in this field!

Branch D, Fisher A and Nugent P (1993) On the relative frequencies of spectroscopically normal and peculiar
Type Ia supernovae. *Astron. J.* **106** 2383–2391
— The authors classify 'normal' and 'peculiar' Type Ia events. Normal supernovae are now usually referred
to as 'Branch-normal'.

Branch D and Tammann G A (1992) Type Ia supernovae as standard candles. *Ann. Rev. Astron. Astrophys.* **30**
359–389
— A thorough review of Type Ia supernovae as distance indicators.

Goldsmith D (1989) *Supernova* (Oxford University Press: Oxford)
— An accessible, though now perhaps slightly dated, account of SN1987A.

Hamuy M *et al.* (1995) A Hubble diagram of distant Type Ia supernovae. *Astron. J.* **109** 1–13
— The authors argue that there is quite a large dispersion in Type Ia peak magnitudes, but that there is a
correlation between the initial decline rate and the peak magnitude that enables one to reduce the dispersion.
Type Ia events can still be used as distance indicators.

Kim A G *et al.* (1977) Implications for the Hubble constant from the first seven supernovae at $z \gtrsim 0.35$.
Astrophys. J. **476** L63–L66
— One of the Supernova Cosmology Project papers; it offers a bound on the Hubble constant of $H_0 <
78\,\mathrm{km\,s^{-1}\,Mpc^{-1}}$.

Kirshner R P and Kwan J (1974) Distances for extragalactic supernovae. *Astrophys. J.* **193** 27–36
— The first application of the expanding photosphere method.

Riess A G, Press W H and Kirshner R P (1995) Using Type Ia supernova light curve shapes to measure the
Hubble constant. *Astrophys. J.* **438** L17–L20
— The authors re-evaluate the use of Type Ia supernovae as standard candles by taking into account a
relation between peak magnitude and rate of decline of brightness.

Riess A G et al (1998) Observational evidence from supernovae for an accelerating universe and a cosmological
constant. *Astron. J.* **116** 1009–1038
— By finding supernovae at very large distances, astronomers are probing the large-scale geometry of the
universe. The High-z Supernova Search team find evidence for a non-zero cosmological constant.

Saha A *et al.* (1997) Cepheid calibration of the peak brightness of Type Ia supernova. VIII. SN1990N in
NGC4639. *Astrophys. J.* **486** 1–20
— The eighth in a series of papers to determine Cepheid distances to galaxies that have hosted Type Ia
supernovae.

Wagoner R V (1977) Determining q_0 from supernovae. *Astrophys. J.* **214** L5–L7
— The author extends the expanding photosphere method of Kirshner and Kwan.

10

Mezzanine: cosmic expansion

There is a limiting distance beyond which even supernovae become too faint to see. How can we determine the distances to the myriad of galaxies that lie beyond this limit? It turns out that there is a powerful method of determining the distance to any cosmological object for which we can take a spectrum. To understand the method we must take one last detour into physics. At the end of the detour we can begin to discuss the size of the universe itself.

10.1 HUBBLE'S LAW

In 1912, eight years before the Great Debate, Vesto Slipher analysed the spectrum of the spiral nebula M 31. He found that its spectral lines were blueshifted, presumably as a result of the Doppler effect. The size of the blueshift indicated that M 31 hurtled towards the Earth at a speed of about $200\,\mathrm{km\,s^{-1}}$. Intrigued by this result, Slipher decided to study the spectra of other spiral nebulae. By 1914, he had data on 15 nebulae. Of these, 13 showed *redshifted* spectral lines and were thus moving *away* from us. By 1922, Slipher knew the radial velocities of 41 spiral nebulae. Almost all of them were rushing away from us at a rate much greater than the typical radial velocity of stars. It was clear that M 31 was very much an exception*.

At about this time Hubble proved that the spiral nebulae were galaxies like our own Galaxy. This made Slipher's results perhaps even more puzzling. Why should such vast masses of stars recede at such high speeds? And why did we not see equal numbers of blueshifted and redshifted galaxies? Hubble resolved to settle the matter, and for this he needed the radial velocities of more galaxies. He enlisted the help of Milton La Salle Humason (1891–1972), a colleague at Mount Wilson. Humason's career path was as tortuous as the paths leading up Mount Wilson. He began as a mule skinner on the mountain. Later, he found a job as a janitor at the telescope complex there. He was insatiably curious, and pestered the astronomers to explain to him what they were doing. He began to help them in their work, and eventually he established his own observing programme.

---

*At the time of writing, astronomers have taken the spectra of many thousands of galaxies. Of these, only a few show a blueshift. Some of these blueshift galaxies are in the Local Group and other nearby groups. Some are in the Virgo cluster. The Virgo galaxy IC 3258, for instance, moves towards us at $517\,\mathrm{km\,s^{-1}}$ — the highest blueshift of any extragalactic object.

Milton La Salle Humason

Hubble asked Humason to use long exposure times to record the spectra of faint — and therefore presumably distant — galaxies. This was not easy. The spectrum of a whole galaxy is a combination of light from all its stars and nebulae and other radiating objects, and is thus close to being a continuous smear. Fortunately, Humason could distinguish the H and K absorbtion lines of ionised calcium, even when other features were lost. (Ionised calcium has an extremely high cross section for light at these wavelengths. When radiation from stars passes through cooler gaseous regions in a galaxy, it is selectively absorbed at the Ca II H and K wavelengths before leaving the galaxy.) For each galaxy he then had to calculate the redshift, z: the fraction by which the wavelengths of the spectral lines were increased over their laboratory values. The z values that Humason found were tiny. Often the redshifts were less than 0.01, which meant that the wavelength shift was less than 1%. But even a small redshift of 0.01 implied a large recession velocity. In 1928, for instance, Humason found the redshift of NGC 7619 to be 0.0126, which corresponded to a recession velocity of about $3800 \, \text{km} \, \text{s}^{-1}$. Humason thus found recession velocities that were much bigger than any determined by Slipher.

When Hubble began to analyse these results in 1929, he decided to plot a graph of the recession velocity of galaxies (as measured by their redshifted spectral lines) against their distance from us (as determined by his observations of Cepheids). On his first graph he compared the redshifts of 18 of Slipher's galaxies with his own estimates of the distances to those galaxies. He *might* have obtained a 'scatter' graph, which would indicate that there was no relationship between the two quantities. What he obtained was a straight line — and thereby one of the greatest scientific discoveries of the century.

(Hubble was not the first to suggest a relation between recession velocity and distance. In 1921, the German astronomer Carl Wirtz (1876–1939) combined the Slipher redshift measurements with distances based on the apparent diameters of galaxies to de-

Vesto Melvin Slipher

duce a tentative correlation between velocity and distance. A few years later the Swedish astronomer Knut Emil Lundmark (1889–1958) did much the same thing. In 1928, the American physicist Howard Percy Robertson (1903–1961) used Slipher's redshifts and Hubble's own published data on galaxian distances to deduce a velocity–distance relation. And as we shall soon see, there were theoretical reasons for supposing that there *should* be a velocity–distance relation. Like so many other discoveries in science, the idea was 'in the air'. Hubble receives the credit, though, because it was his weight of evidence that put the matter beyond doubt.)

Despite what figure 10.1 shows, what Hubble actually discovered was not a linear velocity–distance relation but a linear relationship between redshift, z, and distance, r:

$$z = \text{constant} \times r. \tag{10.1}$$

Hubble *interpreted* the redshift as arising from the Doppler effect. In calculating the recession velocities of the galaxies he essentially multiplied (10.1) throughout by c (since, for small redshifts, the Doppler interpretation of the redshift gives $v = cz$):

$$cz = \text{constant} \times r. \tag{10.2}$$

The constant in the above equation is called the *Hubble constant* and is denoted with the symbol H_o. (It is a pity that we are stuck with the phrase 'Hubble constant'. It would be better to talk of the Hubble *parameter* since, as we shall see, this 'constant' can change with time. In most books and research papers you will see the Hubble constant denoted by H_0 — pronounced 'aitch nought'. I write it with a subscript o rather than 0 to emphasize that we *o*bserve its present-day value.) Equations (10.1) and (10.2) are both statements of *Hubble's law*. A more familiar way of writing it is:

$$cz = H_o r. \tag{10.3}$$

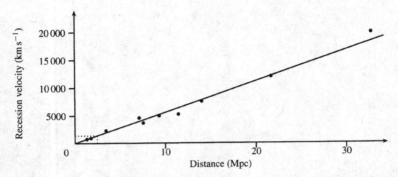

FIGURE 10.1: An early version of the Hubble law (adapted from Hubble (1931)). Recession velocities derive from redshift data; distances derive from a distance ladder built by Hubble, based on the magnitude of brightest stars and the magnitude of galaxies themselves. The slope of the graph is H_0, the Hubble constant. The dotted lines in the lower-left corner show the region referred to in Hubble's 1929 paper. This region contained just 13 data points, with considerable scatter, so Hubble's 1929 extrapolation was brave as well as correct.

Equation (10.3) seems to produce an easy method of calculating the distance to a galaxy. Simply measure its redshift (which nowadays is a matter of routine) and divide by H_0 (the value of which is hotly disputed, but which in principle can be determined by observation). In fact there are some subtleties that we must be aware of when using Hubble's law to calculate distances; and when redshifts are large we must abandon the linear redshift–distance relation altogether. I will return to these complications later, when we take the final step on the distance ladder. In the rest of this chapter, I shall discuss the implications of Hubble's discovery. Hubble's law is more than just a new cosmic yardstick. It tells us something profound about the universe itself.

Why should galaxies rush away from us? What makes us so special? The answer is that there is nothing at all special about us or our particular region of the universe: *all* observers see a linear velocity–distance relation *if the universe is expanding uniformly*. Previously, astronomers had assumed that the universe was a static, unchanging place; the discovery of cosmic expansion changed the way people think about the universe. It was a change similar to that wrought by Copernicus four centuries earlier.

To understand why an expanding universe explains the relation between velocity and distance, consider the following analogy. Imagine sitting in the middle of a large cinema hall where every seat is taken and each person in the audience smokes a cigarette. As the lights dim for the start of the movie, the red tips of the cigarettes glow bright. Now suppose that, for some bizarre reason, the space inside the cinema stretches. Within a few seconds the cinema doubles in size. What do you see as you look around? Your neighbours in front, behind and to either side of you double their distance from you. Since they were nearby to start with they move only a few centimetres. The red tips of their cigarettes therefore move slowly — a few centimetres per second. The people who sit a few seats away also double their distance from you. But because they were farther away to start with they move through a greater distance in the same period of time. Their cigarette tips seem to move faster — a few metres per second. People who are seated at the far end of the cinema from you also double their distance from you, so they move through a much larger distance in the same time period. The most distant points of light thus move the fastest — a few tens of metres per second. So in the darkness of the expanding cinema hall you see a lattice of points of light moving away from you, the nearby lights moving slowly and the more distant lights moving quickly. There is nothing special about you or your cinema seat. The person sitting ten rows back sees the same thing. The effect is due to the expansion of the hall. (Like all analogies, this one is flawed. Patrons sitting next to a wall have a special view: in one direction they see expansion, in the other direction they see the wall. There is no wall for the universe as a whole.)

Similarly, there is nothing special about the Galaxy. Astronomers in a distant galaxy would see *our* Galaxy receding from their own. And no strange force causes distant galaxies to recede faster than nearby galaxies. It is simply that the uniform expansion of the universe causes nearby objects to have a small recession velocity and distant objects to have a large recession velocity. The Hubble constant is just a measure of how quickly the universe is expanding at present. A large value for H_o denotes a rapid expansion; a small value for H_o means the expansion is more sedate.

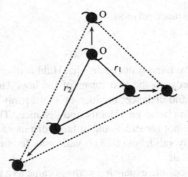

FIGURE 10.2: The positions of three galaxies before and after a twofold expansion of space. An observer in each of the galaxies sees the same thing, namely that the distance to the other galaxies has doubled.

We can also see why a uniform expansion gives rise to a linear velocity–distance law by referring to figure 10.2. Suppose that an observer at O sees two galaxies at distances r_1 and r_2 at some time t. (Note that the distances r_1 and r_2 are *proper* distances, of the tape-measure kind. This is discussed in more detail in the next chapter.) If the triangle formed by the observer and the galaxies remains self-similar upon expansion, the ratio $r_1 : r_2$ is constant. Call the constant H:

$$\frac{r_1}{r_2} = H. \tag{10.4}$$

Because of the expansion, the observer sees that r_1 and r_2 change with time and that the galaxies have a velocity of recession. These velocities at some time t are given by

$v_1(t) = dr_1/dt$ and $v_2(t) = dr_2/dt$. Differentiating (10.4) with respect to time we obtain

$$\frac{v_1}{v_2} = H. \qquad (10.5)$$

If we neglect the time that light takes to travel between the galaxies, we can combine (10.4) and (10.5) to obtain

$$\frac{r_1}{r_2} = \frac{v_1}{v_2} = H \qquad \text{or} \qquad \frac{v_1}{r_1} = \frac{v_2}{r_2} = H$$

or, for a general galaxy i,

$$v_i = H r_i. \qquad (10.6)$$

The expansion thus causes the velocity to increase linearly with distance from any arbitrary point, so long as a geometrical configuration remains self-similar upon expansion. If we use the Doppler relation $v = cz$ then (10.6) becomes $cz = Hr$, which is the Hubble law of (10.3).

The linear velocity–distance relation

$$v = H(t)r \qquad (10.7)$$

is often called Hubble's law, but it is not quite what Hubble discovered. It is worth bearing in mind the distinction between the two forms of the law. Hubble found an empirical relation between redshift and distance; the idea of a uniformly expanding universe leads naturally to a linear relation between velocity and distance. The two forms of Hubble's law, (10.3) and (10.7), are not identical, and at large distances produce different results. Equation (10.7) is generally valid, but (10.3) is valid only for small redshifts. The reasons for this will become clear later.

Hubble's discovery of cosmic expansion perhaps came as a shock to many people, but not to Einstein (at least, not in the same way). A decade earlier, Einstein had come close to making a great prediction. His general theory of relativity seemed to *demand* that the universe expands.

10.2   GENERAL RELATIVITY

Newton's theory of gravity was, for over two centuries, spectacularly successful. It explained all of terrestrial dynamics, the orbits of planets around the Sun, and the movement of binary stars. Each time physicists applied Newton's theory of gravity they obtained excellent results. Despite the theory's success, no one knew what gravity was. It seemed to be some force acting instantaneously at a distance, but exactly what caused the force was unknown. In effect the physicists said: 'here is an equation that governs how massive bodies interact through the force of gravity; the equation works, so let's not ask *why* it works'. Einstein wanted to know why gravity works. He also realised that Newton's theory must be incomplete, since it does not obey the principles of special relativity. These considerations led Einstein to a new theory of gravity: the general theory of relativity.

According to Einstein, gravity is not a force at all. It is an illusion caused by the curvature of spacetime.

The concept of curved spacetime is rather difficult to grasp at first. Even Einstein struggled with the mathematics. The idea of spacetime itself is straightforward, though, so we shall begin the discussion with that.

10.2.1 Spacetime

Throughout this book we have tacitly assumed that the distance between two points in space is a straightforward concept. When we think of space we tend to think of three fixed and unchanging dimensions — a grid of left–right, forward–back, up–down — and distance is just separation measured with reference to this grid. The process of *measuring* a distance may be tricky, but we assume that when we make a measurement the result is something that everyone can agree upon. Our conception of time is more tricky, but most of us perhaps think of it as an infinitely long line: past–present–future. We use time to keep track of changes that occur in space. Space and time thus appear to be quite different and independent entities. Einstein's special theory of relativity, published in 1905, pointed out the need for us to amend our everyday ideas of space and time.

Albert Einstein

The theory of relativity considers *observers* — individuals equipped with a measuring rod and a clock whose function is to record the positions and times at which *events* occur. (An event is something that happens at a particular point in space and at a particular instant of time — the thunk of a particular apple hitting Newton on the head, for instance. An event in relativity is completely characterised by four numbers: its space and time coordinates in a given frame of reference.) An observer in an *inertial reference frame* — a frame in which Newton's first law is valid — is called an *inertial observer*. It is an indication of Einstein's unique genius that he could deduce a whole new system of dynamics from just two postulates about inertial frames.

- All inertial observers are equivalent.

- The speed of light in vacuum, c, is the same in all inertial systems.

From these postulates Einstein derived the *Lorentz transformations*, which relate the space and time coordinates of the same event as measured in different inertial frames S and S'. (The Dutch physicist Hendrik Antoon Lorentz (1853–1928) first derived these transformations without using the assumptions of special relativity.) If S' moves relative to S with speed v along the positive-x axis, then the S coordinates of the event (x, y, z, t) are related to the S' coordinates of the event (x', y', z', t') as follows:

$$x' = \gamma(x - vt) \qquad y' = y \qquad z' = z \qquad t' = \gamma(t - vx/c^2) \tag{10.8}$$

where

$$\gamma = \frac{1}{\sqrt{1 - v^2/c^2}}.$$

One immediate consequence of the postulates is that there is no such thing as absolute simultaneity. Events that are simultaneous in one inertial frame do not occur simultaneously in all other reference frames. For example, imagine that an observer in inertial frame S measures two events, at coordinates x_1 and x_2, to take place at the same time t. (The two events, for example, might be flashbulbs going off from two different cameras. The events are defined to be simultaneous in S if light signals from the two flashes meet at the midpoint between x_1 and x_2.) According to (10.8) an observer in inertial frame S' measures a time interval $t'_1 - t'_2$ between the two events:

$$t'_1 - t'_2 = \gamma(x_2 - x_1)v/c^2.$$

In other words, if $x_1 \neq x_2$ the observer in S' records the two camera flashes as occurring at different times.

If observers in relative motion do not agree about the simultaneity of events, it follows that clocks in relative motion will not stay synchronised. Consider a clock situated at rest in frame S', and suppose that two events occur at this clock. The clock indicates a time t'_1 for the first event and t'_2 for the second event, with $\Delta t' = t'_2 - t'_1$. According to (10.8), an observer in S measures a different time interval Δt between these events:

$$\Delta t = \gamma \Delta t'.$$

The frame in S' is the clock's *rest frame*. In this frame the clock measures the *proper time* between the two events. An identical clock moving relative to the rest frame of the two events always measures a longer time interval between the events. This is the effect of *time dilation* on a moving clock.

Inertial observers in relative motion not only measure time differently: they also measure space differently. Consider a rod placed at rest along the x' axis in S'. One end of the rod is at x'_1 and the other is at x'_2 so, measured in its rest frame, the rod is of length $\Delta x' = x'_1 - x'_2$. An observer in S, who of course must observe both ends of the rod simultaneously to obtain its length, measures a different length for the rod. According to (10.8), an observer in S measures a length Δx for the rod:

$$\Delta x = \frac{\Delta x'}{\gamma}.$$

In other words, an observer in the rest frame of the rod measures its *proper length*. If the rod moves relative to an observer, that observer will measure a shorter length for the rod. This is the effect of *length contraction*.

The effects of time dilation and length contraction seem strange because we have no experience of travelling at high speed. A jet airplane, for instance, moves at about one millionth of the speed of light. At low speeds like this, special relativity reduces to Newtonian physics. But whenever we test the theory with objects that move quickly — subatomic particles, for instance — the predictions of special relativity prove correct.

Different observers may disagree about the time lapse between two events and about the distance separating these events. Is there anything that all observers *can* agree upon? In 1908, Einstein's teacher, the Russian–German mathematician Hermann Minkowski (1864–1909), found that the answer is 'yes'. The quantity ds^2, the square of the *spacetime interval* between two neighbouring events (x, y, z, t) and $(x + dx, y + dy, z + dz, t + dt)$, is an *invariant*: it is the same for all inertial observers. The interval is given by*

$$ds^2 = c^2 dt^2 - (dx^2 + dy^2 + dz^2). \tag{10.9}$$

The total interval Δs between two separated events is obtained by integrating ds along a *worldline* joining the two events. An important result is that, for events connected by a light ray, $ds^2 = 0$. (The expression for the interval is similar to that for dl^2, the square of the distance between two neighbouring points (x, y, z) and $(x + dx, y + dy, z + dz)$. The Pythagorean theorem gives

$$dl^2 = dx^2 + dy^2 + dz^2. \tag{10.10}$$

The quantity dl^2 is invariant for observers in Newtonian physics because, in that framework, simultaneity is absolute. The total distance Δl between two separated points is obtained by integrating dl along a path joining the two points. Clearly the distance between the points depends upon the path; a straight line between the points can be defined as the path for which Δl is a minimum.)

Space and time coordinates get 'mixed up' in a Lorentz transformation. Minkowski argued that if we embrace special relativity then we must abandon the idea of space and time as separate independent entities. Instead, we must talk about a union of the three dimensions of space and the one dimension of time. We live in a universe not of space and time, but of spacetime. From now on, whenever the word 'distance' is encountered in this book, bear in mind that the concept of distance is not straightforward.

10.2.2 Metric equations

In this section I try to give a *flavour* of the mathematics of general relativity. (The 'Further Reading' section has references to some excellent introductory books on general relativity.) It is not necessary to follow all the details of the discussion, but some of the ideas will soon arise again.

We begin by writing the familiar Pythagorean theorem in another way. To do this we introduce g_{ab}, a second-rank tensor. (A *tensor* is a geometrical entity that is independent

---

*Note that ds^2 means $ds \times ds$ and *not* $d(s^2)$; similarly for the other squared differentials.

of a coordinate system. A zero-rank tensor is called a *scalar*: it has a single value at every point in space. A first-rank tensor is called a *vector*: a vector has n components at each point in space, where n is the number of spatial dimensions. A second-rank tensor has n^2 components at each point in space. Temperature is an example of a zero-rank tensor; magnetic field strength is an example of a first-rank tensor; the stresses in a load-bearing beam are described by a second-rank tensor.)

Using g_{ab}, the Pythagorean theorem in three dimensions becomes

$$dl^2 = \sum_{a,b=1}^{3} g_{ab} \, dx^a \, dx^b. \tag{10.11}$$

In this notation $dx^1 \equiv dx$, $dx^2 = dy$ and $dx^3 = dz$. Be careful not to confuse dx^2 with dx^a when $a = 2$, or dx^3 with dx^a when $a = 3$! We can simplify (10.11) if we adopt a *summation convention*: if we see repeated indices in a formula, one superscript and one subscript, we are to sum them over their entire range. Adopting this convention we can write the distance between two neighbouring points in the following way:

$$dl^2 = g_{ab} \, dx^a \, dx^b. \tag{10.12}$$

Expanding the right-hand side of (10.12) we obtain

$$
\begin{aligned}
dl^2 = \quad & g_{11} \, dx \, dx + g_{12} \, dx \, dy + g_{13} \, dx \, dz \\
+ & g_{21} \, dy \, dx + g_{22} \, dy \, dy + g_{23} \, dy \, dz \\
+ & g_{31} \, dz \, dx + g_{32} \, dz \, dy + g_{33} \, dz \, dz.
\end{aligned}
$$

Setting $g_{11} = g_{22} = g_{33} = 1$ and all the other g_{ab} values to zero we recover the relation $dl^2 = dx^2 + dy^2 + dz^2$, which is just Pythagoras in cartesian coordinates. In summary, then, we can calculate the distance between two neighbouring points in space by

$$dl^2 = g_{ab} \, dx^a \, dx^b$$

where

$$g_{ab} = \begin{pmatrix} 1 & 0 & 0 \\ 0 & 1 & 0 \\ 0 & 0 & 1 \end{pmatrix}.$$

Equation (10.12) is a *metric equation*. (Equations (10.9) and (10.10) are also metric equations.) The symmetric second-rank tensor g_{ab} is called the *metric tensor*. It is extremely important because its existence enables us to calculate the distance between two points if we have the coordinate differences between the points. If we know the metric tensor of a space, then we can construct most things we need to know about that space. If a general space (called a *manifold*) does not possess a metric, then 'distance' has no meaning in that space. We saw in the very first step on the distance ladder that the word 'geometry' initially derived from words meaning 'earth' and 'measure'. It is the existence of a metric that makes such distance measurements possible.

Example 10.1 Minkowski spacetime metric

Write down the Minkowski spacetime interval between two neighbouring events in metric form, and give the components of the metric tensor. Use the summation convention and a cartesian coordinate system.

Solution. Proceeding in exactly the same way as in the text, we have $ds^2 = g_{ab}\, dx^a\, dx^b$ where a and b both run from 0 to 3 (0 being the time coordinate). The Minkowski metric tensor is

$$g_{ab} = \begin{pmatrix} c^2 & 0 & 0 & 0 \\ 0 & -1 & 0 & 0 \\ 0 & 0 & -1 & 0 \\ 0 & 0 & 0 & -1 \end{pmatrix}$$

from which we recover $ds^2 = c^2\, dt^2 - (dx^2 + dy^2 + dz^2)$.

We have expressed the components of the metric tensor of three-dimensional space within a cartesian coordinate system. (Example 10.1 does the same for Minkowski spacetime.) But the physically significant quantity we are measuring is the distance between two points, and in Newtonian physics this quantity is invariant: we must obtain the same answer no matter which coordinate mesh we choose to work with. (The same comment applies to the Minkowski spacetime interval: we must obtain the same result whichever coordinate mesh we use.) For instance, we can re-write the Pythagorean theorem using spherical polar coordinates (r, θ, ϕ):

$$dl^2 = dr^2 + r^2(d\theta^2 + \sin^2\theta\, d\phi^2).$$

The equation $dl^2 = g_{ab}\, dx^a\, dx^b$ remains valid, but the components of the metric tensor become

$$g_{ab} = \begin{pmatrix} 1 & 0 & 0 \\ 0 & r^2 & 0 \\ 0 & 0 & r^2\sin^2\theta \end{pmatrix}.$$

We can, if we wish, use a completely arbitrary coordinate mesh. In a three-dimensional space, (10.12) always gives the distance dl between two neighbouring points, but the components of the metric tensor change from their cartesian values.

Equation (10.12) is more elegant and general than the usual form of the Pythagorean theorem, but this is at the expense of introducing a metric tensor with coefficients that depend on the coordinate mesh being used. The whole approach outlined above may seem overly formal and to confer few, if any, advantages. So why go to all this trouble?

The brilliant German mathematician Gauss, whom we met in an earlier chapter for his work on defining the astronomical unit, was the first to show why this approach to geometry is important. In the 1820s, the government of Hanover asked Gauss to take part in a large land survey. Unlike the land surveyors of Ancient Egypt, who worked with small flat areas, Gauss faced a surface that had hills and dales. The standard two-dimensional Pythagorean theorem with $g_{ab} = \left(\begin{smallmatrix} 1 & 0 \\ 0 & 1 \end{smallmatrix}\right)$ was of little use to him. Gauss, though, had a great idea. He proved a result that he called the *theorema egregium:* the 'outstanding theorem'. (Gauss proved a host of outstanding results during his lifetime, so to single out this one shows its importance.) His theorem showed that surveyors could lay down an arbitrary coordinate mesh on top of a surface — different surveyors were free to lay down different meshes — and they could determine the shape of the surface from the manner in which the metric coefficients g_{ab} varied from place to place on the surface. In other words, what was important was not the coordinate mesh (surveyors could use cartesian coordinates, or plane polar coordinates, or any other system) nor the particular values of the metric coefficients (since these were different in different coordinate meshes) but the variation with position of the metric coefficients.

If the two-dimensional space was covered with general coordinates (x^1, x^2) then the *theorema egregium* showed that all surveyors would agree on a quantity K known as the *curvature*. The quantity K is thus an *intrinsic* property of the space. I discuss its physical meaning in the next section. Here — purely for completeness — I give the mathematical form for K. For an orthogonal metric* we have:

$$K = \frac{1}{2g_{11}g_{22}}\left\{-\frac{\partial^2 g_{11}}{\partial(x^2)^2} - \frac{\partial^2 g_{22}}{\partial(x^1)^2} + \frac{1}{2g_{11}}\left[\frac{\partial g_{11}}{\partial x^1}\frac{\partial g_{22}}{\partial x^1} + \left(\frac{\partial g_{11}}{\partial x^2}\right)^2\right]\right.$$
$$\left. + \frac{1}{2g_{22}}\left[\frac{\partial g_{11}}{\partial x^2}\frac{\partial g_{22}}{\partial x^2} + \left(\frac{\partial g_{22}}{\partial x^1}\right)^2\right]\right\}. \tag{10.13}$$

10.2.3   Non-Euclidean geometry

The study of geometry, which began with the rope stretchers of Ancient Egypt and which was developed by Greek philosophers like Thales, showed its full power in the hands of Euclid (*fl.*300 BC). Euclid accepted just five basic geometric *axioms* or *postulates*, along with five so-called *common notions*. The postulates and the common notions were statements whose truth one had to take on faith. He then showed how everything that was then known about geometry followed logically from these few statements. Euclid's treatment was so clear and so elegant that it moved Millay to write: 'Euclid alone has looked on Beauty bare'. For 2000 years, no-one doubted Euclid's system of geometry. His *Elements* is the most successful textbook ever written.

There was just one thing that seemed to mar the beauty of Euclid's work. His five common notions are all obvious and unobjectionable statements — along the lines of 'the whole is greater than the part'. Four of the five postulates are equally bland — for instance, the fourth postulate is 'all right angles are equal'. The problem was the fifth postulate, which can be stated as: 'through any given point there is precisely one straight line that is parallel to a given straight line'. The difficulty is that, while we can accept all

---

*An orthogonal metric is one in which $g_{ab} = g_{ba} = 0$ for $a \neq b$; this implies that the mesh lines for the coordinates x^1 and x^2 cross at right angles. The general expression for K involving non-diagonal g_{ab} is slightly more complicated; since all the metrics we meet in astronomy are orthogonal there is little point in giving the full expression.

the other statements by an appeal to our everyday experience, we cannot do the same with the fifth postulate. None of us has experience about what happens to straight lines when they extend to infinity. Who can tell whether the lines remain parallel all the way?

People have always been uneasy with the fifth postulate. Euclid himself was wary of it, and used it as little as possible. For 20 centuries, mathematicians tried to derive the fifth postulate from the other postulates and common notions, or to find another postulate from which the fifth follows. They could then demote the postulate to the lesser rank of theorem. It says much for Euclid that no one succeeded. The fifth postulate seemed to be as necessary and as fundamental as the other four. (The phrasing of the fifth postulate given above is usually attributed to John Playfair (1748–1819) but a similar version was given long ago by Proclus Diadochus (410–485). The original phrasing of the fifth postulate was: 'if a straight line falling on two straight lines make the interior angles on the same side less than two right angles, the two straight lines if produced indefinitely meet on that side on which the angles are less than two right angles'. Small wonder that from the outset mathematicians worked to prove the postulate — or at least replace it with something more self-evident!)

In the nineteenth century, mathematicians attacked Euclid's fifth postulate from another direction. Suppose that the postulate is false. Suppose that there are *many* parallels (or *no* parallels) to a given straight line. If you combine this supposition with all the other postulates and common notions you can derive a geometry in exactly the same way that Euclid derived his geometry. When you meet a logical contradiction in your geometry it must be due to the initial supposition — you will have proved the supposition to be false. Therefore you will have proved the truth of the fifth postulate. Then, with a sigh, you can drop it from the list of statements that must be accepted without proof. The trouble when you take this approach, as mathematicians quickly found, is that you never seem to come to a logical contradiction. You can have a self-consistent geometry with a fifth postulate that differs from Euclid's statement. You can have a *non-Euclidean geometry*.

In 1829, the Russian mathematician Nikolai Ivanovich Lobachevski (1793–1856) was the first to publish a non-Euclidean geometry. (In fact, he first made public his discovery in a paper he read to the mathematics department of Kazan University on 23 February 1826, but no copy of his paper has survived.) At about the same time, the Hungarian mathematician Janos Bolyai (1802–1860), working independently, developed the same non-Euclidean geometry. He published his work in 1832. Gauss, who without doubt ranks with Newton as one of the greatest mathematicians of all time, developed a non-Euclidean geometry in 1824, before either Lobachevski or Bolyai. Unfortunately he feared the ridicule that might fall upon him if he published work that contradicted Euclid. He wrote to a friend that he would not 'work up my very extensive researches for publication, and perhaps they never will appear in my lifetime, for I fear the hail of the Boeotians if I speak my opinion out loud'. (The phrase 'men of Boeotia' has long since fallen out of fashion, but Gauss used it to refer to dull and unimaginative people.) He did not publish until after Lobachevski and Bolyai.

The geometry of Lobachevski, Bolyai and Gauss is called a *hyperbolic* geometry. It follows from a fifth postulate in which there are many parallel lines to a given straight line. It is the geometry of a shape resembling a saddle, in the same way that Euclidean geometry is the geometry of a plane.

A hyperbolic geometry is not the only possible non-Euclidean geometry. In 1854, the German mathematician Georg Friedrich Riemann (1826–1866) developed *spherical* geometry, which applies to the surface of a sphere. It follows from a fifth postulate in which there are no parallels to a straight line. (In Euclidean geometry we can define a 'straight' line to be one that, as we move along it, does not change direction; or we can define it as the curve of shortest distance between two points. The non-Euclidean generalisation of these two ideas is the *geodesic*. A straight line or geodesic on the surface of a sphere is a *great circle* — a circle whose plane cuts through the centre of the sphere. You can quickly convince yourself that there are no parallel great circles, since they all intersect at two points.) At the same time, Riemann showed how to generalise the work of Gauss to spaces with more than two dimensions. We can thus apply the *theorema egregium* to three-dimensional space or to four-dimensional spacetime.

There are any number of possible geometries. They all follow logically from their axioms, they are all self-consistent, and they are all equally valid. The important question is how well they describe the real world. Of all possible geometries, three are special: Euclidean geometry, hyperbolic geometry and spherical geometry. They are special because they alone, of all the geometries, possess several important features that seem to describe the large-scale structure of the world in which we live. These three geometries are uniform, homogeneous and isotropic, so all points in space and all possible directions are equivalent. They are also invariant under rotations and translations, so if a circle is translated in these spaces it remains the same circle.

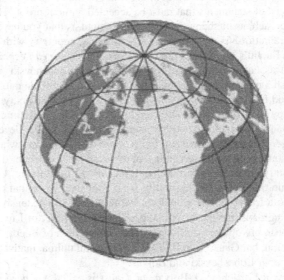

FIGURE 10.3: The shortest path between two cities follows the curved surface of the Earth, but it is not a line of constant latitude. Whenever you take an aircraft, therefore, make sure the airline uses Riemannian geometry when calculating fuel consumption and travel times!

Non-Euclidean geometries are of interest to more than just pure mathematicians. For several everyday applications the spherical geometry of Riemann is more useful than traditional Euclidean geometry. For instance, if you want to travel from Los Angeles (34° N, 118° W) to Tel Aviv (32° N, 35° E) you cannot move in a Euclidean straight line because that would entail boring a hole through the Earth. The shortest path between the two cities follows the curved surface of the Earth, but it is not a line of near-constant latitude. Rather, it is part of a great circle. Figure 10.3 shows the flight path followed by an aircraft; figure 10.4 shows the great circle, of which this path is part, projected onto the plane. A straight line on the Earth's surface thus appears curved on a flat surface.

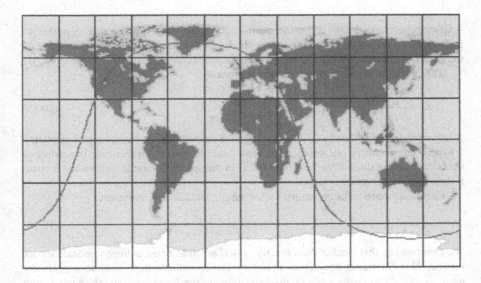

FIGURE 10.4: A great circle on the Earth, when projected onto a flat map, appears curved. The flight path shown in figure 10.3 is part of the great circle in this diagram.

The spherical and hyperbolic geometries differ from Euclidean geometry in several ways. Perhaps the most noticeable difference has to do with the interior angles of a triangle. In a Euclidean space the interior angles of a triangle always add up to exactly 180°; in a spherical space they always add up to more than 180°; in a hyperbolic space they always add up to less than 180°. See figure 10.5.

We can also use the *theorema egregium* of Gauss to distinguish between the three types of geometry. They each have a characteristic K value, or curvature. (In an inhomogeneous space K can vary from point to point. We are interested only in homogeneous spaces, though, which have a single K value.)

The curvature of a spherical surface is given by

$$K = \frac{1}{R^2}.$$

(10.14)

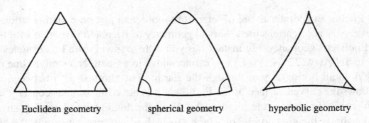

Euclidean geometry   spherical geometry   hyperbolic geometry

FIGURE 10.5: The interior angles of a triangle add up to exactly 180°, more than 180°, or less than 180°, depending upon the geometry.

A spherical surface thus has a constant positive curvature. A spherical geometry possesses an intrinsic scale length R, which we can think of as being the radius of the sphere. As the radius of curvature becomes large, the curvature becomes small.

The curvature of a hyperbolic surface is given by

$$K = -\frac{1}{R^2}. \tag{10.15}$$

A hyperbolic surface thus has a constant negative curvature. Like the spherical geometry, a hyperbolic geometry possesses an intrinsic scale length R. In this case the radius of curvature is more difficult to visualise, but as in the spherical case the curvature is small if the radius of curvature is large.

Euclidean space lacks an intrinsic scale length, and has zero curvature:

$$K = 0. \tag{10.16}$$

We can thus say that Euclidean geometry is *flat,* whereas non-Euclidean geometries are *curved.* Non-Euclidean geometries become indistinguishable from Euclidean geometry as $R \to \infty$. This is why early civilisations thought the Earth was flat: the scale length describing the spherical geometry of the Earth's surface is much bigger than the small patch of land they inhabited.

The curvature of space is potentially an extremely confusing concept, particularly if you try to visualise the curvature. None of us has a problem in imagining the curvature of a two-dimensional space, because we can *embed* a two-dimensional space within our familiar flat three-dimensional space. But if we want to visualise a three-dimensional space curving in a flat higher-dimensional space, then we have to embed the three dimensions in a space of four or more dimensions. And who can visualise four-dimensional space? The important point to appreciate is that we do not have to visualise these higher dimensions; the higher embedding dimensions need not exist. A curved space does not have to be curved 'in' another space. Indeed, we should *not* think in these terms. Curvature is an *intrinsic* geometric property of a space, and it can be determined by measurements made solely *within* that space. This was Gauss' great discovery.

For example, intelligent 'Flatlanders' — infinitely thin creatures who inhabit a two-dimensional surface — can deduce the curvature of their space simply by measuring the

Example 10.2 Curvature of a sphere

Show that $K = 1/R^2$ for the surface of a sphere.

Solution. If we use spherical polar coordinates $(x^1, x^2) \equiv (\theta, \phi)$ to describe a sphere of radius R, then the non-zero metric coefficients are $g_{11} = R^2$ and $g_{22} = R^2 \sin^2 \theta$. We have

$$\frac{\partial g_{11}}{\partial x^1} = \frac{\partial g_{11}}{\partial x^2} = \frac{\partial g_{22}}{\partial x^2} = 0$$

$$\frac{\partial g_{22}}{\partial x^1} = 2R^2 \sin \theta \cos \theta \qquad \frac{\partial^2 g_{22}}{\partial (x^1)^2} = 2R^2 (\cos^2 \theta - \sin^2 \theta).$$

Substituting these values into (10.13) we obtain

$$K = \frac{1}{2R^4 \sin^2 \theta} \left(2R^2 (\sin^2 \theta - \cos^2 \theta) + \frac{1}{2R^2 \sin^2 \theta} \times 4R^4 \sin^2 \theta \cos^2 \theta \right)$$

$$= \frac{1}{2R^4 \sin^2 \theta} (2R^2 \sin^2 \theta - 2R^2 \cos^2 \theta + 2R^2 \cos^2 \theta)$$

$$= 1/R^2.$$

interior angles of a triangle. If the angles add up to exactly $180°$ they know they inhabit a flat Euclidean space. If the angles add up to less than or more than $180°$, they know they live in a hyperbolic or spherical 'curved' space. Another method they could use to determine space curvature would be to plant a large forest of two-dimensional trees with a uniform density of ρ trees per unit area, and then count the number of trees enclosed within a circle of radius r. If they count exactly $\pi r^2 \rho$ trees they may deduce that they live in a Euclidean space. If they count more than $\pi r^2 \rho$ trees they live in a space of negative curvature, and if they count fewer than $\pi r^2 \rho$ trees then their space is positively curved. Since the departure from Euclidean geometry is roughly $(r/R)^2$ they can use these measurements to determine the radius of curvature of their space. (Clearly, since these non-Euclidean effects vary as $(r/R)^2$ it is necessary for Flatland surveyors to work with distances r that are a fair fraction of R.)

Astronomers are, for the most part, intelligent three-dimensional creatures. They can therefore in principle determine the curvature of space using the same types of method that Flatlanders can use. For instance, Gauss tried to determine whether space is curved by measuring the interior angles of a triangle formed by three peaks in the Harz mountains. And Hubble initiated a programme of galaxy counts — the equivalent of counting Flatland trees — to determine the curvature of spacetime. We now know that neither programme could give conclusive results because they sampled a region of space that was too small for non-Euclidean effects to show up. We need a way of probing large distances if we hope to explore the geometry of space. (As proof that we *can* determine the geometry of a surface by making measurements while confined to that surface, recall Eratosthenes. He measured the radius of curvature of the Earth without ever leaving the Earth's surface!)

There is one final point worth making. If we know the curvature K of a surface at all points, we have almost enough information to construct the surface completely. Almost, but not quite. Consider a cylinder; we can easily show that $K = 0$ for a cylinder. In other words a cylinder is flat, like a plane. One manifestation of this is that if you cut through a cylinder and paste it down onto a plane you can do so without causing any distortion of objects on the cylinder's surface. This is not true for a sphere: there are any number of ways of pasting a spherical surface onto a plane and every single way causes distortion. Just look at the Mercator and Peters projections of a world map to convince yourself of this. None of this means that a cylinder is the *same* as a plane. But the differences are *global*, depending on the topology of the object as a whole. Curvature is a *local* property; we cannot determine how space 'links up' globally by making local measurements.

10.2.4   Curved spacetime

Imagine two Flatlanders walking in straight lines. If they live on a plane, and they walk along parallel paths, they never meet. If they live on the surface of a sphere their paths curve and they *do* eventually meet. See figure 10.6. No unseen force bends their paths. The result is simply due to the geometry of the space in which they live.

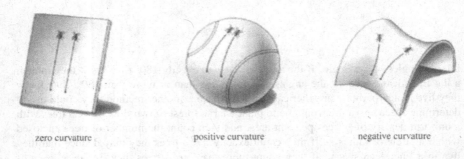

zero curvature   positive curvature   negative curvature

FIGURE 10.6: On the plane, the paths of the two Flatlanders never intersect. On the surface of the sphere, the paths of the two Flatlanders are certain to intersect again. In both cases the paths are the straight lines, or geodesics; on a sphere, though, such paths curve. No force causes the Flatlanders to move towards each other — the effect is due to the curvature of space. The situation on a negatively curved space is more difficult to visualise, as it has many parallel lines to a given line.

Perhaps Einstein's greatest insight was to use non-Euclidean geometry to describe gravity. He argued that any freely falling particle follows a geodesic through spacetime. A massive particle follows a *timelike geodesic*; a photon follows a *null geodesic*. If spacetime is curved, the particle's geodesic will also be curved. Observers not moving with that particle might consider some unseen force to have influenced the motion of the particle — a force they could interpret as gravity. But the particle would be moving as 'straight as possible', subject to the constraints of geometry. For example, a planet that orbits a star moves in as straight a line as possible through spacetime. Spacetime is curved near a star, though, and the planet's geodesic has to follow the curvature of spacetime. So to an observer not moving with the planet its path appears curved. In the language of Newton, the planet moves in response to a gravitational force. In the language of Einstein, the planet moves in the simplest possible way allowed by the curvature of spacetime.

What could make spacetime curve? The obvious answer is that the presence of energy — and, in particular, mass — curves spacetime. The more mass an object has, the more it distorts spacetime. Einstein struggled for years to find a mathematical relationship between the distribution of mass–energy and the curvature of spacetime. In 1915, he finally succeeded in deriving the so-called *field equations* of general relativity, but he was beaten to the punch by the German mathematician David Hilbert (1862–1943). Hilbert was one of the greatest mathematicians of the twentieth century, and after he heard Einstein lecture on the link between gravity, spacetime and non-Euclidean geometry it took him only a few weeks to derive the correct equations. (Hilbert himself awarded all the credit to Einstein. It was Einstein who first understood that gravity arises from the curvature of spacetime, and it was Einstein who worked alone for years to solve the problem. The equations of general relativity, which describe how a given distribution of mass and energy cause spacetime to curve, are quite rightly called the *Einstein* field equations.)

The field equations derived by Einstein take the form

$$R_{ab} - \tfrac{1}{2}g_{ab}\mathcal{R} = \kappa T_{ab} \tag{10.17}$$

where R_{ab} is the Ricci tensor and $\mathcal{R}$ the Ricci scalar, both named after the Italian mathematician Gregorio Ricci-Curbastro (1853–1925); g_{ab} is of course the metric tensor; κ is a constant that contains G, the Newtonian constant of gravitation.

The left-hand side of (10.17) describes the curvature of spacetime. Note that the curvature is described by a tensor, rather than the single function K that Gauss used. Gauss could use K because he considered spaces of only two dimensions; Riemann, and the German mathematician Elwin Bruno Christoffel (1829–1900), showed that a fourth-rank tensor R_{abcd} (the *Riemann curvature tensor*) is required to describe the curvature of higher-dimensional spaces. The Ricci tensor is constructed from the Riemann tensor.

The right-hand side of (10.17) describes the matter content of the theory. It does this through T_{ab}, a second-rank tensor called the *stress–energy* tensor. This is a symmetric tensor, containing ten independent components that describe various forms of energy.

The American physicist John Archibald Wheeler (1911–), the man who coined the phrase 'black hole', famously put the field equations into words: 'mass tells space how to curve; space tells mass how to move'. Wheeler's aphorism makes it sound as if the field equations should be simple to solve: just pick a particular distribution of matter, deduce the curvature, and then calculate geodesics to predict the paths of light rays and particles. Of course, it is not that easy. The field equations are amongst the most difficult equations to solve analytically in the whole of physics. In cases where there is a high degree of symmetry, though, it is possible to obtain exact solutions. In particular, it is possible to find the spacetime metric outside a static, spherically symmetric mass such as the Sun. Einstein therefore first applied his new theory to calculate spacetime curvature near the Sun.

As we saw on page 219, Einstein's theory correctly predicted that a ray of light passing near the Sun's surface would be deflected by 1.75″. His theory also explained a small anomaly in Mercury's orbit. Astronomers had long puzzled over the precession of the perihelion of Mercury's orbit. The perihelion — the point of closest approach to the Sun — was observed to drift at the rate of a few hundred seconds of arc per century, after allowing for the rotation of the Earth. Most of this drift could be explained as the gravitational effects of the other planets, chiefly Venus and Jupiter. However, a tiny residual drift of 43″ per century remained unexplained. General relativity predicted that the precession of Mercury should be 43.03″ per century — exactly what was observed*.

With these successes behind him, Einstein applied his theory to the biggest problem of all: the structure of the universe.

10.2.5   General relativity and cosmology

The curvature of spacetime throughout the universe is undoubtedly complicated. The planets in the solar system all cause small warps in the fabric of spacetime: in the case of the Earth, the components of the metric tensor at the Earth's surface deviate from flat values only by about 1 part in 10^9. The Sun causes a larger warp — about 1 part in 10^6 near its surface — as do all the stars in the Galaxy and all the stars in all the other galaxies in the universe. White dwarfs and neutron stars cause larger warps. (The largest spacetime warp yet recorded is about 1 part in 3; this warping was discovered in 1998 by astronomers who were using the *Rossi X-Ray Timing Explorer* satellite to study the accretion disk around a neutron star.) Black holes puncture the fabric of spacetime completely. But these warps are fine detail. If we look at the universe over large scales, which in this context means over distances larger than about 50 Mpc, all the small-scale irregularities should average out. It is like looking at the surface of the ocean: viewed from a sailboat it looks chaotically complicated, but viewed from a jet plane it looks smooth. Over large enough distances, therefore, we can apply general relativity to study the curvature of the universe as a whole.

In taking this approach we are assuming that the universe is *homogeneous* (i.e. that it is everywhere the same). From what we learned in the previous rung of the distance ladder it is clear that this is a large assumption: the universe is inhomogeneous all the way up to the level of superclusters. Nevertheless, there are good reasons for believing that, on the very largest scales, the universe is homogeneous. (There are certainly excellent reasons for believing that the universe is *isotropic*, i.e. that all directions are alike.) Inhomogeneous cosmologies were popular in the early years of this century. The Swedish astronomer Carl Vilhelm Ludvig Charlier (1862–1906), for example, popularised a multilevel universe with clusters of ever-increasing size. However, it is very difficult to study such models using general relativity. For this reason the first astronomers to apply general relativity to cosmology all assumed that the universe is homogeneous and isotropic; these assumptions are still used by cosmologists.

Einstein was the first to combine the assumptions of homogeneity and isotropy with his general theory of relativity to study the universe as a whole. In 1917, he solved his field equations for an homogeneous and isotropic distribution of stars and found that the universe could not be static. In particular, if the universe had a spherical spacetime geometry then its radius of curvature R was a function of time: the universe expanded or contracted. At the time Einstein published his theory most astronomers believed that

---

*By now, general relativity has been experimentally verified in many different situations — even everyday situations. For example, the global positioning system mentioned on page 22 must take account of general relativistic time dilation.

the universe was static and unchanging. (Remember that in 1917 Hubble had not yet demonstrated the existence of other galaxies, let alone the redshift–distance relation.) In order to fit in with this belief Einstein had to add a 'fudge factor' to his equations: the so-called *cosmological constant* Λ. With this extra term the field equations become

$$R_{ab} - \tfrac{1}{2} g_{ab} \mathcal{R} - g_{ab} \Lambda = \kappa T_{ab}. \tag{10.18}$$

By choosing the value of Λ appropriately, Einstein could obtain a static solution to the equations — a solution in which R did not vary with time.

Einstein was never happy with the extra term. It spoiled the beauty and elegance of his theory. Eventually he disowned it, saying that it was 'the biggest blunder' he had ever made*. Had he kept faith with his original equations he would have made one of the most astounding predictions in the history of science: the universe expands.

The so-called *Einstein universe* had a spherical spacetime. Like the surface of a sphere it was finite, yet unbounded (where, after all, is the 'edge' of the surface of a sphere?) and it thus escaped from the paradox that worried Lucretius. Einstein thought this solution — a static, matter-containing, spherical universe — was the unique cosmological solution to his equations. He was wrong. Other workers soon applied general relativity to the universe as a whole and obtained different solutions.

Willem de Sitter

A few months after Einstein published the field equations, the Dutch mathematician Willem de Sitter (1872–1934) found another cosmological solution to them. The so-called *de Sitter universe* is a flat universe containing a cosmological constant but no matter. The metric of the de Sitter universe is

$$\mathrm{d}s^2 = c^2 \, \mathrm{d}t^2 - \mathrm{e}^{2Ht} \big[\mathrm{d}r^2 + r^2 (\mathrm{d}\theta^2 + \sin^2 \theta \, \mathrm{d}\phi^2) \big] \tag{10.19}$$

---

*We shall see in the final chapter that maybe Einstein was wrong again: perhaps the introduction of the cosmological constant was not a blunder after all!

where $H = \sqrt{\frac{1}{3}\Lambda}$. The term H plays the role of a Hubble constant: in the de Sitter universe, though, it really *is* a constant. We can use the metric to calculate distances and intervals. The radial distance measured at the same time ($dt = 0$) between nearby points on the same radial line ($d\theta = d\phi = 0$) is the proper distance between them: $\sqrt{-ds^2} = e^{Ht}$. In other words, the points move apart from one another: the de Sitter universe expands exponentially. (The idea of an expanding universe that contains no matter seems bizarre. What it means is that if you introduce two small test masses into this empty spacetime, and let them move freely, their separation increases with time. Eddington called the de Sitter universe 'motion without matter'; by contrast, the Einstein universe was 'matter without motion'.)

The de Sitter universe was obviously not a reasonable description of the real universe, but for several years Slipher's redshifts went by the name of the 'de Sitter effect', since the Dutchman's work seemed to have some bearing on the problem. And it proved that the field equations had more than one cosmological solution. On the largest scales space-time could be curved or flat: the field equations alone were not enough to determine the structure of the universe.

In 1922, the Russian physicist Alexander Alexandrovich Friedmann (1888–1925) was the first to state clearly that a matter-containing universe that is homogeneous and isotropic will in general either expand or contract. Friedmann's theory caused little or no impact, but in 1927 the Belgian astronomer Georges Edouard Lemaître (1894–1966) independently retraced Friedmann's calculations and came to the same conclusion. Lemaître's results were widely publicised. So when Hubble published the redshift–distance relation, the world was ready to believe that we live in an expanding universe.

10.3   FRW MODELS

According to general relativity, an expanding universe does not mean that galaxies fly *through* space like the debris from a bomb. Rather, *space itself* expands and the expansion carries the galaxies along with it. Recall the cinema hall analogy of earlier in the chapter. There is a sense in which no-one in the audience actually moved. Everyone stayed seated, and their relative positions were unchanged. The occupant of seat number 22 row 10, for example, kept the same seat throughout the expansion. The grid of seat and row numbers was an example of a *comoving coordinate system* — a system that shares the expansion. The appearance of a recession velocity came about because the distance between seats somehow increased, even though in terms of comoving coordinates no motion was involved.

Figure 10.7 illustrates a more familiar example. It shows two sleeping Flatlanders confined to the surface of an inflating balloon. In terms of the comoving coordinates of latitude and longitude the Flatlanders are stationary: their comoving coordinates do not change with time. Yet when we inflate the balloon the distance between the two Flatlanders increases. It increases not because they have a velocity with respect to the balloon, but because the space in which they live expands. The *coordinate distance* between the Flatlanders is equal to the difference in their comoving coordinates, which is constant as

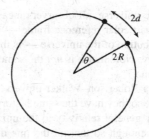

FIGURE 10.7: The Flatlanders do not move, so they keep their comoving coordinates (latitude and longitude in this case). Nevertheless the separation between them doubles because the scale factor of the sphere doubles.

long as they do not move, multiplied by a time-varying scale factor $R(t)$:

$$\text{coordinate distance} = R(t) \times \text{coordinate difference.} \qquad (10.20)$$

The expansion of their two-dimensional space, which is due to the increase in R, stretches the wavelength of radiation just as it stretches other distances. So if the Flatlanders signal each other by sending light rays, the light they receive will be redshifted.

Our own universe has a scale factor associated with it*. How does that scale factor change with time? The answer depends on the particular cosmological model that we choose to describe our universe. We have already seen that general relativity admits at least two cosmological models: the Einstein universe and the de Sitter universe. We also know that neither model can represent the real world (the Einstein contains no expansion and the de Sitter universe contains no matter). There is a host of other models, though, the best studied of which go under the heading of Friedmann–Robertson–Walker (FRW) models. In an FRW universe, there is both expansion and matter, and the distribution of matter is isotropic and homogeneous. These models were first considered by Friedmann, and then studied systematically by Robertson (as part of his work on the redshift–distance law) and the English mathematician Arthur Geoffrey Walker (1909–).

10.3.1   The Robertson–Walker metric

Robertson and Walker were the first to fully appreciate the convenience of comoving coordinates for cosmology. Following their lead, imagine the universe to be covered by some comoving coordinate mesh or grid, and suppose that at each grid point there is a freely floating observer with a clock. Each observer has a particular set of spatial coordinates (r, θ, ϕ) that remain constant for all time. Moreover, all comoving observers measure the same cosmic time t. (In principle, these observers can synchronise their clocks by agreeing to start timing when they determine the temperature of the microwave background radiation to have a particular value. Since we are considering an homogeneous

---

*Note that the scale factor is not necessarily the 'radius of the universe' in the way that R is the radius of the sphere in figure 10.7. We could reasonably call R the radius of the universe if we inhabit a spherical universe, as Einstein originally believed, but not if the universe is flat.

and isotropic universe, all observers measure the same temperature at the same cosmic time. This particular reference frame — the frame that is at rest relative to the average matter distribution of the universe — is the natural one to use when discussing problems in cosmology. Note that it is *not* the same as the Newtonian reference frame of absolute space and time.)

In such a Robertson–Walker universe all observers see the same thing at the same cosmic time, so they have the same history. Clearly this is an idealisation; but remember that to apply general relativity to the universe as a whole we have to consider distance scales large enough so that all the fine detail — like people and planets and stars and galaxies — average out. If we make the idealisation that the universe is the same at all distance scales then the spacetime of Robertson and Walker separates nicely into cosmic time, which is common to all observers, and curved three-dimensional space. The metric of this spacetime is thus

$$ds^2 = c^2 \, dt^2 - R^2(t) \, dl^2 \tag{10.21}$$

where $R(t)$ is the universal scale factor, which changes with time. Purely from the assumptions of homogeneity and isotropy, and making no reference to the field equations, Robertson and Walker showed that the general form for the spatial part of the metric is

$$dl^2 = \frac{dr^2}{1 - kr^2} + r^2 \, d\theta^2 + r^2 \sin^2 \theta \, d\phi^2 \tag{10.22}$$

where $k = +1, 0$ or -1. The k term denotes the curvature of space: $k = +1$ corresponds to positive curvature; $k = -1$ corresponds to negative curvature; $k = 0$ corresponds to flat space.

The general form for the Robertson–Walker metric is thus

$$ds^2 = c^2 \, dt^2 - R^2(t) \left(\frac{dr^2}{1 - kr^2} + r^2 \, d\theta^2 + r^2 \sin^2 \theta \, d\phi^2 \right) \quad k = \pm 1, 0. \tag{10.23}$$

Notice how the simplifying assumptions of homogeneity and isotropy reduce the number of unknowns in the metric tensor from ten to just two: k and $R(t)$.

It is useful to write the metric explicitly for the three different cases:

$$ds^2 = c^2 \, dt^2 - R^2(t) \left(\frac{dr^2}{1 - r^2} + r^2 \, d\theta^2 + r^2 \sin^2 \theta \, d\phi^2 \right) \quad \text{spherical} \tag{10.24}$$

$$ds^2 = c^2 \, dt^2 - R^2(t) \left(dr^2 + r^2 \, d\theta^2 + r^2 \sin^2 \theta \, d\phi^2 \right) \quad \text{flat} \tag{10.25}$$

$$ds^2 = c^2 \, dt^2 - R^2(t) \left(\frac{dr^2}{1 + r^2} + r^2 \, d\theta^2 + r^2 \sin^2 \theta \, d\phi^2 \right) \quad \text{hyperbolic.} \tag{10.26}$$

The Robertson–Walker metric says nothing about which particular k describes our universe, or of the form that $R(t)$ can take. To learn more about the curvature and the scale factor we must use general relativity and a particular cosmological model. Before considering this further, it is worth looking at (10.23) in more detail.

First, it is important to remember that r in (10.23) is a comoving coordinate. If an observer here on Earth is at $r = 0$ and a distant galaxy is at $r = r_e$, then the observer remains at $r = 0$ and the distant galaxy remains at $r = r_e$. The term r is thus better thought of as a label than as a distance. The coordinate distance, d_C, is given by (10.20). If the light emitted by a galaxy with comoving radial coordinate r_e is observed by us at the present time t_o, then the present coordinate distance to the galaxy is given by

$$d_C(t_o) = R(t_o)r_e \tag{10.27}$$

where $R(t_o)$ is the present value of the scale factor. The coordinate distance to the galaxy changes because $R(t)$ changes, not because the galaxy has a large velocity through space away from us.

What is the actual distance to the galaxy? We have to be very careful about what we mean by cosmological distance in an expanding universe, where galaxies are separated not just in space but also in time. The next chapter discusses various definitions of distance we could use. In the present context it is easiest to use the proper distance. To measure the proper distance to a galaxy, imagine that there is a chain of observers between us and the galaxy. Each observer measures the distance between himself and his immediate neighbour in the direction of the galaxy at the same cosmic time t. If we then add up all these small distance elements the result is the proper distance d_P to the galaxy at cosmic time t. Taking the radial part of (10.23) we see that

$$d_P(t) = R(t) \int_0^{r_e} \frac{dr}{\sqrt{1 - kr^2}}. \tag{10.28}$$

There are thus three expressions for the proper distance to an object, depending on the curvature of the universe:

$$d_P(t) = R(t) \sin^{-1} r \qquad \text{spherical} \tag{10.29}$$
$$d_P(t) = R(t)r \qquad \text{flat} \tag{10.30}$$
$$d_P(t) = R(t) \sinh^{-1} r \qquad \text{hyperbolic.} \tag{10.31}$$

Note that the proper distance is equal to the coordinate distance only in the case of a flat (i.e. $k = 0$) space. In a spherical geometry (i.e. $k > 0$) the proper distance to an object is always greater than its coordinate distance; in a hyperbolic geometry (i.e. $k < 0$) the proper distance is always less than the coordinate distance. Note also that proper distance is *not* the same as measuring the light-travel time from $r = r_e$ to $r = 0$ and multiplying by c. It is not the distance to the galaxy as we see it, because we can only 'see' it where it was when the light that we receive left it. We 'see' into the past. The proper distance given by (10.28) is the distance to where the galaxy will be at the cosmic time we consider as 'now'. Furthermore, the proper distance is not something that we can measure in practice (since we are unlikely to ever arrange the required chain of observers). Nevertheless, as I describe in the next chapter, we can relate the proper distance to quantities that we *can* measure. For the rest of this chapter I will consider only proper distance, and not worry about how we measure it.

Second, the Robertson–Walker metric clearly accounts for cosmological redshifts. Imagine light waves of wavelength λ_e leaving a distant galaxy which has comoving radial coordinate r_e; one crest is emitted at cosmic time t_e and the next crest is emitted at cosmic time $t_e + \Delta t_e$. We receive the crests 'now', in the cosmic time interval t_o to $t_o + \Delta t_o$. Light always propagates along a null geodesic, i.e. $ds^2 = 0$. Also, since the light travels radially towards us, we have $d\theta = d\phi = 0$. We can then use the Robertson–Walker metric to relate the various cosmic times in the problem to r_e. Putting $ds^2 = d\theta = d\phi = 0$ into (10.23) we obtain

$$0 = c^2 \, dt^2 - \frac{R^2(t) \, dr^2}{1 - kr^2}.$$

For the first crest we therefore have

$$\int_{t_e}^{t_o} \frac{dt}{R(t)} = \frac{1}{c} \int_0^{r_e} \frac{dr}{\sqrt{1 - kr^2}} \tag{10.32}$$

and for the second crest we have

$$\int_{t_e + \Delta t_e}^{t_o + \Delta t_o} \frac{dt}{R(t)} = \frac{1}{c} \int_0^{r_e} \frac{dr}{\sqrt{1 - kr^2}}. \tag{10.33}$$

Subtracting (10.32) from (10.33), and manipulating the limits of the integrals, produces

$$\int_{t_o}^{t_o + \Delta t_o} \frac{dt}{R(t)} - \int_{t_e}^{t_e + \Delta t_e} \frac{dt}{R(t)} = 0.$$

For electromagnetic radiation the frequencies are such that Δt_o and Δt_e are small fractions of a second; the scale factor of the universe $R(t)$ cannot change substantially over such small timescales. We can treat $R(t)$ as a constant in the above equations, so that

$$\frac{\Delta t_o}{R(t_o)} - \frac{\Delta t_e}{R(t_e)} = 0 \quad \text{or} \quad \frac{\Delta t_o}{\Delta t_e} = \frac{R(t_o)}{R(t_e)}.$$

Since $\lambda_o = c \Delta t_o$ and $\lambda_e = c \Delta t_e$ we can write this as

$$\frac{\lambda_o}{\lambda_e} = \frac{R(t_o)}{R(t_e)}.$$

In (6.2) we defined redshift to be

$$z = \frac{\Delta \lambda}{\lambda_s} \quad \text{i.e.} \quad z = \frac{\lambda_o - \lambda_e}{\lambda_e}$$

so finally we obtain the beautifully simple relation

$$1 + z = \frac{R(t_o)}{R(t_e)}. \tag{10.34}$$

Equation (10.34) is extremely important since it tells us the nature of the cosmological redshift: it is simply a measure of how much the universe has expanded during the time that light from a galaxy has been travelling towards us. *Cosmological redshift is thus an expansion effect rather than a velocity effect.* (We shall see in the next chapter how the Hubble law arises.)

Example 10.3 Redshift and expansion

You observe a galaxy to have a redshift of 0.37. By how much has the universe expanded since the observed light left the galaxy?

Solution. The ratio $R(t_o)/R(t_e)$ tells us how much the universe has expanded during the light travel time. We see from (10.34) that $R(t_o)/R(t_e)$ for this case is 1.37. In other words, the universe is 1.37 times larger now than it was when the galaxy emitted the light: it has expanded by 37%.   In general, if you multiply the redshift by 100 you obtain the percentage increase in the size of the universe since the light was emitted by the galaxy.

10.3.2   The Hubble flow

Not all of a galaxy's redshift will be the result of cosmic expansion. Some of the redshift will be due to the traditional Doppler effect, because galaxies in general have a random motion: they change their comoving coordinates. In our Local Group, for instance, the light from M 31 and M 33 is blueshifted because these galaxies have a component of velocity towards us. The gravitational interaction between Local Group galaxies cause these peculiar motions. The same thing happens in distant groups of galaxies. Local gravitational interactions will cause some galaxies to move towards us and be Doppler shifted to the blue, and others to move away from us and be Doppler shifted to the red. Groups of galaxies resemble flocks of birds flying away from each other. An individual bird has two velocities: a recession velocity, which is the velocity of the flock, and a peculiar velocity, which is its velocity relative to the flock.

Even as far away as the Virgo cluster, the Doppler shift due to peculiar velocity can exceed the redshift due to cosmic expansion. Several Virgo galaxies have blueshifts. This is one reason why it is difficult to decide upon the value of the Hubble constant from local studies: two nearby galaxies at the same distance can have quite different redshifts. At large enough distances, though, redshifts due to expansion become much greater than Doppler shifts due to peculiar velocities. Far enough away, essentially all of the redshift that we observe is a result of cosmic expansion. We enter the region of the *Hubble flow*.

The expansion of space is one of those concepts for which we have no direct experience to aid our understanding. In the time it takes you to read this sentence, something like 40 000 miles of space will have opened up between our Galaxy and the galaxies of the Coma cluster. Loosely (and perhaps incorrectly) speaking, this is 'new' space: space that the expansion in some sense created. It is difficult to appreciate what this means when there are no everyday examples of the phenomenon.

You might object that, because we do not see the effects of the cosmic expansion in everyday life, our interpretation of the redshift must be wrong. In the Woody Allen film *Annie Hall*, a young boy has read something that has made him depressed, with the result that he has stopped doing his homework. His mother takes him to a psychiatrist and prods him to explain the source of his depression. The boy says: 'Well the universe is everything, and if it's expanding some day it will break apart and that will be the end of everything ... What's the point?' His mother shouts: 'What has the universe got to do with it? You're here in Brooklyn! Brooklyn is not expanding!' His mother is correct. Matter sticks together through atomic forces that are strong enough to overcome the cosmic expansion. Thus objects like bacteria or human bodies or planets do not take part in the expansion. Similarly, the gravitational effects that make the planets orbit the Sun, the Sun orbit the centre of the Galaxy, and the Galaxy interact with other Local Group galaxies, are strong enough overcome the cosmic expansion. Thus planetary systems, galaxies and groups of galaxies do not take part in the expansion. But when we consider the distances involved between clusters of galaxies, the expansion of space takes over.

10.3.3   The Einstein–de Sitter universe

Friedmann showed that the field equations for an homogeneous and isotropically expanding universe produce the equations

$$\ddot{R}(t) = -\frac{4\pi G R(t)}{3}\left(\rho(t) + \frac{3p}{c^2}\right) + \tfrac{1}{3}\Lambda R(t) \qquad (10.35)$$

$$\dot{R}^2(t) = \frac{8\pi G\rho(t)}{3}R^2(t) + \tfrac{1}{3}\Lambda R^2(t) - k. \qquad (10.36)$$

In these equations, G is the gravitational constant, R is the scale factor and Λ is the cosmological constant; k is a curvature term that can take the value $+1$, -1 or 0; ρ is the density of all the mass and energy in the universe, and p is its pressure.

Using these equations we can generate any number of possible models of the universe, each one with different values of ρ, p, Λ and k, and each one evolving in a different way. The usual analyses of these equations assume that galaxies behave like *dust*; in other words, there is smooth motion with no pressure ($p = 0$). We can also assume that $\Lambda = 0$. (Although there are recent indications that the cosmological constant may be non-zero, all we know for certain is that it is small; taking $\Lambda = 0$ is the simplest option.) Finally, what value should we take for k? Again, the simplest option is to take $k = 0$, and assume that on large scales the universe is flat. A Friedmann model with $p = \Lambda = k = 0$ was first studied in 1932 by Einstein and de Sitter; it is called the *Einstein–de Sitter universe*.

Substituting $p = \Lambda = k = 0$ into (10.36) we obtain

$$\dot{R}^2(t) = \tfrac{8}{3}\pi G\rho(t)R^2(t). \qquad (10.37)$$

It is easy to show that

$$H(t) = \frac{\dot{R}(t)}{R(t)}. \qquad (10.38)$$

Therefore

$$8\pi G\rho(t) = 3H^2(t). \qquad (10.39)$$

Equation (10.39) essentially defines the Einstein–de Sitter universe.

We can easily solve (10.37) for the scale factor if we can eliminate the density $\rho(t)$. We do this by using the conservation of matter: the mass within a comoving sphere remains constant. The volume of a sphere is proportional to $R^3(t)$, so $\rho(t)R^3(t)$ is constant:

$$\rho(t) = \frac{\rho(t_0)R^3(t_0)}{R^3(t)}$$

where, as usual, a subscript 'o' denotes the observed present-day value. We thus have

$$\left(\frac{dR(t)}{dt}\right)^2 = \tfrac{8}{3}\pi G\rho(t_0)\frac{R^3(t_0)}{R(t)}$$

and using (10.39) this becomes

$$\left(\frac{dR(t)}{dt}\right)^2 = H_0^2\frac{R^3(t_0)}{R(t)}.$$

If we write this as

$$\frac{dR(t)}{dt} = H_0 R^{3/2}(t_0) R^{-1/2}(t)$$

then, assuming that $R = 0$ at $t = 0$, we can integrate:

$$R(t) = R(t_0)\left(\tfrac{3}{2}H_0 t\right)^{2/3}. \tag{10.40}$$

Equation (10.40) represents the scale factor of the Einstein–de Sitter universe as a function of time.

In the rest of the book we will deal mainly with the Einstein–de Sitter universe since it is the simplest of all possible Friedmann models. It is important to remember, though, that all of our arguments can be extended to other Friedmann models at the cost of some extra mathematical manipulation. Whether any of the Friedmann models describe the real world is a question that we can answer only by observation.

10.4   THE BIG BANG

Imagine that some super-intelligence made a movie of the history of the universe. As we watched the movie we would see the universe expanding, with groups of galaxies increasing their separation from each other. Now imagine that we played the movie in reverse. We would see the universe contract, with groups of galaxies *approaching* one another. Presumably, if we rewound the movie long enough, we would see a time when everything was very close together. In the Einstein–de Sitter universe this conclusion follows directly from (10.40). For example, (10.40) tells us that when the universe was 10% of its present age the scale factor was 21.5% of its present size. The universe was therefore much denser then than it is now. It is easy to see that the density of the Einstein–de Sitter universe as a function of time is given by

$$\rho(t) = \rho(t_0)\left(\frac{t_0}{t}\right)^2. \tag{10.41}$$

Equation (10.41) tells us that the density of the Einstein–de Sitter universe becomes very large as $t \to 0$.

Lemaître was the first to suggest that there was a time when our universe was very much smaller and denser than it is now. His work seemed to imply that the universe had a definite origin. In the beginning there was a 'primeval atom', resting in a static state for an indefinite period. Then, for some reason, the universe began to expand.

In deriving (10.40) we assumed that $R = 0$ at $t = 0$. How much time has elapsed since that instant? In other words, how old is the universe? We can determine the age of the Einstein–de Sitter universe directly from (10.40). But before calculating an exact answer, it is useful to first obtain a rough estimate of the age of the universe by using Hubble's law.

Pick a galaxy at distance d and with recession velocity v. At a time $t = d/v$ in the past the galaxy was on top of us. Now pick a galaxy that is at a distance $2d$. Hubble's law tells us that it recedes with recession velocity $2v$. Therefore, it would have been on top of us at a time $t = 2d/2v = d/v$ in the past: exactly the same time as the first galaxy. The same argument applies to any other galaxy taking part in this idealised constant cosmic expansion. At some time in the past, given by a galaxy's distance from us divided by its recession velocity, everything was close together.

According to Hubble's law, the distance to a galaxy divided by its recession velocity is $1/H_0$. In other words, the age of such a constant-expansion universe is $1/H_0$. The reciprocal of the Hubble constant is called the *Hubble time*, H_T:

$$H_T = \frac{1}{H_0}. \tag{10.42}$$

The Einstein–de Sitter universe, which we are using as a model of the real universe, does not expand at a constant rate. In this model, the mutual gravitational attraction of all the mass in the universe slows the expansion. Gravity acts as a brake. If we measure a particular expansion rate *now*, we can be sure that the rate of expansion was faster in the past (and will be slower in the future as gravity continues to brake the expansion). Since the universe expanded more quickly in the past, the universe must be *younger* than the Hubble time.

We determine the age of the Einstein–de Sitter universe from (10.40). At the present time $R(t) = R(t_0)$, so

$$t_0 = \frac{2}{3H_0} \tag{10.43}$$

or equivalently

$$t_0 = \tfrac{2}{3} H_T. \tag{10.44}$$

The age of an Einstein–de Sitter universe is thus exactly two-thirds of the Hubble time.

Figure 10.8 shows how the scale factor of the universe changes with time in the case of constant expansion (straight line) and in the case of an Einstein–de Sitter universe (curve).

The Russian–American physicist George Gamow (1904–1968) developed Lemaître's ideas. In 1948 he published a paper with his student Ralph Asher Alpher (1921–) and

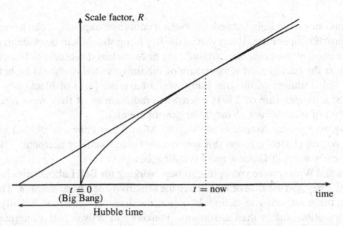

FIGURE 10.8: The diagram shows how the scale factor of the universe changes with time for two cases. If the expansion is constant, the scale factor was zero a Hubble time in the past. For universes that contain matter, and in the absence of a cosmological constant, then the expansion was faster in the past than we now measure it. In the Einstein–de Sitter universe the Big Bang happened two-thirds of a Hubble time ago.

his colleague Hans Bethe* in which he described the universe as beginning in a highly compressed state of extremely hot radiation. The expansion started in a moment called the *Big Bang*. (Hoyle coined the phrase 'Big Bang' in a 1950 radio interview. Perhaps he used it in a disparaging sense; he preferred a different cosmology, one in which the universe has no unique origin. In any case, the term 'Big Bang' stuck. This was unfortunate because the phrase carries with it connotations of a chemical explosion, and this does not properly convey what happened at the start of the universe. A chemical explosion occurs at a single point and creates pressure differences that hurl material outward into pre-existing space. The Big Bang, on the other hand, occurred *everywhere*; and there were no pressure differences causing an expansion — the whole of spacetime began in some initial singular state, and then the scale factor expanded according to the equations of general relativity. Picturing the Big Bang as a large bomb leads to confusion and does not do justice to the strangeness of the event.)

As the universe expanded, the radiation that filled the universe cooled, and hydrogen and helium formed. Gamow and Alpher showed how light elements could form just after the Big Bang, and their calculation of the relative abundances of hydrogen and helium closely matched the observed abundances. It was the first piece of evidence that the universe really did start in a highly condensed state. It was also the first inkling that workers could use well-understood physics to investigate the early universe.

Ralph Alpher and Robert Herman (1914–) went on to make a specific prediction. The

---

*Although Bethe is a great physicist, and won the 1967 Nobel Prize for his work on energy production in stars, he did no work on this paper and was surprised to find himself listed as co-author. Gamow was a joker. He thought it would be appropriate to have a paper on the creation of the universe to have authors whose names sounded like alpha, beta and gamma — the first three letters of the greek alphabet. To Gamow's delight, the paper appeared on April 1.

universe was once incredibly hot and, although it cools as it expands, it can never reach the absolute zero of temperature. Energy from the Big Bang should still be evident; dissipated by the expansion of the universe, certainly, but nevertheless detectable in principle. They predicted that the background temperature of the universe today should be between 5 K and 50 K. This remnant of the Big Bang would have the form of blackbody radiation. Radiation at a temperature of 5 K is microwave radiation so, if they were right, the sky should be full of microwaves — the 'afterglow of creation'.

In 1964, two young American physicists, Arno Allan Penzias (1933–) and Robert Woodrow Wilson (1936–), found this microwave background by accident. They had no idea of the early work of Gamow and his colleagues.

Penzias and Wilson were young researchers working for Bell Laboratories in Holmdel, New Jersey. They had the task of measuring the sensitivity of a horn antenna. The antenna was among the most sensitive radio telescopes in existence, but it was initially designed for communications rather than astronomy. Penzias and Wilson had an agreement with their bosses at Bell: once they had made the appropriate calibrations of the antenna, they could use it to do some radio astronomy. Their results were good enough for satellite communications, but not for astronomical observations. The trouble was that a low, steady noise persisted in the receiver at microwave frequencies. Penzias and Wilson had no idea what this noise was, but until they identified it they would not use the horn for astronomy.

Penzias and Wilson took the equipment apart. They found a pigeon's nest inside the horn, and the horn itself was coated with pigeon droppings. Heat from such droppings radiates in the microwave region, so Penzias and Wilson removed the pigeons and scrubbed the horn clean. They noted a small drop in the background radiation, but most of the noise was still there. They took the equipment apart again, eliminated anything that could conceivably add to the background noise, and reassembled the antenna. Nothing worked. The hiss remained. The noise was there, and it seemed to come from all parts of the sky. Having accounted for all possible natural and man-made sources of microwave radiation, Penzias and Wilson realised they were observing the Big Bang. They estimated the temperature of the radiation to be about 3.5 K.

These two successful predictions — the relative abundance of light elements and the cosmic microwave background radiation — helped to convince most astronomers that the universe began as a Big Bang. Not surprisingly, immense effort has gone into developing the Big Bang model. We now have a good picture of how the universe has evolved since it was just one millionth of a second old. (In fact, it is much more than just a picture. Physicists can make detailed and specific predictions about the universe on the basis of the Big Bang model[*]. Here, though, there is room only for a sketch.) We need to know something of the evolution of the universe when we take the final step on the distance ladder and discuss the size of the universe.

One millionth of a second after the Big Bang.   At this point the universe contains a dense mix of particles. Protons and neutrons — plus their antiparticles — dominate this era, although a host of other particles (such as electrons, neutrinos and photons) also exist.

---

[*]This is not to say that the Big Bang model is correct in all its details, or even that its central tenets are certainly correct. But it *is* the most successful model of the birth of the universe that mankind has ever had.

left: Robert Woodrow Wilson
right: Arno Allan Penzias

The universe is hot: a few trillion degrees! Protons and neutrons continually annihilate with their antiparticles, creating pure energy from which form new proton–antiproton and neutron–antineutron pairs.

One ten thousandth of a second after the Big Bang. The universe expands, and as it does so it cools. It is now at 10^{12} K. As it cools, the rate at which protons and neutrons annihilate outstrips their rate of creation. Almost all the heavy particles vanish in a flash of energy. At $t = 10^{-4}$ s the universe contains mainly photons, electrons and neutrinos. For every proton there are roughly a billion electrons, a billion neutrinos and a billion photons. Compared to their numbers before the annihilation, the residue of protons and neutrons is insignificant. (It is significant for us, of course. All the matter that we see today, from the matter that makes up galaxies to the matter that makes up people, can ultimately be traced back to this point in the very early universe.)

One second after the Big Bang. The temperature is down to 10^{10} K. Just prior to this time, particles like photons, electrons and neutrinos form a soup with all their antiparticles. They share the energy of the universe between them. Electrons annihilate with antielectrons, creating pure energy from which form new electron–antielectron pairs. As the universe cools, though, the annihilation rate begins to outstrip the creation rate. Electrons begin to disappear and pass their energy to the photons. The neutrinos, which interacted closely with the electrons, *decouple*. They essentially never again interact with the rest of the universe. To this day, they travel at — or close to — the speed of light, passing through matter without hindrance. There is thus a cosmic neutrino background in the same way that there is a cosmic microwave background. While you have been reading this paragraph, 10^{20} cosmic neutrinos — relics of an event that happened when the universe was about one second old — have passed through your body. (The detection of this cosmic neutrino background is a surefire way of winning the Nobel Prize.)

Two hundred seconds after the Big Bang.   Most of the energy of the universe is in the form of radiation, but there is a thin gas of protons, neutrons and electrons. At $t = 100$ s the temperature is roughly 10^9 K, cool enough for neutrons and protons to form deuterons through the reaction $n + p \rightarrow {}^2H + \gamma$. (The reaction was taking place prior to this time, but the reverse reaction ${}^2H + \gamma \rightarrow n + p$ took place equally quickly.) The deuterons quickly burn to form helium nuclei. Two hundred seconds after the Big Bang, a quarter of all matter is helium. The remainder is hydrogen.

One hundred thousand years after the Big Bang, and later.   At $t = 10^5$ years, a bright yellow light fills the universe: the temperature is now about 3000 K. It is still too hot for atoms to form. Matter is ionised: electrons move freely, unattached to any nucleus. Photons cannot travel far before being scattered by these electrons.

By the time the universe is about a million years old it has cooled so that it glows only red hot — cool enough for atoms to form. This is known as the *recombination epoch*, since the protons and electrons combine to form stable hydrogen atoms. Once neutral atoms form, radiation can travel through space without hindrance. So, about 10^6 years after the Big Bang, the photons decouple — just as the neutrinos had decoupled at an earlier stage in the universe. Most photons never again interact with matter. They form the cosmic microwave background radiation.

As the universe expands and cools, the cosmic microwave background radiation cools with it. Ten million years after the Big Bang, the background radiation is at room temperature — 300 K. At the present time, the radiation has a temperature of 2.728 ± 0.04 K.

After the recombination epoch, matter began to dominate the universe. Perhaps a few million years after the Big Bang, in a process that is still mysterious, the matter that now forms clusters of galaxies began to condense out of the universal expansion. After more than 10^9 years, stars and planets began to form. Eventually, 10^{10} years after the Big Bang, intelligent life developed on at least one planet in the universe. It is surely a tribute to that intelligent species that it can piece together what happened all the way back to when the universe was just a millionth of a second old.

10.5 ALTERNATIVE THEORIES

Historically, the redshift–distance relation fuelled much of the development of Big Bang cosmology. But what if our interpretation of the redshift is wrong? We would need to develop a whole new theory of cosmology.

Many workers have suggested alternative cosmologies. For example, the great pioneer of quantum physics, Paul Adrien Maurice Dirac (1902–1984), proposed a model in which the gravitational constant G varies with time. Zwicky favoured the *tired light* hypothesis in which a photon loses energy in collisions with particles in the intergalactic medium, thus causing a redshift. This hypothesis is ruled out by observation; a version of the hypothesis developed by the French physicist Jean-Pierre Vigier (1920–) is less easy to rule out. Vigier adds a new term to the equations of quantum mechanics, a term that allows for the vacuum itself to absorb the photon's energy. A few physicists favour *plasma cosmology*, which suggests that plasmas and magnetic fields play the dominant role in the universe and that redshifts arise from a process called the Wolf effect.

One of the strongest attacks on the standard Big Bang model comes from an eminent group of scientists including Fred Hoyle, the English-born astronomer Geoffrey Burbidge (1925–), the Indian theoretician Jayant Vishnu Narlikar (1938–), the Sri Lankan-born British theoretician Nalin Chandra Wickramasinghe (1939–) and the American observational astronomer Halton Christian Arp (1927–).

Arp is a gifted observational astronomer, well known for his *Atlas of Peculiar Galaxies* — a beautiful collection of strange objects that defy the simple Hubble classification scheme. In 1966, he noticed that several compact radio sources lie close in the sky to several of the galaxies in his *Atlas*. He soon found about 30 high-redshift quasars lying close to peculiar galaxies with a much lower redshift. There were two ways to explain these occurrences: they could be line-of-sight coincidences, or the galaxy and the quasar could be physically associated with one another. Arp thought that the second explanation was more likely, even though this destroyed the basis of the redshift–distance relation.

In 1971, Arp found what he thought was evidence of a quasar (Markarian 205) interacting with a galaxy (NGC 4319). His photographs seemed to show a faint wispy 'bridge' between the spiral shape of the galaxy and the small fuzzy blob that was the quasar*. If the bridge were real, and it connected Markarian 205 to NGC 4319, then both objects had to be at the same distance from us. But redshift measurements indicated that the quasar was 12 times farther away than the galaxy. Arp's bridge, if real, would prove that not all redshifts are cosmological in origin. It would show that our standard ideas about the Big Bang are wrong (and an explanation of it would almost certainly require completely new physics).

The 'bridge' between Markarian 205 and NGC 4319 was the first of several similar 'structures' connecting objects of different redshift. Arp and his colleagues argue that there are now too many such examples to explain away as line-of-sight coincidences. But is this true? There are perhaps a 100 billion galaxies in the sky, so it would be strange indeed if we did *not* see some striking coincidences. The whole argument becomes bogged down in a dispute about statistics. Unless we can directly measure the distance to Markarian 205 and NGC 4319, or to similar pairings, we have to accept what the Hubble law tells us: the two objects are at different distances.

Arp, Burbidge, Hoyle, Narlikar and Wickramasinghe continue to question the Big Bang model. In 1990 they met at Cardiff, and proposed an alternative cosmological

---

*Many of Halton Arp's photographs of peculiar galaxies and quasar–galaxy pairs bear an uncanny resemblance to the paintings of his uncle, Jean Arp (1887–1966). Jean was an abstract painter who produced his best work 40 years *before* Halton photographed similar abstract shapes in the deep sky.

model. They published an account of their discussions, called the *Cardiff Manifesto*, in *Nature*. We should always welcome such an iconoclastic approach to accepted notions. (The role of iconoclast is one that Hoyle in particular has filled wonderfully over the years.) Nevertheless, most astronomers believe that the *Cardiff Manifesto* is wrong.

Yet another attack on the conventional interpretation of redshift comes from the American radio astronomer William Grant Tifft (1932–). Tifft sees quantisation effects in redshift data. According to him, galaxies may have only certain allowed values of redshift. Tifft and his colleagues have developed a rival cosmological theory to explain redshift quantisation; a theory that can also accommodate the discordant redshifts of Arp. Again, though, few astronomers accept Tifft's argument. They believe that the perceived quantisation will go away when more data become available.

Conventional opinion is that the blackbody background radiation and the abundance of light elements are compelling pieces of evidence in favour of the Big Bang model. The universe *does* expand. Redshift *does* indicate distance, and we can use it on the final rung of the distance ladder.

CHAPTER SUMMARY

- Hubble's law is an empirical relation between the redshifts and distances of nearby galaxies:

$$cz = H_o r.$$

H_o is the present-day Hubble constant. Its value is a matter of dispute.

- A more general form of Hubble's law, which applies in an expanding homogeneous and isotropic universe, relates velocity and distance:

$$v = H(t)r.$$

- Hubble's law implies that the universe expands. That the universe expands can be explained by Einstein's theory of general relativity.

- The field equations of general relativity relate the curvature of spacetime to the presence of matter and energy. In turn, the motion of matter is determined by the precise form of the curvature.

- We can solve the field equations as applied to the universe as a whole if we assume that, on a large enough scale, the universe is homogeneous and isotropic. Such assumptions give rise to Friedmann–Robertson–Walker (FRW) models of the universe.

- The Robertson–Walker metric in comoving coordinates (r, θ, ϕ) is

$$ds^2 = c^2 \, dt^2 - R^2(t)\left(\frac{dr^2}{1 - kr^2} + r^2 \, d\theta^2 + r^2 \sin^2 \theta \, d\phi^2 \right)$$

where $R(t)$ is the universal scale factor at cosmic time t and k is a curvature term.

- In FRW models, $k = +1$ corresponds to a universe with spherical geometry; $k = 0$ corresponds to a flat universe; $k = -1$ corresponds to a universe with hyperbolic geometry.

- The proper distance d_P to a galaxy with comoving coordinate r depends upon the curvature of the universe:

$$d_P = R(t) \sin^{-1} r \qquad k = +1$$
$$d_P = R(t) r \qquad k = 0$$
$$d_P = R(t) \sinh^{-1} r \qquad k = -1.$$

The proper distance to a galaxy increases because $R(t)$ increases.

- The cosmological redshift, z, of a galaxy is a measure of how much the universe has expanded between t_e (when it emitted the light) and t_o (the time we observe that light):

$$1 + z = \frac{R(t_o)}{R(t_e)}.$$

- The way in which the universe scale factor varies with time depends upon the details of a particular cosmological model. A popular model is the Einstein–de Sitter universe, which has $p = \Lambda = k = 0$. Here, p is the pressure of matter and Λ is the cosmological constant. In the Einstein–de Sitter universe the scale factor varies with time as follows:

$$R(t) = R(t_o) \left(\frac{t}{t_o} \right)^{2/3}.$$

- The density of the Einstein–de Sitter universe varies with time as follows:

$$\rho(t) = \rho(t_o) \left(\frac{t_o}{t} \right)^2.$$

As $t \to 0$, the density becomes very large. At $t = 0$ the universe began in a Big Bang.

- In the Einstein–de Sitter universe, the age of the universe is given by

$$t_o = \tfrac{2}{3} H_T$$

where

$$H_T = \frac{1}{H_o}$$

is the Hubble time. The Hubble time is the age of a constant-expansion universe; the Einstein–de Sitter age is less than H_T because the expansion slows with time.

- The cosmic microwave background radiation, discovered by Penzias and Wilson, is a relic of a time when the universe was much hotter and denser.

- A few astronomers question the interpretation that redshift indicates distance.

QUESTIONS AND PROBLEMS

10.1 Who deserves the credit for discovering the 'expansion of the universe'? (Refer to the papers in the 'Further Reading' section below to disentangle this complicated story.)

10.2 Some people prefer the phrase 'big stretch' when referring to the start of the universe. Devise your own alternative name for the Big Bang.

10.3 The 'tired light' explanation of the redshift supposes that photons somehow lose energy on their way to us from distant galaxies. If the energy loss is due to scattering of particles, as Zwicky supposed, propose a simple observational test of the tired light hypothesis.

10.4 A photon from the cosmic microwave background has much less energy now than it had just after the recombination epoch. Where does this energy go?

10.5 Use (10.13) and suitable metric coefficients to show that $K = -1/R^2$ for a hyperbolic surface and $K = 0$ for a plane.

10.6 If, every second, 40 000 miles of space is created between the Galaxy and the Coma cluster, where does that space make its entry? Discuss whether this is a meaningful question. If it is not meaningful, why is it not? If it *is* meaningful — explain it!

10.7 Explain why there is no centre to the universe which originated from the Big Bang.

10.8 Derive (10.38).

10.9 The universe was hotter in the past, so the 2.7 K background radiation should be warmer at moderate redshifts. What would be the background temperature at $z = 2$? Suggest observations that could search for this increased temperature.

10.10 What are the components of the Robertson–Walker metric tensor (for the coordinate system used in the text)?

$g_{tt} = -1$, $g_{rr} = R^2(t)/(1 - kr^2)$, $g_{\theta\theta} = R^2(t)r^2$, $g_{\phi\phi} = R^2(t)r^2 \sin^2\theta$, other components zero; $k = \mp 1$ or 0.

10.11 What is the Minkowski metric in terms of cylindrical coordinates? $ds^2 = c^2 dt^2 - dr^2 - r^2 d\phi^2 - dz^2$

10.12 You observe a galaxy to have a redshift of 0.75. By how much has the universe expanded since the observed light left the galaxy?

75%

10.13 What is the Hubble time if $H_0 = 100 \, \mathrm{km \, s^{-1} \, Mpc^{-1}}$?

9.8×10^9 years

FURTHER READING

GENERAL
Abbott E A (1952) *Flatland* (New York: Dover)
    — A reprint of a classic book. Most authors use his idea of 'Flatlanders' when discussing spacetime.

Chown M (1996) *Afterglow of Creation* (University Science Books: Sausalito, CA)
    — A popular account of the COBE discoveries and their significance.

Einstein A (1923) *The Principle of Relativity* (Methuen: London)
    — A classic book, reprinted most recently by Dover.

Fauvel J and Gray J (1987) *The History of Mathematics: A Reader* (Milton Keynes: Open University)
    — Contains an interesting account of the role of Gauss in the development of non-Euclidean geometry.

Geroch R (1978) *General Relativity from A to B* (University of Chicago Press: Chicago)
    — A non-mathematical account of general relativity.

Harrison E R (1981) *Cosmology: The Science of the Universe* (Cambridge University Press: Cambridge)
— Perhaps the best introduction to cosmology.

Harrison E R (1993) The redshift–distance and velocity–distance laws. *Astrophys. J.* **403** 28–31
— Why the relations $cz = H_0 r$ and $v = H(t)r$ are not equivalent.

North J D (1965) *The Measure of the Universe* (Clarendon Press: Oxford)
— An authoritative history of general relativistic cosmology.

Silk J (1989) *The Big Bang* (Freeman: New York)
— An accessible introduction to the Big Bang model written by a leading scientist.

Thorne K S (1994) *Black Holes and Time Warps: Einstein's Outrageous Legacy* (New York: Norton)
— A wonderful popular account of general relativity.

Weinberg S (1977) *The First Three Minutes* (André Deutsch: London)
— Now rather dated; nevertheless, this book remains a classic popular account of the early universe.

Wheeler J A (1990) *Gravity and Spacetime* (Freeman: New York)
— A beautifully illustrated book (one of the *Scientific American Library* series) written by one of the world's leading theoretical physicists.

Will C M (1986) *Was Einstein Right?* (Basic Books: New York)
— A clear, non-mathematical account of experimental tests of general relativity.

TEXTBOOKS ON GENERAL RELATIVITY AND COSMOLOGY
Berry M V (1989) *Principles of Cosmology and Gravitation* (IOPP: Bristol)
— A clear introduction to the subject written by a physicist who is famous for work in other fields.

D'Inverno R (1992) *Introducing Einstein's Relativity* (Clarendon Press: Oxford)
— A thorough coverage of special and general relativity for the undergraduate physicist.

Misner C W, Thorne K S and Wheeler J A (1973) *Gravitation* (Freeman: New York)
— A weighty text on gravitation. Full of insights.

Narlikar J V (1983) *Introduction to Cosmology* (Jones and Bartlett: Boston)
— A very clear account of 'standard' cosmology, and a discussion of some non-standard cosmologies.

Peebles P J E (1993) *Principles of Physical Cosmology* (Princeton University Press: Princeton, NJ)
— A standard text for postgraduates, covering all aspects of physical cosmology.

Rindler W (1977) *Essential Relativity* (Springer: Berlin)
— An excellent text that covers both the special and general theories.

Sandage A R, Kron R G and Longair M S (1995) *The Deep Universe* (Berlin: Springer)
— Three beautiful sets of lecture notes on cosmology, for the graduate student.

Schutz B F (1985) *A First Course in General Relativity* (Cambridge: Cambridge University Press)
— A standard undergraduate textbook.

Weinberg S (1972) *Gravitation and Cosmology* (Wiley: New York)
— A superb postgraduate-level account of the principles and applications of general relativity, written by a Nobel Prize-winning particle physicist.

CLASSIC PAPERS IN COSMOLOGY
Alpher R A, Bethe H A and Gamow G (1948) The origin of chemical elements. *Phys. Rev.* **73** 803–804
— The first suggestion of the cosmic microwave background and of how the birth of the universe might give rise to the elements.

Alpher R A and Herman R C (1948) On the relative abundance of the elements. *Phys. Rev.* **74** 1737–1742
— A companion paper to the original Alpher/Bethe/Gamow paper.

Alpher R A and Herman R C (1950) Theory of the origin and relative abundance distribution of the elements. *Rev. Mod. Phys.* **22** 153–212
— A detailed early review of big-bang nucleosynthesis; now known to be incorrect in its details.

Alpher R A, Follin J W Jr and Herman R C (1953) Physical conditions in the initial stages of the expanding universe. *Phys. Rev.* **92** 1347–1361
— Another early paper that applies basic physics to the birth of the universe.

de Sitter W (1917) On the curvature of space. *Proc. Kon. Ned. Akad. Wet.* **20** 229–243
— de Sitter produced the second cosmological solution to Einstein's field equations.

Einstein A and de Sitter W (1932) On the relation between the expansion and the mean density of the universe. *Proc. Nat. Acad. Sci.* **18** 213–214
— This paper introduced the Einstein–de Sitter model, which is the simplest of the Friedmann cosmologies.

Friedmann A A (1922) Über die Krümmung des Raumes. *Z. Phys.* **10** 377–386
— This revolutionary paper, 'On the curvature of space', was published in a leading journal and yet for some reason it had little effect on contemporary cosmologists. A translation can be found in: Lang K R and Gingerich O (1979) *A Source Book in Astronomy and Astrophysics 1900–1975* (Harvard University Press: Cambridge, MA)

Hubble E (1929) A relation between the distance and radial velocity among extra-galactic nebulae. *Proc. Nat. Acad. Sci.* **15** 168–173
— A landmark paper in twentieth century science: the discovery of a redshift–distance relation for galaxies.

Hubble E and Humason M L (1931) The velocity–distance relation among extra-galactic nebulae. *Astrophys. J.* **74** 43–80
— This confirmed Hubble's 1929 result; it settled the case for a linear redshift–distance relation for galaxies.

Lundmark K E (1925) The motions and distances of spiral nebulae. *MNRAS* **85** 865–894
— A very early paper on the velocity–distance relation. Unfortunately, Lundmark chose to fit his data to a polynomial of the form $v = a + br + cr^2$ and obtained a negative value for c.

Penzias A A and Wilson R W (1965) A measurement of excess antenna temperature of 4080 Mc/s. *Astrophys. J.* **142** 419
— The discovery of the cosmic microwave background radiation.

Robertson H P (1928) On relativistic cosmology. *Phil. Mag.* **5** 835–848
— An early discussion of some of the observational consequences of relativistic cosmology.

Wirtz C (1921) Einiges zur Statistik der Radialbewegungen von Spiralnebeln und Kugelsternhaufen. *Astron. Nachr.* **215** 349–354
— An early indication of the possibility of a linear velocity–distance relation for galaxies.

NON-STANDARD COSMOLOGIES

Arp H C (1987) *Quasars, Redshifts and Controversies* (Interstellar Media: Berkeley, CA)
— A fascinating defence of the idea that redshift does not indicate distance, written by a leading observational astronomer.

Arp H C *et al.* (1990) The extragalactic universe: an alternative view. *Nature* **346** 807–812
— The *Cardiff Manifesto*: a thought-provoking article on alternatives to the Big Bang, by Arp, Burbidge, Hoyle, Narlikar and Wickramasinghe.

Field G B, Arp H and Bahcall J N (1973) *The Redshift Controversy* (Benjamin: New York)
— Contains reprints of several classic papers on redshift, and a discussion of whether redshift indicates distance.

Peratt A L (1992) Plasma cosmology. *Sky & Telescope* **83** (2) 136–140
— A non-mathematical account of a cosmological model favoured by some physicists.

11

Eighth step: the universe

Of all the methods of measuring large distances, surely the easiest is to use the Hubble law. Take the spectrum of a galaxy that is in the Hubble flow, measure the redshift of the spectral lines, and multiply the redshift by a number that depends on the Hubble constant: the result is the distance to the galaxy. It sounds simple, and it is. There are just three caveats.

First, we cannot always use the linear redshift–distance relation found by Hubble, since this applies only for small redshifts. For large redshifts the global curvature of the universe becomes important, and in general the redshift–distance relation is nonlinear. The exact form of the relation depends upon the cosmological model under consideration, and must be derived separately for each particular model. So we have to specify which cosmological model we are working with.

Second, we must be careful about what we mean by distance in an expanding universe. When we convert from redshift to distance we must specify our definition of distance as well as our cosmological model.

Third, we need to know the value of the Hubble constant.

In the next three sections I discuss these three points. Once we agree on a cosmological model, a definition of distance and the value of H_o, we can climb to the top of the distance ladder.

11.1   DO WE LIVE IN AN EINSTEIN–DE SITTER UNIVERSE?

We saw in the previous chapter that (10.35) and (10.36) are the Friedmann equations for a zero-pressure homogeneously and isotropically expanding universe. Cosmologists usually write the Friedmann equations in terms of three parameters given by the following equations:

$$\text{curvature parameter} \qquad K = \frac{k}{R^2} \qquad\qquad (11.1)$$

$$\text{Hubble parameter} \qquad H = \frac{\dot{R}}{R} \qquad\qquad (11.2)$$

$$\text{deceleration parameter} \qquad q = -\frac{\ddot{R}}{RH^2}. \qquad\qquad (11.3)$$

The quantities K and H are familiar; q, the *deceleration parameter*, is a measure of how much the cosmic expansion is decelerating due to the mass of the universe. In terms of the measurable quantities K, H, q, Λ and ρ the Friedmann equations become

$$K = 4\pi G\rho - H^2(q+1) \tag{11.4}$$

$$\Lambda = 4\pi G\rho - 3H^2 q. \tag{11.5}$$

There are many possible combinations of parameters in (11.4) and (11.5), each combination representing a different homogeneous and isotropically expanding universe. The empty de Sitter universe, for instance, has $\rho = K = 0, q = -1, \Lambda = 3H^2$. The models we chose to focus on in the last chapter have $\Lambda = 0$, so that

$$K = H^2(2q-1) \tag{11.6}$$

$$4\pi G\rho = 3H^2 q. \tag{11.7}$$

The Einstein–de Sitter universe is flat (i.e. $K = 0$) so that

$$q = \tfrac{1}{2} \tag{11.8}$$

$$\rho = \frac{3H^2}{8\pi G}. \tag{11.9}$$

In models with positive curvature (i.e. $K > 0$) then $q > \tfrac{1}{2}$. In models with negative curvature (i.e. $K < 0$) then $q < \tfrac{1}{2}$. For all models that have the same value for H, the density increases with q. Therefore the Einstein–de Sitter model is a *critical density* universe:

$$\rho_c = \frac{3H^2}{8\pi G}.$$

To see why the Einstein–de Sitter universe is said to have a critical density, write the density of all other models in terms of ρ_c through the relation $\rho = 2q\rho_c$:

$$K = H^2\left(\frac{\rho}{\rho_c} - 1\right). \tag{11.10}$$

If $\rho > \rho_c$ the curvature is positive and the universe is closed. One day the expansion will stop, the universe will collapse back in on itself and, as Frost wrote in his famous poem *Fire and Ice*, 'the world will end in fire'. If $\rho < \rho_c$ the curvature is negative and the universe is open. The expansion will never stop and in this case 'the world will end in ice'. The Einstein–de Sitter universe, with $\rho = \rho_c$, sits on the boundary between open and closed universes; the expansion continues forever, but at an ever-slowing pace.

The ratio of average density to critical density is such a useful parameter for distinguishing between model universes that it has its own symbol. The *density parameter*, Ω, is defined through

$$\Omega = \frac{\rho}{\rho_c} = 2q. \tag{11.11}$$

The density parameter is a function of time, since both ρ and ρ_c are functions of time; we denote its observed present-day value by Ω_o. In terms of presently-observed values, (11.10) becomes

$$K = H_o^2(\Omega_o - 1). \tag{11.12}$$

If $\Omega_o = 1$ then the universe is flat and we live in an Einstein–de Sitter universe; if $\Omega_o < 1$ there is insufficient mass to halt the expansion and the universe is open; if $\Omega_o > 1$ the universe is closed.

Astronomers can estimate the value of Ω_o in various ways. The simplest is to estimate the amount of luminous matter in the universe. It turns out that the amount of material in the universe that shines brightly enough for us to see is not nearly enough to reach the critical density: $\Omega_o(\text{luminous}) \approx 0.01$. We know that the universe contains matter that we do not see: so-called *dark matter*. The rotation curves of galaxies, for instance, immediately tell us that galaxian halos contain a great deal of dark matter, even though we do not know the form the dark matter takes. Taking into account galaxian rotation curves we find $\Omega_o(\text{galaxies}) \approx 0.1$. If we look at the motion of galaxy clusters on the largest scales then perhaps $\Omega_o(\text{clusters}) \approx 0.3$. But it is difficult to find enough material, dark matter or luminous, to ensure that $\Omega_o = 1$.

There are good theoretical reasons (which I explain at the end of the chapter) for supposing that the universe is flat. But if there is not enough mass to reach the critical density, how can we get a flat universe? Einstein's 'blunder', the cosmological constant, may turn out to be not such a blunder after all. The cosmological constant has an energy density, and this can contribute to the density parameter. We can have a flat universe with $\Omega_o^M + \Omega_o^\Lambda = 1$, where Ω_o^M and Ω_o^Λ are the contributions to the critical density from mass and the cosmological constant.

Preliminary results from the Supernova Cosmology Project and the High-z Supernova Search suggest that if the universe is flat then $\Omega_o^M \approx 0.3$ and $\Omega_o^\Lambda \approx 0.7$. Alternatively, if the cosmological constant is zero, then there is insufficient mass to close the universe.

Although the evidence hints that we do *not* live in an Einstein–de Sitter universe, there is sufficient uncertainty in the results that we cannot rule it out. Since the mathematics simplifies so much if we work with an Einstein–de Sitter universe, and since an Einstein–de Sitter universe is not yet ruled out by observation, I will give distances based on this model. In other words, I take

$$\Lambda = 0$$
$$2q = \Omega = 1.$$

This is still the 'standard model'. An interesting exercise for readers would be to examine distances in models with non-zero cosmological constant.

11.2 FIVE TYPES OF DISTANCE

When we study distances in the solar system, or in the Galaxy, or even in the Local Group, we can safely use our everyday notions of Newtonian distance. In these terms, the distance to a celestial object is the Euclidean distance to the point it occupies in absolute space at a particular absolute time. Relativity of course tells us that such a simple view of distance is incorrect, but we can easily shift to the correct view if necessary. The concept of distance is much more complicated when the expansion of the universe is involved: we must be *very* careful about what we mean by the distance to a cosmological object. For instance, when the light from a remote galaxy set out on its journey to us the galaxy might have been nearby. During the light's journey time, the intervening space is stretched. By the time we receive the light, the galaxy is farther away. Thus an object's *reception distance* is always greater than its *emission distance* — at least in a universe that expands constantly, like the Einstein–de Sitter universe. We can use the Robertson–Walker metric to calculate both a reception distance and an emission distance for any given redshift. In this chapter I discuss only reception distances.

Even when we agree to discuss only reception distances, there are at least five types of distance we could use: proper distance, luminosity distance, angular diameter distance, proper motion distance and light travel-time distance. Fortunately, it is fairly easy to relate these to what we observe: redshift.

11.2.1 Proper distance

The proper distance to an object is not something we can measure, but it is what most of us have in mind when we talk about distance. It is a tape-measure kind of distance. In the previous chapter we saw that the proper distance, d_P, is equivalent to coordinate distance (but only in an Einstein–de Sitter universe, which is flat and thus has $K = 0$). We have from (10.30) that

$$d_P(t_o) = R(t_o)r_e \tag{11.13}$$

where r_e is the comoving radial coordinate of an object and $R(t_o)$ is the scale factor at the present cosmic time t_o. Suppose that a light ray leaves an object of redshift z at cosmic time t_e. What is the proper distance to the object? Over a small time interval the light ray travels a small distance — a distance depending upon the scale factor of the universe at that instant. A different scale factor applies at each stage in the history of the light ray. So for each instant in the light ray's history, from the time it set off at t_e to the time we observe it at t_o, we must calculate the distance element. The result of adding all these small distance elements is the proper distance to the object.

A light ray always propagates along a null geodesic so, using the same argument we used in the previous chapter when discussing the redshift, we have

$$r_e = \int_{t_e}^{t_o} \frac{c \, dt}{R(t)}.$$

But from (10.40) and (10.43) we have

$$R(t) = R(t_o)\left(\frac{t}{t_o}\right)^{2/3} \tag{11.14}$$

so the expression for r_e becomes

$$r_e = \frac{c}{R(t_o)}t_o^{2/3}\int_{t_e}^{t_o}t^{-2/3}\,dt = \frac{3t_o^{2/3}c}{R(t_o)}\left(t_o^{1/3} - t_e^{1/3}\right)$$

$$= \frac{3t_o c}{R(t_o)}\left[1 - \left(\frac{t_e}{t_o}\right)^{1/3}\right].$$

From (10.43) we know that

$$t_o = \frac{2}{3H_o}$$

so we can rewrite the expression for r_e as

$$r_e = \frac{2c}{H_o R(t_o)}\left[1 - \left(\frac{t_e}{t_o}\right)^{1/3}\right].$$

From (10.34) and (11.14) we have

$$1 + z = \frac{R(t_o)}{R(t_e)} = \left(\frac{t_o}{t_e}\right)^{2/3}$$

so

$$r_e = \frac{2c}{H_o R(t_o)}\left(1 - \frac{1}{\sqrt{1+z}}\right).$$

Finally, we can define the *Hubble length*, H_L:

$$H_L = \frac{c}{H_o}. \tag{11.15}$$

The Hubble length provides a rough estimate of the size of the observable universe. (Clearly, the numerical value of H_L depends upon the numerical value of H_o. I discuss this in more detail soon.) In these units we have

$$r_e = \frac{2H_L}{R(t_o)}\left(1 - \frac{1}{\sqrt{1+z}}\right)$$

and so the proper reception distance of an object with redshift z is

$$d_P = 2H_L\left(1 - \frac{1}{\sqrt{1+z}}\right). \tag{11.16}$$

Remember that this expression only holds for an Einstein–de Sitter universe. Expressions for proper distance in other cosmological models are more complicated.

11.2.2   Luminosity distance

Since a proper distance is unobservable, it makes sense to relate redshift with a distance that astronomers actually measure. As we have seen, the commonest way that astronomers measure distance is to identify a standard candle and then calculate its distance from its apparent luminosity. In effect they calculate a luminosity distance, d_L.

When we consider stars and nearby galaxies, where the cosmological redshift plays no role, the observed radiation flux, f, is simply given by the inverse-square law:

$$f = \frac{L}{4\pi d_L^2} \tag{11.17}$$

where L is the intrinsic luminosity of the object. Now consider a galaxy at a cosmological distance. It has comoving radial coordinate r_e, and it emits radiation at cosmic time t_e that we now see at t_o. The radiation is now crossing a surface of area $4\pi r_e^2 R^2(t_o)$.

Two quite separate effects reduce the flux. First, the radiation is redshifted and thus arrives with its energy reduced by a factor $1 + z$. (Recall that the energy of a photon is given by hc/λ, where h is the Planck constant. The increase in wavelength reduces the photon's energy.) This factor of $1 + z$ arises because of the *fact* of the redshift — the *cause* of the redshift is unimportant. If the redshift were due to some cause other than cosmic expansion, the radiation would still be reduced in energy by a factor $1 + z$. The second effect is due to the expansion itself: the rate of reception of photons is less than the rate of emission. This effect causes the flux to be reduced by another factor of $1 + z$. In total, therefore, the flux is reduced by a factor $(1 + z)^2$ and we have

$$f(t_o) = \frac{L(t_e)}{4\pi r_e^2 R^2(t_o)(1 + z)^2}$$

and from (11.17) we obtain

$$\begin{aligned}
d_L &= (1 + z)r_e R(t_o) \\
&= (1 + z)d_P \\
&= 2H_L\left(1 + z - \sqrt{1+z}\right).
\end{aligned} \tag{11.18}$$

The luminosity distance is what Hubble measured when he worked with Humason on developing his redshift–distance relation.

If we expand the right-hand side of (11.18) in a Taylor series about $z = 0$, and keep terms up to z^2, we obtain

$$d_L = H_L(z + \tfrac{1}{4}z^2) = \frac{cz}{H_o}\left(1 + \frac{z}{4}\right).$$

Compare this with (10.3). We see that the expansion generates the redshift–distance relation discovered by Hubble, but also that Hubble law is an approximation. When $z = 0.4$, the second term in the bracket is 10% of the first. Thus for large redshifts we need to abandon the straight-line Hubble law.

11.2.3 Angular diameter distance

Some of the distance indicators described in this book have been standard rods rather than standard candles, so we also need to define an angular diameter distance, d_A, and relate this to redshift.

Consider a galaxy of proper linear diameter D that subtends an angle $\Delta\theta$ at our telescopes. As usual, the galaxy is at redshift z and has comoving radial coordinate r_e; at time t_e it emits the light that we now observe. Consider two light rays (null geodesics) from the edges of the galaxy that are directed towards Earth. (See figure 11.1 for the geometry of this argument.)

FIGURE 11.1: The geometry for calculating an angular diameter distance. (Not to scale.)

The Robertson–Walker metric enables us to calculate the proper diameter D. If we put $dt = dr = d\phi = 0$ and $d\theta = \Delta\theta$ into (10.25) we obtain

$$ds^2 = -r_e^2 R^2(t_e)\Delta\theta^2 = -D^2.$$

We therefore have

$$D = r_e R(t_e)\Delta\theta.$$

If we define the angular diameter distance in the same way that surveyors do, namely $d_A = D/\Delta\theta$, then we see that

$$d_A = \frac{r_e R(t_o)}{1+z} = \frac{d_P}{1+z} = \frac{d_L}{(1+z)^2} \tag{11.19}$$

$$= 2H_L\left(\frac{1}{1+z} - \frac{1}{(1+z)^{3/2}}\right). \tag{11.20}$$

Equation (11.20) contains a very surprising result. We can use it to write the angular diameter of a galaxy, $\Delta\theta$, as

$$\Delta\theta = \frac{D}{2H_L}\frac{(1+z)^{3/2}}{\sqrt{1+z}-1}.$$

The surprise is that this expression for angular diameter is not a smoothly decreasing function of redshift. It has a minimum value at $z = 1.25$. Consider two identical galaxies, one at redshift $z = 1.25$ and one at redshift $z = 2.25$; the distant galaxy appears larger than the nearby galaxy! This effect is due to the matter in the universe — which is homogeneously distributed in this model — acting as a gravitational lens.

Example 11.1 Angular diameter of the Local Group

If the Local Group has a proper diameter of 2 Mpc, what is the *minimum* angle it can subtend to any alien astronomer who may be observing it somewhere in the distant universe?
(Assume that we live in an Einstein–de Sitter universe, and that $H_L = 6000$ Mpc.)

Solution. It will subtend a minimum angle for an astronomer who measures the redshift of
the Local Group to be $z = 1.25$. In this case

$$\Delta\theta = \frac{2\,\text{Mpc}}{2H_L} \times 6.75 = 1.125 \times 10^{-3}\,\text{rad} = 232''.$$

11.2.4   Proper motion distance

If we somehow know the transverse velocity of an object, and if we can measure its proper
motion, then we can deduce its distance. We have used this technique several times
on earlier rungs of the distance ladder. The derivation of the relation of proper motion
distance, d_M, to redshift is similar to that for the other types of distance; I therefore just
give the result:

$$d_M = 2H_L\left(1 - \frac{1}{\sqrt{1+z}}\right). \tag{11.21}$$

The proper motion distance thus produces the same answer as the proper distance.

11.2.5   Light travel-time distance

When we say that the distance to Proxima Centauri is about 4 ly, we are using the constancy of the speed of light to convert from a time to a distance. The time that light takes
to reach us from an object is called the *lookback time*, since when we look out in the
universe we look back in time. We always see Proxima Centauri as it was four years ago,
never as it is now.

If we can calculate the lookback time in years then we immediately have a distance in
terms of light years. This is the lookback distance, or the light travel-time distance, d_T. In
a Friedmann universe we can calculate the lookback time as a function of redshift; here,
as above, we calculate it for the flat $K = 0$ case.

The lookback time, t_T, is simply the difference between the present time, t_o, and the
time when the light was emitted, t_e:

$$t_T = t_o - t_e.$$

We saw in (10.44) that $t_o = \frac{2}{3}H_T$, where the Hubble time is just the reciprocal of the
Hubble constant. We can obtain an expression for t_e from (11.14) and (10.34):

$$t_e = \frac{2H_T}{3(1+z)^{3/2}}.$$

The lookback time is thus expressed by

$$t_T = \tfrac{2}{3} H_T \left(1 - \frac{1}{(1+z)^{3/2}} \right).$$

If we multiply both sides of this expression by c, and note that $H_L = cH_T$, we obtain the light travel-time distance:

$$d_T = \tfrac{2}{3} H_L \left(1 - \frac{1}{(1+z)^{3/2}} \right). \tag{11.22}$$

11.2.6   Which distance should we use?

A glance at (11.16), (11.18), (11.20), (11.21) and (11.22) shows that the different distances have a variety of redshift dependences. For small z, though, they all agree:

$$d_P \approx d_L \approx d_A \approx d_M \approx d_T \approx \frac{cz}{H_0}.$$

In our cosmic neighbourhood we observe a linear Hubble relationship, no matter which definition of distance we use. It is only when we go to higher redshifts that the differences between distance measures become noticeable. Figure 11.2 shows a graph of distance (in Hubble lengths) against redshift (up to $z = 2$) for the five different distance measures.

So which distance should we use? When astronomers attempt to measure H_0, perhaps by drawing a Hubble diagram for Type Ia supernovae, they are most likely to use luminosity distance. For some applications the angular diameter distance is more appropriate. But perhaps the most straightforward definition of distance — and the one that I will for the most part use — is the light travel-time distance. When the popular press writes a statement like 'galaxy A is ten billion light years away' they are using light travel-time distance. This is a good choice, since light is our messenger from the universe. Always remember, though, that what we *measure* is a redshift: converting a redshift into a particular number of light years requires that we specify the relevant parameters of our model, and in particular the Hubble constant.

Finally, for any given cosmological model we can convert a redshift into a proper distance — which is what most of us mean by 'distance' in an everyday sense. As the final step on the distance ladder we examine proper distances in our universe.

11.3   THE HUBBLE CONSTANT

The Hubble constant is just the slope of the line in figure 10.1 on page 242. If we measure recession velocities (the vertical axis of the graph) in units of $km\,s^{-1}$, and distances (the horizontal axis of the graph) in units of Mpc, then H_0 has units of $km\,s^{-1}\,Mpc^{-1}$. This is convoluted, but all it says is that — in the Hubble flow — a galaxy gains an extra $|H_0|\,km\,s^{-1}$ in recession velocity for every million parsecs in distance. Hubble's 1929 estimate was that $H_0 = 530\,km\,s^{-1}\,Mpc^{-1}$. So, using this first estimate, for every million parsecs in distance a galaxy gains an extra $530\,km\,s^{-1}$ in recession velocity.

The recession velocities that Hubble used were reasonably secure, since redshifts can be measured accurately. (Of course, he had to correct for peculiar velocities, which have

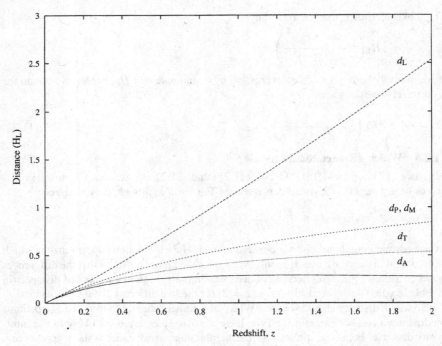

FIGURE 11.2: Distance against redshift for five different distance measures in an Einstein–de Sitter universe. (Distance is expressed in terms of the Hubble length, H_L.)

nothing to do with cosmic expansion; so he had to correct for the motion of the Sun around the Galaxy and the peculiar velocity of the Galaxy itself.) The main problem with Hubble's estimate for H_o was that his galaxian distance estimates were wrong. For nearby galaxies they depended upon Cepheids and, as we saw on page 174, in 1952 Baade revised this step of the distance ladder by a factor of two. The distances to remote galaxies depended upon the brightest stars in those galaxies. In 1956, Hubble's protégé Allan Sandage showed that what Hubble believed were bright stars were actually huge regions of ionised hydrogen. The net effect of all these revisions, according to Sandage, was that galaxies were three times farther away than Hubble had thought. The slope of Hubble's original diagram was therefore three times too steep, and H_o was more like $180 \, \text{km s}^{-1} \, \text{Mpc}^{-1}$. By 1958, Sandage had reduced this even further, to $75 \, \text{km s}^{-1} \, \text{Mpc}^{-1}$. By 1974, Sandage and his colleague Gustav Tammann declared that $H_o = 55 \pm 10 \, \text{km s}^{-1} \, \text{Mpc}^{-1}$, a value they have adhered to ever since. Such a value for the Hubble constant is known as the *long distance scale*, since it means that the universe is much larger than initially believed.

In the late 1970s, Gérard de Vaucouleurs wrote a scathing attack on the methods used by Sandage and Tammann. de Vaucouleurs took a different approach to measuring distances. Rather than rely on just a few well known indicators, such as Cepheids, in effect he used any method he could think of to estimate a galaxian distance, and then averaged the results. He argued that by using a variety of distance indicators it would not matter if one or two of them were wrong: the averaging would smooth out the errors.

Allan Rex Sandage

Few astronomers agreed with all of de Vaucouleurs' distance indicators, but they agreed that the overall approach had some validity. Over-reliance on a few key indicators is dangerous, since if one of the indicators turns out to be faulty the whole carefully constructed distance ladder comes crashing down. (This almost happened with one of the indicators used by Sandage: the size of H II regions. When a high-mass star forms from a dense cloud of hydrogen it emits ultraviolet photons that ionize the hydrogen. The Danish astronomer Bengt Georg Daniel Strömgren (1908–1987) showed that there is a sharp boundary between the ionised hydrogen near the star and the neutral hydrogen farther away. This boundary marks the limits of the *Strömgren sphere* around the star. In 1960, the Argentinean astronomer José Luis Sérsic (1933–1993) suggested that the diameter of Strömgren spheres could be used as a standard rod. Sandage and Tammann developed a method based on this suggestion, but it turned out to be a poor distance indicator. They later dropped the technique.)

When de Vaucouleurs applied his rather individualistic approach to distance measurement he obtained $H_o = 100 \, \text{km s}^{-1} \, \text{Mpc}^{-1}$, which was twice the value favoured by Sandage. The de Vaucouleurs universe was thus half the size (and therefore half the age) of the Sandage universe. A value for H_o of around $100 \, \text{km s}^{-1} \, \text{Mpc}^{-1}$ is known, for obvious reasons, as the *short distance scale*. The error bars for the two scales did not overlap; a

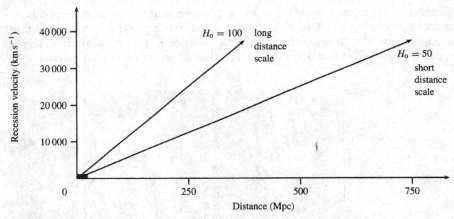

FIGURE 11.3: The Hubble constant is the slope of a diagram like this one. Every time a step on the distance ladder is revised, the slope of the line changes. Recent years have seen the line jump about like the speedometer needle in a joyrider's car. Present thinking is that the Hubble constant almost certainly lies somewhere in the range 45–$100\,\mathrm{km\,s^{-1}\,Mpc^{-1}}$, and probably somewhere in the range 50–$80\,\mathrm{km\,s^{-1}\,Mpc^{-1}}$. The small grey box at the bottom corner of the graph shows the region of the universe first charted by Hubble in 1929. (Note that at larger distances than shown here, the straight lines begin to curve. The exact manner in which the lines curve depends upon the particular cosmological model being studied.)

compromise figure somewhere between the two scales satisfied neither camp. Figure 11.3 shows two slopes, which correspond to $50\ \mathrm{km\,s^{-1}\,Mpc^{-1}}$ and $100\ \mathrm{km\,s^{-1}\,Mpc^{-1}}$, on a modern version of Hubble's graph.

11.3.1   The short and long distance scales

The argument raged between the two camps for over a decade. The dispute revolved around the distance to the Virgo cluster, which is the closest large cluster but which is far enough away to be taking part in the Hubble flow. The *ratio* of the distance of a remote cluster to that of the Virgo cluster was usually not in dispute. What was controversial was the absolute distance to Virgo. Supporters of the short distance scale favoured a distance of about 16 Mpc to the centre of the Virgo cluster. Supporters of the long distance scale favoured a distance of about 22 Mpc. None of the arguments presented by either side were compelling, but in the mid-1980s several new distance indicators were brought to bear on the problem. The Tully–Fisher and Faber–Jackson relations appeared first, and then astronomers calibrated the luminosity functions of planetary nebulae and globular clusters. More recently, the SBF method began to yield distances. Most estimates for H_0 based on these methods were on the high side. In 1992, an influential review by Jacoby and several of his colleagues collated the various Virgo-distance results from these new indicators (see table 11.1). There was good agreement between the various methods, and the results clearly seemed to favour the short distance scale. Perhaps the general consensus among the astronomical community during the 1990s has been that the short distance scale, with $H_0 = 80 \pm 10\,\mathrm{km\,s^{-1}\,Mpc^{-1}}$, is correct.

TABLE 11.1: The distance to the Virgo cluster (adapted from Jacoby et al. (1992)). The distances assume a distance to M 31 of 770 kpc. Based on these distances the authors backed the short distance scale, and obtained $H_o = 80 \pm 11 \, \mathrm{km\,s^{-1}\,Mpc^{-1}}$.

Method	Distance (Mpc)
GCLF	18.8 ± 3.8
Novae	21.1 ± 3.9
SN Ia	19.4 ± 5.0
Tully–Fisher	15.8 ± 1.5
PNLF	15.4 ± 1.1
SBF	15.9 ± 0.9
D_n–σ	16.8 ± 2.4
weighted average	16.0 ± 1.7

TABLE 11.2: The distance to the Virgo cluster according to Sandage and Tammann. They use these distances to derive $H_o = 55 \pm 5 \, \mathrm{km\,s^{-1}\,Mpc^{-1}}$.

Method	Distance (Mpc)
GCLF	21.6 ± 1.5
Novae	19.6 ± 4.0
SN Ia	21.0 ± 1.6
Tully–Fisher	21.8 ± 1.6
Cepheids	20.2 ± 2.0
D_n–σ	23.4 ± 2.2
average	21.5 ± 0.8

Sandage was unconvinced. He has consistently supported the long distance scale with $H_o = 55 \pm 10 \, \mathrm{km\,s^{-1}\,Mpc^{-1}}$. Table 11.2 shows the distance to the Virgo cluster as determined recently by Sandage and Tammann.

There are several reasons why Sandage and Tammann disagree with the short distance scale. Without going into great depth there is room here to mention only a few. Firstly, and not just facetiously, Sandage has argued that the proponents of the short distance scale are victims of the lemming effect. In other words, he accuses them of playing 'follow my leader', and obtaining a high value for H_o simply because everybody else does. The lemming effect is not unknown in astronomy. The South African–Canadian astronomer John Donald Fernie (1933–) wrote of it in entertaining fashion when discussing why it took so long for workers to uncover Shapley's miscalibration of the Cepheid zero point:

'The definitive study of the herd instincts of astronomers has yet to be written, but there are times when we resemble nothing so much as a herd of antelope, heads down in tight parallel formation, thundering with firm determination in a particular direction across the plain. At a given signal from the leader, we whirl about, and, with equally firm determination, thunder off in a quite different direction, still in tight parallel formation.'

FIGURE 11.4: An image of the Virgo cluster of galaxies. The lenticular galaxy M 86 (near the centre of the image) lies close to the heart of the Virgo cluster. The mean distance to the cluster is a matter of dispute.

Such an effect probably occurs in all branches of science, not just astronomy. It does not mean that scientists fraudulently publish work that they know to be incorrect. It has much more subtle origins in psychology. For example, if you obtain an unexpected experimental result it is relatively easy to come up with, say, a dozen ways in which you can refine the experiment. But if the result is exactly what you expected, it is harder to criticise the experiment and generate the same number of refinements. It is impossible to quantify this effect, and in most cases if you disagree with majority scientific opinion it is probably *you* who is wrong. In astronomy though, because measuring distances is so difficult, it is important that workers should question the majority opinion.

The second reason is more substantial. It has to do with the relative motion of the Local Group and the Virgo cluster: the so-called *Virgocentric infall*. Although the Virgo cluster is in the Hubble flow, its vast mass retards the pure cosmic expansion. Astronomers express this retardation as a velocity — an infall velocity of the Local Group towards the Virgo cluster. Just as astronomers involved in this work must correct for the motion of the Sun around the Galaxy, they must also correct for the Virgocentric infall. Sandage takes $v_{in} \approx 220\,\text{km s}^{-1}$, whereas much higher values have often been used by supporters of the short distance scale. Unfortunately, the question of the correct value for the Virgocentric infall generates its own labyrinthine debates.

Third, Sandage and Tammann were unconvinced by the calibrations used by proponents of the short distance scale, and of the effectiveness of some of the new distance indicators. They argued that the PNLF method is flawed in supposing there is a universal sharp dispersionless cut-off in the luminosity function, and that our current understanding of the SBF method is imperfect because it can produce distances in conflict with Cepheid distances. Supporters of the short distance scale have strong counter-arguments to these criticisms, and once more the debate becomes tangled in a morass of technical detail.

Fourth, and perhaps most importantly, they accused proponents of the short distance scale of neglecting, or insufficiently accounting for, *Malmquist bias*. Since Malmquist-like biases are important in setting the extragalactic distance scale, particularly through the Tully–Fisher and Faber–Jackson methods, it is worth considering it in a little detail.

11.3.2   Malmquist bias

The method of using standard candles to calculate a distance seems straightforward. In the real world, though, a standard candle does not have a single luminosity; even our best standard candles have a spread of luminosities. This in itself is not a problem. If we assume that the candle obeys a luminosity function with a mean absolute magnitude M_0 and a dispersion σ about the mean, then we can use the usual theory of errors to calculate the uncertainty in our distance estimates. This is what all physical scientists must do when treating the results of their experiments. But when we use standard candles as a distance indicator the dispersion can cause an extra problem.

When we look out at distant galaxies there is inevitably an apparent magnitude threshold beyond which we cannot detect our standard candle. Unfortunately, if we use our standard candle on a magnitude-limited sample of galaxies, we introduce a bias. (There are several types of bias that can creep into astronomical observations. We have already seen that the Lutz–Kelker bias exists in parallax measurements, and there are others.) Figure 11.5 illustrates a magnitude threshold of $m = 13$. Well below the threshold, when the candles are nearby, there is no problem; we see essentially all of the candles. Near the threshold, though, the luminosity function is progressively cut away. We see only the brightest examples of the candle. There is thus a tendency to underestimate the absolute magnitude of distant candles, which causes us to underestimate their distance. In turn, this causes us to overestimate the value of H_o. The magnitude of this bias was calculated in 1920 by the Swedish astronomer Karl Gunnar Malmquist (1893–1982). Malmquist argued as follows. (If you are not interested in the derivation, skip straight to (11.25).)

Suppose that the standard candle has a Gaussian luminosity function $\phi(M)$ with mean M_0 and dispersion σ:

$$\phi(M)\,dM = \exp\left(-\frac{(M - M_0)^2}{2\sigma^2}\right).$$

Suppose further that this function is the same at all distances and that the candles are uniformly distributed throughout space. Now consider a thin shell at distance r whose inner boundary is at $r - \frac{1}{2}dr$ and whose outer boundary is at $r + \frac{1}{2}dr$; the shell has volume $4\pi r^2\,dr$. Within that shell, only candles with absolute magnitude M in the range $M + \frac{1}{2}dM$ to $M - \frac{1}{2}dM$ are observed at apparent magnitude m in the range $m + \frac{1}{2}dm$ to $m - \frac{1}{2}dm$. The number of candles with this absolute magnitude is $4\pi r^2\phi(M)\,dM\,dr$. If we consider a shell at larger distance r then a brighter absolute magnitude M is needed for

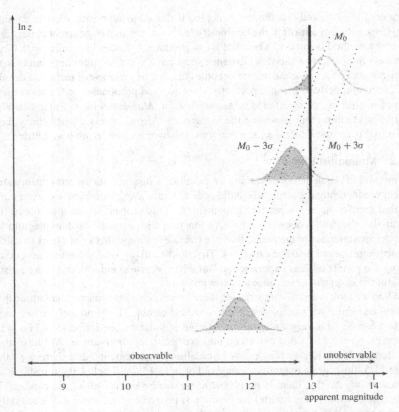

FIGURE 11.5: The origin of the Malmquist bias. A standard candle always has a spread of luminosities; in this diagram the luminosity function is Gaussian with mean absolute magnitude M_0 and dispersion σ. The vertical line denotes a magnitude threshold: we cannot see objects fainter than $m = 13$. This introduces a bias because, as we approach the threshold, more and more of the luminosity function is cut away.

us to observe the same apparent magnitude m; if we assume that no light is absorbed on its way to us, then the three quantities m, M and r, are related through the usual expression $m - M = 5 \log r - 5$. If we integrate the contribution of each shell over all distances we obtain $N(m)$, the total number of candles over the whole sky with an apparent magnitude in the range $m + \frac{1}{2}\mathrm{d}m$ to $m - \frac{1}{2}\mathrm{d}m$:

$$N(m) = 4\pi \int_0^\infty r^2 \phi(M)\,\mathrm{d}M\,\mathrm{d}r$$

$$= 4\pi \int_0^\infty r^2 \exp\left(-\frac{(m + 5 - 5\log r - M_0)^2}{2\sigma^2}\right)\mathrm{d}r. \tag{11.23}$$

Suppose we can observe all the candles up to a limiting apparent magnitude m. Then M_m, the mean absolute magnitude of this magnitude-limited sample, is given by

$$M_m = \frac{\int_0^\infty M r^2 \phi(M) \, dM \, dr}{\int_0^\infty r^2 \phi(M) \, dM \, dr}. \tag{11.24}$$

Differentiating (11.23) with respect to m gives us

$$\frac{dN(m)}{dm} = -\frac{4\pi}{\sigma^2} \int_0^\infty r^2 (m + 5 - 5\log r - M_0)$$
$$\times \exp\left(-\frac{(m + 5 - 5\log r - M_0)^2}{2\sigma^2}\right) dr.$$

Using the definition $m + 5 - 5\log r = M$, and (11.23) and (11.24), this becomes

$$\frac{dN(m)}{dm} = -\frac{4\pi}{\sigma^2}\left[\int_0^\infty M r^2 \exp\left(-\frac{(m + 5 - 5\log r - M_0)^2}{2\sigma^2}\right) dr\right.$$
$$\left. - \int_0^\infty M_0 r^2 \exp\left(-\frac{(m + 5 - 5\log r - M_0)^2}{2\sigma^2}\right) dr\right]$$
$$= -\frac{1}{\sigma^2}\left(4\pi\int_0^\infty M r^2 \phi(M) \, dM \, dr - 4\pi\int_0^\infty M_0 r^2 \phi(M) \, dM \, dr\right)$$
$$= \frac{1}{\sigma^2}[M_0 N(m) - M_m N(m)].$$

Upon rearranging, this becomes

$$M_m = M_0 - \frac{\sigma^2}{N(m)}\frac{dN(m)}{dm} = M_0 - \frac{\sigma^2}{\log e}\frac{d\log N(m)}{dm}.$$

Finally, as Herschel showed (see (7.3)), for a uniform distribution of candles we have

$$\log N(m) = 0.6m + K$$

where K is a constant. Therefore, since $0.6/\log e = 1.382$, we have

$$M_m = M_0 - 1.382\sigma^2. \tag{11.25}$$

This is the equation that Malmquist derived. It shows that the mean absolute magnitude of a magnitude-limited sample is brightened by an amount $1.382\sigma^2$. If we had an ideal dispersionless standard candle, with $\sigma = 0$, we could ignore the effect. Similarly, if we restrict ourselves to short distances, where we can detect the full sample of galaxies, we need not worry. In practice, since our standard candles may have a significant dispersion and we wish to probe deep into the universe, we have to take account of the bias.

Many studies in the mid-1980s used the Tully–Fisher relation on galaxies drawn from magnitude-limited catalogues. The results of these studies always gave high values of H_0. Sandage and Tammann argued that the authors of these studies had failed to account for the Malmquist bias, and thus that the results of such Tully–Fisher studies were suspect.

Example 11.2 A Malmquist correction

We know from local calibrations that a standard candle obeys a Gaussian luminosity function with mean absolute magnitude M_0 and with an intrinsic dispersion of 0.6 mag. We use a magnitude-limited survey of distant candles to deduce a value of H_0. Potentially, what is the effect of ignoring the Malmquist bias on our value of H_0?

Solution. If we ignore the bias then we potentially introduce an error of 0.5 mag. The bias always acts in such a way that we *underestimate* distance, and therefore *overestimate* the value of H_0. A 0.5 mag error in brightness corresponds to an overestimate in H_0 of 25%; so if $H_0(\text{true}) = 50\,\text{km s}^{-1}\,\text{Mpc}^{-1}$ we would observe $H_0(\text{observed}) = 62.5\,\text{km s}^{-1}\,\text{Mpc}^{-1}$.

(Indeed, Sandage has called the Tully–Fisher method the 'fishy Tuller method'.) It is scarcely worth mentioning that the authors of the original studies deny that their work is subject to the Malmquist bias; or, if it is, that appropriate corrections were taken to arrive at an unbiased answer.

If we have large samples of data then there are well-defined procedures for recognising the presence of Malmquist bias; if the bias is present, there are techniques to remove them and produce an unbiased answer. The moral, though, is that we must exercise care when deriving distances — otherwise we risk being fooled.

11.3.3   The *Hubble Space Telescope* and the Hubble constant

There seemed to be little hope of reconciling the two distance scales, and deciding upon the correct value for H_0, until the *Hubble Space Telescope* became operational. The *Hubble Space Telescope* can detect Cepheids in the Virgo cluster, and other clusters. Potentially it can detect Cepheids out to 30 Mpc, which encompasses the whole local supercluster. All workers agree that Cepheids are the most reliable extragalactic distance indicators, so there was hope that the *Hubble Space Telescope* would produce a value of H_0 upon which everyone could agree.

Three teams have used the *HST* to search for extragalactic Cepheids: a team led by Sandage, a team led by Tanvir and the *Hubble Space Telescope* Key Project team.

The Sandage team.   Sandage uses the *Hubble Space Telescope* to search for Cepheids in galaxies that have hosted Type Ia supernovae. This approach leads to the Hubble constant in four steps. In the first step, the Cepheid period–luminosity law gives the distance to the host galaxy. The second step combines this information with the observed peak brightness of the supernova to obtain the absolute peak magnitude of the supernova. The third step uses supernovae as standard candles. Supernovae can provide the distances to galaxies that are so remote they must take part in the Hubble flow. At large enough distances the recession velocity dwarfs any peculiar velocity a galaxy might have. The final step — the recession velocity divided by the distance — produces H_0.

As we saw on page 231, Sandage found Cepheids in the Virgo spiral galaxy NGC 4639 and deduced that the galaxy was 25 ± 2.5 Mpc distant. He then used data from SN1990N, a well-studied supernova in that galaxy, to deduce an absolute peak magnitude for supernovae of about -19.5. Using these as a standard candle, he estimated H_0. His result? He found $H_0 = 57 \pm 4$ km s^{-1} Mpc^{-1}, consistent with what he had been saying since 1974.

The Tanvir team.   Virgo is perhaps not the ideal route to H_0. It has a complex structure and, as the Sandage and Key Project teams have demonstrated, it is extended along our line of sight. Perhaps a better route to H_0 is through a different cluster. The team led by Tanvir decided to search for Cepheids in galaxies in the Leo I group. They chose this group because it is much more compact than the Virgo cluster, and it is the nearest group with a good mix of both spiral and elliptical galaxies.

In 1995, the Tanvir team used the *Hubble Space Telescope* to detect eight Cepheids in M 96, a large spiral in the Leo I group, and calculated its distance to be 11.6 ± 0.8 Mpc. This can be taken to be the distance to the centre of the group. The group is too close to be useful for a direct calculation of H_0: peculiar velocities are a sizeable fraction of the redshift. But because the *relative* distance of Leo I and Coma clusters is quite well established, Tanvir's team could determine the distance to the latter. They calculated a distance to Coma of 105 ± 11 Mpc. Peculiar velocities are less important at this distance and they could thus derive a value for H_0. Their result? They obtained $H_0 = 69 \pm 8$ km s^{-1} Mpc^{-1}, a value between the short and long distance scales!

The Hubble Space Telescope Key Project team.   In 1987, scientists adopted three *Key Projects* as the primary objectives of the *Hubble Space Telescope*. One of them was the calibration of the extragalactic distance scale, with the goal of determining H_0 to an accuracy of 10%.

This particular Key Project has three Principal Investigators — Wendy Freedman, the American astronomer Robert C. Kennicutt Jr (1951–) and the Australian astronomer Jeremy Richard Mould (1949–) — but it is in truth a large collaborative effort. As well as the three Principal Investigators, it employs 11 Investigators and numerous postdoctoral workers and students.

Unlike the Sandage and Tanvir teams, which target the calibration of just one distance indicator, the Key Project team aims to improve the calibration of five secondary distance indicators and use them as steps to the Hubble constant. We have already met the five distance indicators: the infrared Tully–Fisher method, the surface brightness fluctuation method, the PNLF method, the expanding photosphere method for Type II supernovae and the peak luminosity of Type Ia supernovae. One problem with these indicators is that they have been calibrated with only a very few galaxies of known distance. So although the various methods seem to produce quite accurate *relative* distances, there is a chance that the local calibrations of the various methods are systematically in error. The Key Project team therefore set out to extend the number of galaxies with a known Cepheid distance, and thereby improve the calibration of the secondary distance indicators.

Their strategy is to measure the precise Cepheid distances to about 25 galaxies, of which 18 are particularly important. These are spiral galaxies that are ideal for calibrating the infrared Tully–Fisher relation, as well as the other distance indicators. Observations began in 1994, and the Cepheid distances to most of the Key Project galaxies have already been published. The result that made headlines around the world was their first result: the distance to the Virgo spiral M 100. In 1994, the team found 20 Cepheids in M 100, which gave a distance of 17.2 ± 1.8 Mpc (56 ± 5.9 Mly). Until Sandage found Cepheids in NGC 4639 a few months later, this was the most distant galaxy in which Cepheids had been seen.

The Key Project team used this result to calculate H_0. They did this by assuming that M 100 is close to the Virgo cluster core, so that its distance is the distance to the centre of the cluster. Combining this with the average redshift of the Virgo cluster produces an estimate of H_0. Their result? They obtained $H_0 = 80 \pm 17$ km s^{-1} Mpc^{-1}, consistent with the short distance scale and inconsistent with the Sandage result!

The distance to NGC 4639, which is also a Virgo galaxy, cast doubt on this high value for H_0: since NGC 4639 is 1.5 times as distant as M 100, it is clear that the Virgo cluster has a huge line-of-sight depth. It is unlikely that M 100 defines the centre of the cluster. Nevertheless, since that initial result the Key Project team has produced other estimates of H_0 based upon quite different observations. For instance, they have detected Cepheids in the Fornax cluster. Fornax lies at about the same distance as Virgo, but it is more compact and so there is less uncertainty in the distance of member galaxies from its centre. The Key Project estimates of H_0 so far remain in the range ≈ 70–80 km s^{-1} Mpc^{-1}. Table 11.3 shows recent Key Project estimates from various secondary indicators.

TABLE 11.3: Key Project estimates for H_0 using five different secondary indicators. (Adapted from Mould, Kennicutt and Freedman (1999).)

Method	H_0 (km s^{-1} Mpc^{-1})
Tully–Fisher	73 ± 10
D_n–σ	73 ± 10
SBF	78 ± 8
SN Ia	68 ± 10
SN II	73 ± 12

Their present (January 1999) adopted value for H_0 is 73 ± 10 km s^{-1} Mpc^{-1}. Note that this is now very close to the value obtained by Tanvir, although it is still incompatible with the value preferred by Sandage. Their final estimate for H_0 should be available some time in 1999.

The values of H_0 from the three teams do not take into account the Feast–Catchpole calibration of the Cepheid period–luminosity relation. If this calibration is correct, it reduces the derived value of H_0 by several per cent.

11.3.4   Future prospects for measuring the Hubble constant

Figure 11.6 shows a form of the cosmological distance ladder. With each arrow on this diagram there is the possibility of introducing some uncertainty. So it is perhaps not surprising that there should be a dispute over the value of H_0. Twenty years ago, the debate was whether $H_0 = 50$ or 100 km s^{-1} Mpc^{-1}. There has at least been some progress since then, because now the debate is at the level of whether $H_0 = 50$ km s^{-1} Mpc^{-1} or $H_0 = 75$ km s^{-1} Mpc^{-1}. Maybe before another two decades have passed we will know

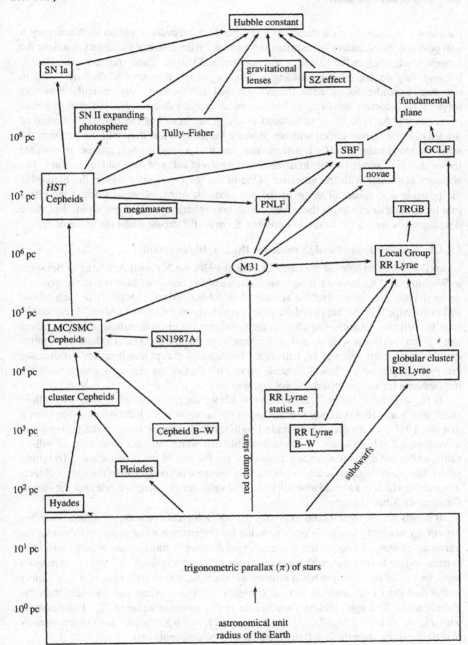

FIGURE 11.6: Routes to the Hubble constant: the cosmological distance ladder.

the Hubble constant to an accuracy of a few per cent. In order to obtain such accuracy it will probably be necessary for astronomers to develop the distance indicators that have the longest reach, such as the SZ effect and gravitational lenses. These methods are valuable because they do not rely on building a distance ladder; they take us directly to H_0 in one step. Consider, for instance, the gravitational lens method. Any particular lens may produce an incorrect distance, and hence an incorrect value for H_0, because of errors in measuring the time delay or through errors in the model of the mass distribution of the lens. But different lenses will not produce the *same* error; they will not produce the *same* incorrect value for H_0. If astronomers can find a large enough number of suitable lenses they will surely be able to determine the correct value of the Hubble constant. The situation is similar with the SZ effect. One of the major uncertainties in the method is the possible non-spherical shape of the gas cloud; if astronomers can detect the effect in a large number of clouds the errors due to non-sphericity will average out. But these developments are for the future. At present, there is still debate about the value of H_0.

11.3.5   Implications of a high value for the Hubble constant

In April 1996, in an echo of the Great Debate in 1920, the National Academy of Sciences in Washington, DC, hosted a debate on the scale of the universe. In essence, the question under debate was: is the Hubble constant low (about $50 \, km \, s^{-1} \, Mpc^{-1}$) or high (about $80 \, km \, s^{-1} \, Mpc^{-1}$)? As happened 76 years previously, nothing was settled. Presumably, as in the original debate, some clues to the solution were given insufficient weight, others were given too much weight, and still others were misunderstood. To date, the controversy regarding the value of H_0 still rages. But why — except that it makes our distance estimates imprecise — should it matter so much? Part of the reason has to do with the implications for our preferred cosmological model.

If $H_0 = 75 \, km \, s^{-1} \, Mpc^{-1}$, then from (10.42) we see that the Hubble time is 13 billion years, and from (10.44) we see that the age of the universe is 8.7 billion years. The reason that the 1994 Key Project result made headlines is that many stars in the universe are older than 8.7 billion years. It seems probable that some stars are older than 10 billion years, and some extreme estimates have even put the age of the oldest stars at 16 billion years. The short distance scale means that the universe is younger than the oldest objects it contains, which is a logical absurdity. A large value for H_0 therefore rules out the simple Einstein–de Sitter universe.

It seems we are then forced to invoke the cosmological constant, Λ, which describes the energy of empty space. A positive value for the cosmological constant increases the universal expansion rate: as the universe expands there is more space, which causes the expansion rate to increase, which creates space even more quickly ... and the expansion 'runs away'. If the universe has a cosmological constant then the expansion was *slower* in the past than it is now. In turn, that means that the universe can be *older* than the Hubble time. The age problem vanishes for any reasonable value of H_0. For example, with $H_0 = 75 \, km \, s^{-1} \, Mpc^{-1}$, $\Omega_0^M = 0.15$ and $\Omega_0^\Lambda = 0.85$, which are observationally plausible values, then the Big Bang happened 15 billion years ago.

Cosmologists have been reluctant to invoke the cosmological constant, primarily for aesthetic reasons. It is somehow more 'elegant' to have $\Lambda = 0$. Letting aesthetics be your guide often works well in theoretical physics, but it can lead you astray. For instance, for many centuries people believed that planetary orbits were circular because circles are somehow the 'purest' geometrical shape. It was not until the work of Kepler and Newton that it was finally understood why planets should follow elliptical orbits. We may be in the same position in our thinking regarding the cosmological constant. Perhaps someone will one day develop a theory that explains why Λ should have a small but non-zero value.

On the other hand, if $H_0 = 50\,\mathrm{km\,s^{-1}\,Mpc^{-1}}$ then the Hubble time is 19.5 billion years and the universe is 13 billion years old. An Einstein–de Sitter universe would be old enough to contain the oldest stars. (If correct, the Feast–Catchpole Cepheid calibration, and the work of Reid on the distance and age of subdwarf stars, indicate that the oldest stars are 11–13 billion years old.) The Einstein–de Sitter universe is tenable, at least regarding the age problem.

So which value should we take for H_0? I take $H_0 = 50\,\mathrm{km\,s^{-1}\,Mpc^{-1}}$ in the rest of this book. Partly this is because the distance indicators that work over the largest distances (the SZ effect, gravitational lenses, Type Ia supernovae), well into the regions where we can ignore peculiar motions and the complications of our local environment, have tended to support the long distance scale. But mainly it is because a low value of H_0 is required if we wish to make honest use of the simple Einstein–de Sitter model.

For the rest of the chapter, therefore, I will calculate distances in an Einstein–de Sitter universe with the following parameters:

$$H_0 = 50\,\mathrm{km\,s^{-1}\,Mpc^{-1}}$$
$$\Omega_0^M = 2q_0 = 1$$
$$\Lambda = 0$$
$$k = 0.$$

Making allowance for different values of H_0 is easy. For instance, if you believe that $H_0 = 100\,\mathrm{km\,s^{-1}\,Mpc^{-1}}$ then just divide any distances I give by a factor of 2. It is more difficult to make allowance for $k \neq 0$ or $\Lambda \neq 0$; this is left to the reader.

11.4   REDSHIFT SURVEYS

Once we know H_0 we can calculate H_L, the Hubble length. The Hubble length is the distance that light can travel in a Hubble time, so it serves as a measure of the extent of the observable universe. For $H_0 = 50\,\mathrm{km\,s^{-1}\,Mpc^{-1}}$, $H_L = c/H_0 = 6000\,\mathrm{Mpc}$. If we substitute this value for H_L into (11.22) we can convert redshift into light travel-time distance in an Einstein–de Sitter universe:

$$d_T(\mathrm{Mpc}) = 4000\left(1 - \frac{1}{(1+z)^{3/2}}\right). \tag{11.26}$$

Table 11.4 gives the equivalent distances for several redshifts. (Some estimates in the table are for specific objects. For instance, the galaxy that appears to have hosted the γ-ray burst of 8 May 1997 was the first such host galaxy to have its spectrum taken. Its spectral lines were redshifted by $z = 0.8$. The γ-ray burst of 14 December 1997 took

TABLE 11.4: The light travel-time distance for various values of redshift in an Einstein–de Sitter universe. (The conversions assume $H_0 = 50 \, \text{km s}^{-1} \, \text{Mpc}^{-1}$, $q_0 = \frac{1}{2}$, $\Lambda = 0$.)

Redshift, z	Light travel-time distance (Mpc)	Light travel-time distance (Bly)
0.01	60	0.20
0.05	275	0.90
0.1	530	1.73
0.2	950	3.10
0.5	1810	5.90
0.8	2345	7.64
1	2585	8.43
1.25	2815	9.18
2	3230	10.53
3	3500	11.41
3.42	3570	11.64
4	3640	11.87
5	3730	12.16
5.64	3766	12.28
10	3890	12.68
20	3960	12.91
∞	4000	13.04

place in a galaxy with $z = 3.42$. These redshifts seem to offer conclusive proof that γ-ray bursts are cosmological in origin. The burst at $z = 3.42$ was so distant that for us to see it now it must have been the most powerful explosion ever recorded, releasing more energy in a few seconds than the total output of a hundred supernovae.)

By measuring redshifts we can map the distribution of galaxies in three dimensions. Proceeding in this way we should be able to complete the programme started by Herschel, and gain 'a knowledge of the construction of the heavens'. Unfortunately, there was until recently a major hindrance to charting the heavens in this way: measuring the redshift of a distant galaxy took a long time. Even with the largest telescopes it often took several hours to collect enough photons to produce a spectrum. A reliable redshift survey, of the kind Herschel would surely have attempted, was impossible. It would simply be too expensive in telescope time. The result was that, four decades after Hubble's discovery, the number of published galaxian redshifts was numbered in the hundreds.

During the 1970s the situation began to improve. The development of charge coupled devices, with their superior light-gathering capability, enabled astronomers to take a spectrum in minutes rather than hours. More recent has been the introduction of multi-fibre spectrographs, in which many fibre optic cables carry light from different galaxies into the spectrograph for analysis. This means that every object in a telescope's field of view can have its spectrum taken at the same time. With these and other technological advances, astronomers moved from custom-made to mass-produced redshift measurements. They began to map the universe using redshifts.

One of the first to carry out a redshift survey was Kirshner, the Harvard supernova expert. Kirshner chose to study a small patch of sky in the direction of the constellation of Boötes. He found lots of galaxies with a redshift of up to 0.03, and lots of galaxies with a redshift greater than 0.06, but none with redshifts between 0.03 and 0.06. Using Hubble's law, Kirshner deduced that there is a region of space some 200 Mpc (650 Mly) deep in which there are no galaxies. Astronomers call it the *Boötes Void*.

Kirshner's was not the only redshift survey to find voids. The American astronomers John Peter Huchra (1948–), Margaret Joan Geller (1947–) and their colleagues, were enthusiastic pioneers of redshift surveys. As well as voids they also discovered filaments — chains of clusters of galaxies that stretch for 1000 Mpc or more. Such filaments are the largest structures in the universe. The more data we obtain from redshift surveys, the more it seems that the distribution of galaxies resembles a bowl of soapsuds. The surfaces of soap bubbles make a similar pattern to that made by galaxies, while the interiors of the bubbles resemble the voids.

The early redshift surveys covered only a small area of sky and a limited number of galaxies. Two recent surveys — the Las Campanas Redshift Survey in Chile and the Two Degree Field Survey in Australia — have measured several hundred thousand redshifts. And the Sloan Digital Sky Survey is now underway; its first images of galaxies were obtained in June 1998. It uses a dedicated 2.5-m telescope at Apache Point Observatory, New Mexico, in order to study galaxies over one quarter of the sky. For five years the Sloan Survey will take the spectra of about 900 000 galaxies out to a redshift of about 0.2. I wrote on page 196 that we had to give up the hope of measuring the distance to every single galaxy. The Hubble law revives this hope. The Sloan Survey will provide us with our best understanding yet of the 'construction of the heavens'.

Now that we can measure redshifts as a matter of routine, we can ask the obvious question: which object has the highest known redshift? This is the same as asking: what is the farthest known object.

11.5 HIGH-REDSHIFT GALAXIES

By the end of the 1950s the radio telescope had become an important tool for observing the universe, and one of the most skilful of the new breed of observers was the English astronomer Martin Ryle (1918–1984). By the end of the decade, Ryle and his colleagues had compiled a collection of 450 bright radio sources, which was eventually published as the *Third Cambridge Catalogue* (3C). Sandage was asked by a colleague to use the 200-inch Hale telescope to find the visible counterpart of 3C 48 — the 48th object in the catalogue. In September 1960, Sandage found a small blue dot at the coordinates of 3C 48. It appeared that 3C 48 was something unique: a radio star.

Sandage obtained the spectrum of 3C 48 and found that the object was unique in another way. It had a spectrum quite unlike any other star. He consulted the American astrophysicist Jesse Leonard Greenstein (1909–), who was an expert in decoding intricate stellar spectra. This spectrum stumped even Greenstein. He could invent no plausible physical model of a star that could produce such a strange set of lines.

For the next three years Sandage continued to amass data on 3C 48. He found that it could vary in brightness on a timescale of a few weeks, which in turn implied that it could not possibly be bigger than a few light weeks in diameter — not much bigger than the solar system. This discovery seemed to rule out the possibility, as some had suggested, that 3C 48 was a distant galaxy. It seemed to confirm the idea that 3C 48 was a nearby star with most peculiar properties.

Astronomers soon identified the optical counterparts of several other compact radio sources from the Cambridge catalogue. The spectra of these objects were as strange as the spectrum of 3C 48. They contained clear spectral lines, but the lines were in places where no element had its characteristic 'fingerprint'. Just as odd was the presence of bright emission lines in the spectra: radio galaxies were known to have emission lines in their spectra, but stars characteristically had dark absorbtion lines. In 1963, the Dutch–American astronomer Maarten Schmidt (1929–) took the spectrum of 3C 273, and found the usual strange assortment of spectral lines. He was as puzzled as everyone else. Then one day, sitting in his office, it occurred to him that the spectrum made perfect sense if the lines were redshifted by 16%. He told Greenstein, who instantly realised that the spectral lines in 3C 48 were redshifted by 37%.

Redshifts of $z = 0.16$ and $z = 0.37$ were large but not unprecedented; several faint galaxies had larger measured redshifts. On the other hand, 3C 48 and 3C 273 were quite unlike conventional galaxies: they were a hundred times brighter than other galaxies at the same distance, yet the region of energy emission was much smaller than any known galaxy. They were given the name 'quasi-stellar radio sources'. The name was not particularly appropriate, since most quasars are poor radio sources. In 1964, the Chinese–American astronomer Hong-Yee Chiu (1932–) suggested that this term should be shortened to *quasar*, and the name stuck.

We now know that quasars have nothing to do with stars. It seems likely that they are the violent nuclei of active galaxies. The outskirts of the galaxies become lost in the glare, in much the same way that light from a car's headlamps drowns out light from the car itself. The details of how so much energy can originate from such a small volume remain unknown, but most models of quasars have a supermassive black hole as the probable power source.

The quasars 3C 273 and 3C 48 were the first of many to be discovered. Astronomers have catalogued about 6000, and one estimate is that about 100 thousand quasars were created. But when we set this number against the 100 *billion* galaxies that may exist, it becomes clear that quasars are extremely rare. Understanding their extreme rarity, and their extreme luminosity, are just two reasons why the study of quasars is still an active area of research. Another reason is that we find most quasars at high redshifts (3C 273 is the closest quasar), so when we study them we see the universe at an earlier stage of its development. They may help us understand how the universe evolved. Yet another reason for studying quasars is that they can help us determine distances in the universe. I have already discussed them in connection with gravitational lenses (and with Arp's more controversial work on discordant redshifts).

Since quasars are so bright, we can see them over vast reaches of space. Astronomers eventually found quasars with redshifts greater than 2; later, they began to find a few quasars with redshifts above 3. In 1991, astronomers discovered that the quasar PC 1247+3406 has a redshift of 4.897. Until recently, PC 1247+3406 was the farthest quasar — and the farthest object — known to mankind.

For many years, quasars were the most distant objects astronomers could see. Now, improved techniques mean that astronomers can see galaxies at distances beyond the realm of quasars. In July 1997, the quasar PC 1247+3406 lost its record as the most distant known object* to a galaxy with $z = 4.92$. This galaxy (shown in figure 9.15) is at a light travel-time distance of 3720 Mpc (12.13 Bly). Light from the galaxy set off on its journey to us less than a billion years after the Big Bang. We can see it only because a cluster of galaxies called CL 1358+62 gravitationally lensed the light from the distant galaxy, and acted as a cosmic telescope.

Whereas quasar PC 1247+3406 held the record for the most distant object for six years, the lensed galaxy held the record for just over six months. In 1998, astronomers used the Keck telescopes to detect a galaxy called 0140+326RD1 at $z = 5.34$, the first time the $z = 5$ barrier had been broken. (See figure 11.7.) This galaxy held the record for just six weeks! The American astronomer Esther Ming Hu (1953–) along with the British-born astronomers Lennox Lauchlan Cowie (1950–) and Richard Gerard McMahon (1959–) used the Keck telescopes to detect a galaxy at $z = 5.64$. This galaxy is at a light travel-time distance of 3766 Mpc (12.28 Bly). Light from the galaxy set off on its journey to us just 720 million years after the Big Bang.

In 1995, the *Hubble Space Telescope* took a ten-day exposure of a small patch of sky in the constellation of Ursa Major. The resulting 'deep' look into space — the famous Hubble Deep Field (see figure 11.8) — recorded some of the faintest and therefore presumably most distant galaxies ever seen. A similar observation of a small region of space in the southern constellation of Tucana, made in October 1998, doubled the number of these very distant galaxies available for study by astronomers. In January 1998, the *HST* peered even deeper into space. A team led by the American astronomer Rodger Irwin Thompson (1944–) took a 36-hour exposure of a part of the Hubble Deep Field, a region of sky just above the handle of the Plough. Thomson used the Near Infrared Camera and Multi-Object Spectrometer (NICMOS), which can see beyond the limits of the visible Deep Field. (An infrared camera like NICMOS was needed because cosmic expansion stretches the light from such very distant galaxies out to infrared wavelengths.) Some of the galaxies in the Hubble Deep Field and NICMOS images may have redshifts as high as 7. Work is already in progress to take the spectra of these objects.

By the time this book appears in print, the redshift record will almost certainly have been broken again. At the time of writing, though, the most distant known object in the universe is the galaxy at $z = 5.64$. (The search for high-redshift objects is not a game, with astronomers vying with each other for the record. When we see high-redshift objects we see the universe as it was when it was young. We can learn much about how galaxies were born, and how they evolve.)

We now have a possible answer to a question posed throughout the book: how big is the universe? If we define the size of the known universe to be the light travel-time distance to the farthest known object, then the size of the universe — assuming the model parameters given above — is 3766 Mpc. But we can push out to higher redshifts than that.

---

*In December 1998, astronomers working on the Sloan Digital Sky Survey announced that an analysis of the first few months' data had uncovered three very high-redshift quasars. Two of the quasars had redshifts of 5.0 and 4.9. So PC 1247+3406 no longer even holds the record for the quasar with the highest redshift. The Sloan Survey is still at a very early stage, so further discoveries are likely.

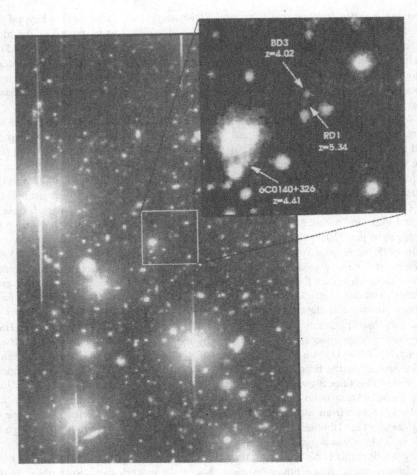

FIGURE 11.7: An area of sky containing a redshift 5.34 galaxy. The inset shows a magnified view of the area within the white box (30″ × 30″ in size). Some of the distant galaxies within this field are labelled with their redshifts. The galaxy at $z = 5.34$ briefly held the record for the most distant known object in the universe.

11.6   THE EDGE OF THE UNIVERSE

If we see an object at $z = 5.64$ we see the universe as it was 12.28 billion years ago. Presumably, if we look out to even greater distances in space, and thus further back in time, we will eventually reach an era when the galaxies began to condense out of the universal expansion. Astronomers do not yet understand the processes by which galaxies and quasars form, but it seems reasonable to suppose that quasars 'switch on' at a redshift of about 10 while the oldest stars in galaxies and globular clusters form at a redshift of about 20. Although it would be exceedingly difficult, in principle it may be possible to

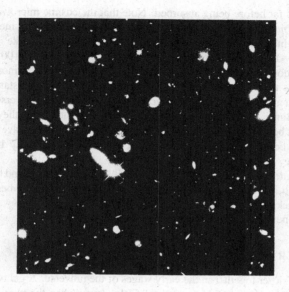

FIGURE 11.8: Part of the Hubble Deep Field. Almost every object on this image is a distant galaxy.

look back to objects with $z = 20$. This corresponds to a light travel-time distance of
3960 Mpc (12.91 Bly).

If we relax the condition that we should look for luminous *objects* like quasars and
galaxies, then there is a sense in which we have already discussed distances much larger
than $z = 20$. The recombination epoch (see page 272) occurred about one million years
after the Big Bang, when expansion caused the cosmic temperature to drop to about
3000 K. Before the recombination epoch, when the temperature was greater than 3000 K,
photons had enough energy to ionize hydrogen atoms. After the recombination epoch,
photons no longer had enough energy to prevent the combination of electrons and protons
into hydrogen atoms. Like a morning mist lifting with sunrise, the universe became transparent to photons. Since most of the photons from the recombination epoch never again
interact with matter, they still exist. This radiation is what Penzias and Wilson discovered. The present temperature of the radiation is about 2.73 K, so it has cooled by a factor
of $\frac{3000}{2.73} \approx 1100$ since the recombination epoch. This cooling is caused by the cosmic
expansion redshifting the radiation. In other words, the cosmic microwave background
radiation is simply the photons from the recombination epoch that have been cooled by a
redshift of $z = 1100$.

Cosmic microwave background photons, with their redshift of 1100, in some ways
represent an 'edge' to the universe. At higher redshifts, and therefore greater lookback
times, stars and galaxies could not exist, because even the simplest atom was unstable.
Even if an object did exist before the recombination epoch we can never see it: to see an
object we need to receive photons from it, but before the recombination epoch photons

could not travel far before being absorbed. Note that the cosmic microwave background is emphatically *not* an 'edge' to the universe in the sense that Lucretius meant. Indeed, the word 'edge' is inappropriate. The microwave background is more like an impenetrable shroud of fog — a blanket that prevents us from seeing beyond $z = 1100$.

If we abandon photons as our messenger, and instead use neutrinos, then one day we may be able to probe to even higher redshifts and thus larger distances. Neutrinos decoupled from matter at a very early stage in the evolution of the universe — one second after the Big Bang, when the temperature was 10^{10} K. Since then, they have travelled freely at or just below the speed of light. Like the photons, they too have cooled with the expansion: the cosmic neutrino background has a temperature of only 1.9 K. And, like the photons, they bathe the entire universe.

Neutrinos interact so weakly with matter that it is difficult to stop and then study them. Perhaps, though, our technology will one day advance to the point where we can study the neutrino background in the same way that we now study the microwave background. If this ever happens, we will be investigating the universe at $z = 10^{10}$.

11.7 PROPER DISTANCES

Galaxies were closer together in the early stages of the universe. A galaxy at $z = 20$, for instance, was only about 450 Mpc (about 1.5 Bly) away when the photons that we now see left it. So why did the photons take so long — about 12.9 billion years according to table 11.4 — to get here? The reason is that the universe expanded and, early on in its evolution, it did so much faster than the speed of light. Note that there is no violation of special relativity here, because nothing actually moved *through* space. Space itself expanded in such a way that the distance between objects increased faster than light speed*. Initially, the photons found that the distance between them and us increased, even though they moved towards us at the speed of light. (It is like trying to run up the down escalator: unless you are quicker than the escalator you move down, not up.) The expansion slowed, though, and eventually the photons began to make progress. At the present time, 12.9 billion years after they set off towards us, we see them.

The light travel-time distance to a galaxy does not tell us how close the galaxy was to us when it emitted its light. Nor does it tell us how far away the galaxy is now. It simply tells us how long it has taken light to reach us from the galaxy. Light travel-time distance is thus not really a distance at all.

When we talk of distance, most of us want to know the distance between two objects *now*, at one instant in time, rather than the distance between us 'here and now' and an object 'there and then'. For example, when we say that the Sun is 8.3 light minutes away we probably do not refer to the time light has taken to reach us. Instead, we probably have a mental picture of somehow stretching a long enough tape measure between Earth and Sun and measuring, *now*, a distance of 93 million miles — which is equivalent to 8.3 light minutes. According to special relativity, it makes no sense to talk of a single 'now' in this way. When discussing the universe as a whole, though, we can use the temperature of the

---

*Remember that R, the scale factor of the universe, is just a number. According to general relativity, this number can increase at an arbitrarily fast rate. Special relativity has nothing to say about this situation.

cosmic background radiation to define what we mean by 'now'. The distance to an object 'right now', with all observations made at the same instant in time, is its proper distance (see page 282 for a mathematical treatment).

It is important to remember that we never *measure* a proper distance. Perhaps, at this very instant, malevolent aliens have destroyed the Sun; the Sun then has no proper distance since it no longer exists. Nevertheless, it will continue to shine peacefully in our skies for 8.3 minutes after its destruction. The finite speed of light means that we never see things as they are 'now'; when we look out at the universe we look back into the past. Light travel-time distance is what is physically important. Although we never measure a proper distance, if we specify a particular cosmological model we can always *calculate* a proper distance from a redshift. Equation (11.16) gives the expression for proper reception distance in an Einstein–de Sitter universe. For $H_o = 50\,\text{km}\,\text{s}^{-1}\,\text{Mpc}^{-1}$, the expression becomes

$$d_\text{P}(\text{Mpc}) = 12\,000\left(1 - \frac{1}{\sqrt{1+z}}\right).$$

For nearby objects like the Sun or Andromeda, proper distance and light travel-time distance are effectively equivalent: during the time that light takes to travel to us from these objects, the objects themselves do not move far. The situation is different for objects taking part in the Hubble flow. Cosmic expansion increases the distance to these objects during the light travel time. Galaxies taking part in the Hubble flow are thus farther away, right now, than their light travel-time distance would indicate.

Table 11.5 gives the proper reception distances for several values of z in an Einstein–de Sitter universe.

From table 11.5 we see that the most distant object known, with a redshift of 5.64, lies at a proper distance of 7340 Mpc (23.94 Bly). The proper distance of 7340 Mpc is the distance to the galaxy *now*, at the present cosmic time, but we cannot see it as it is now. Instead, we see it as it was in its youth, 12.28 billion years ago. How has the galaxy evolved? It may have settled down into a quiet and uneventful middle-age, or perhaps it no longer exists as a distinct physical system. We cannot possibly know. To find out, we must wait several billion years for light from the galaxy to reach us.

11.8 THE PARTICLE HORIZON

Tables 11.4 and 11.5 each contain an entry for infinite redshift, which corresponds to the Big Bang itself. Perhaps we will never probe directly beyond the neutrino decoupling time, and the Big Bang will always be unobservable, but that does not stop us calculating. Table 11.4 shows that an infinite redshift corresponds to a light travel-time distance of 4000 Mpc (13.04 Bly). The equivalent lookback time — the age of the universe — is of course 13.04 billion years. Table 11.5 shows that this maximum light travel-time distance corresponds to a proper distance of 12 000 Mpc (39.12 Bly). This, then, is the present radius of our Einstein–de Sitter universe: 12 000 Mpc.

There is a boundary to the universe, and it lies at a distance of about 12 000 Mpc in all directions. It is a strange kind of boundary, though: more of an information barrier than

TABLE 11.5: The proper distance for various values of redshift in an Einstein–de Sitter universe. (The conversions assume $H_0 = 50\,\mathrm{km\,s^{-1}\,Mpc^{-1}}$, $q_0 = \frac{1}{2}$, $\Lambda = 0$.)

Redshift, z	Proper distance (Mpc)	Proper distance (Bly)
0.01	60	0.20
0.05	290	0.95
0.1	560	1.83
0.2	1050	3.42
0.5	2200	7.17
0.8	3055	9.96
1	3515	11.46
1.25	4000	13.04
2	5070	16.53
3	6000	19.56
3.42	6290	20.51
4	6630	21.61
5	7100	23.15
5.64	7340	23.94
10	8380	27.32
20	9380	30.58
∞	12000	39.12

an edge. To fully understand the nature of the boundary, it is easiest to first consider a completely different universe to the one in which it seems we live.

Imagine a universe that is static. In other words, suppose there is no cosmic expansion, so that the galaxies do not move away from one another. Suppose this imaginary universe is infinite in extent, and that it contains an infinite number of galaxies scattered throughout space; the space density of galaxies is the same as in the real universe. The origin of this imaginary universe was not a hot, dense Big Bang. Rather, a single creation event some 13 billion years in the past somehow 'switched on' every galaxy at the same time. What would the night sky look like on an Earth in such a universe? Would we find it unbearably bright, our eyes blinded by light from an infinite number of galaxies? No! Remarkably, the sky would look pretty much the same as our own night sky. We could see only galaxies within a sphere of radius 13 billion light years. None of the very distant galaxies would be visible. There simply would not have been time for light to reach us from sources farther away than 13 billion light years. Even in such an infinite universe, strewn throughout with stars and galaxies, the night sky would be dark.

The crucial point is that although this imaginary universe is infinite in extent, every observer sees only a finite portion of it. Every observer is at the centre of a sphere that contains his or her *observable* universe. The surface of the sphere is a boundary, and it is impossible to see beyond the boundary because insufficient time has elapsed for light from such regions to reach the observer. The boundary is similar to the horizon we see when we are on a ship in the middle of an ocean. For that reason, the boundary of the observable universe is called a *horizon*. More precisely, it is a *particle horizon*. (The word 'particle' in this context refers to an object, like a star or a galaxy, that endures. Thus it is different from an event, which happens at a single spacetime point. The particle horizon is the boundary between particles that can at present be observed and those that cannot. One can also talk of an *event horizon*: the boundary between events that we can at some time observe, and those we can never observe. The Einstein–de Sitter universe does not possess an event horizon, so I will concentrate on particle horizons. The terminology is due to the German physicist Wolfgang Rindler (1924–), who first clarified these issues.)

Figure 11.9 shows the particle horizon for three observers — A, B and C — who live in this imaginary static universe. Notice that the horizons for A and B overlap. A and B see the same things in the overlapping region, but they cannot see each other and they each see things that the other cannot. The observable universe of C has nothing in common with the observable universe of A or B, since C's particle horizon does not overlap with that of A or B. We say that the three observers are *causally disconnected*. No action that any of them has ever taken can possibly cause an effect on the others (at least, not yet).

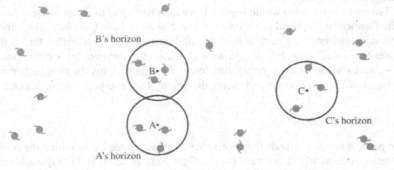

FIGURE 11.9: Particle horizons for three observers A, B and C. The observers are not yet in causal contact.

Unlike terrestrial horizons, the particle horizon is not static. One year after the galaxies 'switched on', the particle horizon for each observer was one light year away. Ten years after the creation event, it was ten light years away. Clearly, the particle horizon sweeps outward at the speed of light, and the observable universe increases in size (even though the universe itself does not expand). The three observers in figure 11.9, if they wait long enough, will find that their particle horizons will completely overlap and they will be visible to each other. They will come into *causal contact*.

A particle horizon exists in this static infinite universe because it is not infinitely old. We know that the real universe is not infinitely old, and that it began with a Big Bang about 13 billion years ago. Our universe therefore has a particle horizon. We have already calculated the proper distance to the horizon: from (11.16) it is at a distance of $2H_L$, which is 12 000 Mpc (39.12 Bly) for $H_o = 50 \, \mathrm{km \, s^{-1} \, Mpc^{-1}}$.

(The concept of particle horizons is trickier in an expanding universe than it is in a static universe. It can sometimes cause confusion to learn that the radius of the observable universe is three times larger than the distance light can possibly have travelled. How can the observable universe extend beyond the maximum light travel-time distance? There are two points to remember. First, the observable universe contains particles that are visible to us *at some stage in their evolution*, not necessarily as they are at their *present* stage of evolution. Particles beyond the horizon are not yet visible to us at *any* stage of their evolution. Second, the proper distance to the particle horizon is something we calculate rather than measure. The expansion of the universe carries the horizon away from us, and we can calculate, within any particular cosmological model, the rate at which it does so.)

Do not overstretch the analogy between terrestrial and cosmological horizons. A terrestrial horizon is not a real barrier. We can communicate with someone who is beyond the horizon using a variety of methods. Lord Nelson, for instance, could signal his entire fleet with flags; there is nothing in principle to stop a message going right round the world using this method. Nowadays we use communications satellites. A particle horizon is different. It marks a definite boundary. No amount of clever signalling with other observers can provide us with any information about particles that may exist beyond the boundary. Objects beyond our horizon do not affect us in any way: not by their light, nor by their gravity, nor by anything else.

Lucretius asked what would happen if we marched up to the edge of the universe and hurled something outwards. We can see now that there is no paradox. The edge of the observable universe is a particle horizon: an information barrier, rather than a physical barrier like a wall or a cliff. If you want to see what is beyond the horizon you can do so — but by using patience rather than force. As time goes by, the particle horizon ever increases. Wait long enough and eventually you will see everything there is to see.

11.9 INFLATION

The particle horizon bounds the observable universe. Since by definition the observable universe contains all that we can observe at present, we could end our quest here. There are strong indications, though, that the universe must extend far beyond the horizon. To see why this should be so, we need to consider where the particle horizon was in the past.

At time t after the Big Bang the horizon distance, d_h, is always at twice the Hubble length:

$$d_h(t) = 2H_L = 3ct. \tag{11.27}$$

(Strictly speaking, this holds only in a *matter-dominated* Einstein–de Sitter universe. For a few hundred years after the Big Bang the universe was dominated by radiation rather than matter. The properties of the Einstein–de Sitter universe during a *radiation-dominated* era were computed by the American physicist Richard Chace Tolman (1881–1948). Tolman showed that the radiation-dominated universe behaves much like a matter-dominated universe, except that the scale factor evolves slightly differently: $R(t) \propto t^{1/2}$ instead of $t^{2/3}$. In the early stage of the universe, $d_h(t) = 2ct$.)

We can use (11.27) to calculate the distance to the horizon at some time in the past. At the recombination epoch, for instance, when the universe was about 3×10^5 years old, $d_h \approx 9 \times 10^5$ ly.

The horizon distance increases with time, but it is important to remember that cosmic expansion — the increase in the universal scale factor, which causes space itself to stretch — is something entirely different. (We saw earlier that, even in a static universe, the

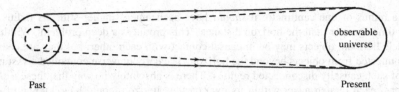

FIGURE 11.10: In the past, the distance to the particle horizon (the broken circle) was smaller than the region of spacetime that expanded to form our present observable universe (full circle). The further back in time we look, the greater the discrepancy becomes. This leads to the so-called horizon problem. The solution must be that, at some time in the past, the universe was inside the broken circle.

horizon recedes at the speed of light and so the observable universe steadily increases in size.) We can use (11.14) to calculate at any earlier time the size R of the spacetime region that expanded into our present observable universe. At the recombination epoch, for instance, $R \approx 3 \times 10^7$ ly.

By comparing the numbers from the above two paragraphs we see that at the recombination epoch the radius of the universe was about 30 times larger than the radius of the *observable* universe (i.e. the horizon distance). The mismatch between the horizon distance and the radius of the universe increases the further back in time we look. A few seconds after the Big Bang, the horizon distance was a few light seconds; what would later become our observable universe had a radius of a few light *years*. When the universe was 10^{-35} s old, the particle horizon was about 10^{-27} m away; but the universe itself — the material that would later expand to form our present visible universe — had a radius of about 10^{-2} m. Figure 11.10 shows a sketch of these ideas.

Example 11.3 Expansion and horizon distances

Calculate the distance to the horizon when the universe was 10^9 years old. Also calculate the radius of the universe at that time. Why are the two distances different?

Solution. We are in the matter-dominated era at 10^9 years, so use (11.27) to determine the horizon distance: $d_h = 3 \times 10^9$ ly. From (11.14),

$$R(t = 10^9 \text{ yr}) = R(t_0)\left(\frac{10^9}{1.3 \times 10^{10}}\right)^{2/3} = 0.18 R(t_0) = 7 \times 10^9 \text{ ly.}$$

In other words, one billion years after the Big Bang the particle horizon is about half the size of the universe. The difference arises because the distance to the particle horizon increases more quickly than the cosmic expansion. The horizon distance is proportional to t, whereas the radius of the universe is proportion to $t^{2/3}$ (or $t^{1/2}$ at very early times).

It is difficult to comprehend that everything we know — the Earth, the Galaxy, clusters of galaxies — would once have fitted inside a thimble. And yet, when the universe had a radius of one centimetre it *should* have been much smaller still: its radius was 10^{25} times greater than the horizon distance. This produces a deep problem. Within the particle horizon, objects may be in causal contact with each other; but they are causally disconnected from objects beyond the horizon. The early universe contained a vast number of such causally disconnected regions. There is absolutely no way that these isolated regions, each hidden away within its own particle horizon, could have known of each others' existence. And yet, for some reason, they all began their cosmic expansion at precisely the same time and in precisely the same way. We know they did, because our present observable universe, into which all these regions expanded, is so isotropic: the cosmic microwave background, for instance, looks the same in all directions*. This is the deep problem. Cosmologists call it the *horizon problem*.

There seem to be only two solutions to the horizon problem. The first is to rely upon coincidence, and say simply that the isotropy of the universe is just one of those things that happens. This is a profoundly unsatisfying solution. If you were a teacher marking a test, and you found that not only had every student in your class scored 98.63% but that they had all made exactly the same mistakes — mistakes, moreover, that were in your model answers — you would suspect a case of cheating. Someone must have obtained your model answers before the test and given them to each student. You would not consider it to be coincidence. The likelihood is vanishingly small that vast numbers of causally disconnected regions — thousands at the recombination epoch, trillions when the universe was a few seconds old — all behave in the same way by chance. There *had* to be some 'cheating' or prior communication.

The second solution to the horizon problem explains how the different causally disconnected regions managed to 'cheat'. Figure 11.10 makes the solution clear. It shows that the horizon problem arises because the present observable universe seems to have expanded from a region of space that, in the past, was much bigger than the particle horizon. Therefore, our universe could *not* have expanded from such a large region. The problem is solved if, immediately after the creation, the universe was *within* the particle horizon. The spacetime region that expanded to form our present observable universe would then have been small enough for all its parts to be in causal contact. This is when the 'cheating' took place: all regions could 'talk' to each other, so any differences between regions could smooth out. Then, in a fleeting moment, a period of superfast expansion must have occurred. This super-expansion, which is called *inflation*, outpaced the particle horizon: it put causally connected regions of spacetime far beyond the horizon. After inflation ended, the universe expanded more sedately — in the cosmic expansion we see today — and the particle horizon began to catch up.

There seem to be no other plausible explanations of the horizon problem. (Other proposed solutions, such as the so-called *mixmaster universe*, met with no success.) So we are driven to the idea of inflation. The description given above, though, is just a sketch of an idea. Why should the universe expand so quickly? We know that, if the conditions are right, the scale factor $R(t)$ of a universe can increase at arbitrarily high rates. We have already seen an example of a model universe (the de Sitter universe) in which the scale

---

*When we study the cosmic microwave background in patches of sky separated by more than about 2°, we study parts of the universe that were not in causal contact when radiation decoupled from matter. There is no way these regions can 'know' that they should be at the same temperature; and yet they are.

factor increases exponentially with time. But what could make our real universe undergo such an extremely rapid period of expansion?

11.9.1   Cosmology and particle physics

With hindsight it is clear that many physicists and cosmologists anticipated some of the key features of inflation. It seems, though, that in 1979 the Russian physicist Alexei Alexandrovich Starobinsky (1948–) was the first to develop a working model of inflation. Starobinsky's model, which was based on a theory of quantum gravity, was very complicated. Furthermore, although it provoked great interest amongst cosmologists in the Soviet Union, the political situation at that time prevented his model from being discussed by workers in the rest of the world. Thus the concept of inflation did not become generally popular until 1981, when the American particle physicist Alan Harvey Guth (1947–) published a key paper.

Alan Harvey Guth

As a particle physicist, Guth was interested not in the large-scale structure of the universe but in the details of the universe at the very smallest scales. His research involved the way in which subatomic particles may interact with each other. He was particularly interested in *grand unified theories* of particle interactions.

We know of four interactions, or forces, in nature. By far the strongest interaction is the unimaginatively titled *strong interaction*. The strong interaction has a short range — about the size of an atomic nucleus — but within its domain it governs the behaviour of particles like protons and neutrons. According to quantum theory, the strong force is mediated by the exchange of particles called gluons. The next strongest interaction is electromagnetism, which takes place through the exchange of photons. Electromagnetism can act over large distances. However, because electric and magnetic sources exist in two 'opposite' types (namely positive and negative charges, north and south poles) the tendency is for large amounts of matter to be neutral. Next in strength is the *weak interaction*, whose range is limited to a distance roughly 100 times smaller than the diameter of an atomic nucleus. The weak interaction is responsible for the radioactive decay of certain nuclei, and for the occasional interaction of neutrinos with matter. It is mediated by the exchange of very massive particles called W and Z particles. By far the weakest interaction is gravity. This may seem odd, since gravity plays the dominant role in astronomy. Gravity wins by default, though: the strong and weak interactions do not make themselves felt outside the atomic realm, while the two types of electric and magnetic charge tend to cancel each other when large amounts of matter are present.

For many years physicists have tried to show that different forces are different manifestations of some underlying, unified force. This endeavour began in earnest with the Scottish physicist James Clerk Maxwell (1831–1879), who showed that electricity and magnetism are just different aspects of the electromagnetic interaction. During the 1960s and 1970s, the American physicists Steven Weinberg (1933–) and Sheldon Lee Glashow (1932–), and the Pakistani physicist Abdus Salam (1926–1996), showed that at high enough energies the weak and electromagnetic interactions become unified. They are different aspects of a single *electroweak* interaction.

The electroweak unification of Glashow, Salam and Weinberg, implies that a scalar *Higgs field*, named after the British physicist Peter Ware Higgs (1929–), permeates the universe. (A scalar field — temperature, for instance — is simply a single number at each point in space.) Above the electroweak unification temperature the mean value of the Higgs field is zero, so there is symmetry between the weak and electromagnetic interactions: the photon, W and Z particles are all massless, and there is no fundamental difference between the interactions. Below the unification temperature the Higgs field 'freezes' to a non-zero value in the vacuum. The W and Z particles interact with this constant scalar field and become massive; the photon does not interact with this field, and remains light. The symmetry of the theory has been broken.

The energy at which electroweak unification takes place is not now found in nature. (The electroweak interaction split into the two distinct interactions we see today when the universe was only 10^{-12} s.) That is why physicists who wish to study the electroweak force must build large particle accelerators, such as those at CERN in Geneva and Fermilab in Chicago.

The success of the Glashow–Salam–Weinberg theory provoked the thought that at even higher energies, and with the addition of more Higgs fields, the same symmetry-breaking mechanism might unify the electroweak interaction with the strong interaction*. This is the idea behind grand unified theories. The energies needed to study grand unification are beyond those that any man-made particle accelerator can produce. But such energies were commonplace in the early universe (about 10^{-35} s after the Big Bang — a time when, as we have seen, the universe had a radius of about one centimetre). This was why Guth, a particle physicist, became interested in cosmology. It seemed the only way to study what might happen at grand unification energies.

Guth wanted to know what happened as the universe made the transition from a grand unified phase, when the Higgs fields were zero and there was just the single grand unified interaction, to a phase in which the Higgs fields had non-zero values and the strong

---

*The unification of gravity with the strong and electroweak interactions is a profoundly difficult problem, which requires a quantum theory of gravity. Particle physicists are currently very excited about *M theory*, which may hold the clue to a 'theory of everything'. Hopes have been high before, but have always been dashed.

interaction became separate from the electroweak interaction. He therefore investigated what the various candidate theories predicted would happen as the universe cooled past the grand unification point. (The idea of modelling the development of the early universe as a series of phase transitions was first proposed in 1972 by the Russian physicists David Abramovich Kirzhnits (1926–) and Andrei Linde (1948–).)

There were two possibilities. The transition might have occurred at the instant the temperature reached the critical point, or it might have been delayed. Guth did not know the correct form of the grand unified theory (physicists *still* do not know the correct form of the theory), but all the best candidates suffered a problem with the first possibility. If the transition occurred immediately, a host of exotic particles called *magnetic monopoles* would form. Magnetic monopoles are precisely what their name suggests: isolated north and south magnetic poles. You cannot make a monopole. If you cut a bar magnet in half you do not produce isolated poles: you produce two smaller magnets, each with a north and south magnetic pole. Although we cannot make monopoles today, the huge number generated in the early universe would still exist today. Furthermore, they would be easily observed. Since we do not see monopoles in the universe, Guth ruled out the idea of an instantaneous phase transition at the grand unified temperature.

Guth therefore concentrated on the second possibility. He determined the consequences of a transition that was delayed while the universe cooled below the transition temperature. (Delayed transitions can happen in the more common phase transitions studied by physicists. For example, if water is cooled very rapidly it can remain in the liquid phase up to 20 degrees below the freezing point.) Guth's calculations suggested that the universe would be trapped in a *false vacuum* state that could fuel inflation.

We probably all have a mental picture of a vacuum as a state of nothingness. To a particle physicist, though, a vacuum is simply the state of least energy. Once the universe had cooled sufficiently, the vacuum should have resembled the familiar state of broken symmetry that physicists can probe today in high-energy particle experiments. But because the transition from grand unification was delayed, the state of least energy — the vacuum — became trapped in the 'wrong' phase: a phase where the symmetry between the strong and electroweak forces had not yet broken. (By way of analogy, suppose you lived in a valley surrounded on all sides by mountains: you would consider yourself to be in a state of 'lowest altitude'. But you could be wrong. Out of sight on the other side of the mountains there might be an even lower valley, the 'true' state of lowest altitude. To reach this 'true' state requires a great deal of energy expenditure, in the form of a trek to a mountain summit. Or — if this situation were governed by the bizarre rules of quantum physics — you might be quantum tunnelled to the 'true' state! In either case, you might be trapped in the 'false' valley for some time.)

Guth found that all grand unified theories predicted that the false vacuum had a large and *constant* energy density ρ_{fv}. Moreover, $p_{\text{fv}} = -\rho_{\text{fv}} c^2$, so the false vacuum had the seemingly bizarre property of *negative* pressure. The false vacuum lasted for only a tiny fraction of a second, but while it lasted its effect upon the universe was extraordinary.

To see why this situation should fuel inflation, imagine that a small region of spacetime quantum tunnels into the state of true vacuum. Although this tiny bubble of 'normal' universe has zero pressure, it has higher pressure than the false vacuum that surrounds it. The bubble of true vacuum therefore expands. The expansion of the bubble is governed by the Friedmann equations. For a flat universe we can write (10.35), the first of the Friedmann equations, as

$$\ddot{R} = \tfrac{8}{3}\pi G \rho_{\text{fv}}\, R. \tag{11.28}$$

This differential equation has a general solution that is the sum of exponentially increasing and decreasing terms. Since the decreasing term rapidly tends to zero we can ignore it. If R_0 is the value of the scale factor when inflation begins, then $R(t)$ — the scale factor of the universe at time t — is given by

$$R(t) = R_0 e^{Ht} \qquad \text{where} \quad H = \sqrt{\tfrac{8}{3}\pi G \rho_{\text{fv}}}. \tag{11.29}$$

In other words there is a constant Hubble parameter, and the universe grows exponentially quickly. It is like the strange de Sitter universe; the false vacuum plays the role of a very large positive cosmological constant, and drives the expansion ever faster. At the end of inflation the energy of the false vacuum is dumped into the universe, and the universe re-heats to the temperature it had before inflation. From this point onwards the universe evolves in exactly the same way as in the standard Big Bang picture.

Different models of inflation produce different values for the duration of the false vacuum phase and for H. We can see the effect of inflation, though, by choosing some conservative values. A value of 100 for the exponent Ht would be on the low side. If we use this estimate we can see that the universe was $e^{100} \approx 2.7 \times 10^{43}$ times larger at the end of the inflationary period than it was at the start. According to some grand unified theories, the phase transition took place at about 10^{-34} s. The horizon distance at this time was 3×10^{-26} m, and we can take this as the radius of the initial bubble. At the end of the inflationary period — at about 10^{-32} s — the radius of the bubble had inflated to

$$3 \times 10^{-26} \times 2.7 \times 10^{43}\ \text{m} \approx 10^{18}\ \text{m} \approx 100\,\text{ly}.$$

By comparison, the region that later expanded to become our present-day observable universe was only about 3 m in radius!

Thanks to inflation the horizon problem is solved: at very early times the spacetime region that became today's observable universe occupied only a minuscule fraction of the causally connected region. Inflation solves the monopole problem: the phase transition created monopoles in abundance, but inflation swept them beyond the horizon, out of our sight. Inflation also explains why we observe the universe to be very nearly flat*: any curvature is flattened if space stretches exponentially.

The original formulation of inflation proved to be unworkable; after a dose of Guth's inflation the universe would look nothing like the universe we actually see. Nevertheless, the idea of inflation seemed so appealing, and it seemed to avoid so many of the pitfalls of Big Bang cosmology, that it became a hot topic of research. Soon after Guth published

---

*The so-called *flatness* problem is just as puzzling as the horizon and monopole problems. That the present universe is flat (or very nearly flat) can be expressed by saying that Ω is close to 1; we measure Ω to be about 0.3, but it may well be larger. However, cosmic expansion drives Ω away from 1. The puzzle is that for Ω to be so close to 1 now, it must have been *incredibly* close to 1 at the time of grand unification: it was unity to within one part in 10^{60}. Where does this 'fine tuning' originate?

his model, Linde — who was then still based in Russia — showed how it could be improved. Quite independently, the American physicists Paul Joseph Steinhardt (1952–) and Andreas Johann Albrecht (1957–) introduced the same idea a few months later. The key to the *new inflationary scenario,* as it became called (to distinguish it from Guth's 'old' inflationary model), was the way in which the phase transition from false vacuum to true vacuum took place. In old inflation the phase transition, when it finally happened, took place almost instantaneously. In new inflation, the transition took place much more slowly. While the universe gently rolled to its true vacuum it would still be in the false vacuum state, and it would continue to inflate. The inflation would stop automatically when the true vacuum was reached, and the rapid oscillations of the Higgs fields at this point would produce the elementary particles and reheat the universe. Linde showed that this new model of inflation not only kept all the successes of Guth's original model, but also explained the homogeneity and isotropy of the universe as well as predicting the form of the density perturbations that would eventually evolve into the large-scale structure that we see today. (In the 1970s, Zel'dovich and the English-born cosmologist Edward Robert Harrison (1919–) independently studied models of galaxy formation from primordial density fluctuations. They argued on mathematical grounds that the spectrum of density perturbations is 'scale free'. In other words, if the average matter density ρ has a small density perturbation $\delta\rho$ then the ratio $\delta\rho/\rho$ does not single out any particular linear scale. This is the so-called Harrison–Zel'dovich spectrum. Inflation provided a physical explanation for the spectrum: the inflating field is subject to quantum fluctuations; as the scale factor increases, these tiny perturbations stretch and become large, generating density perturbations with a scale-free spectrum.)

The success of new inflation led to an explosion of interest in the topic. The amount of research activity devoted to inflation underwent its own period of inflationary expansion. And yet still the various scenarios seemed somewhat artificial and unrealistic; the scalar fields had to be chosen with care in order to make the models work. In 1983, though, Linde realised that there was no need to invoke complicated quantum gravity effects, as Starobinsky had done. There was no need to base the model on high-temperature phase transitions, as Guth had done. There was even no need to make the usual assumption that the very early universe was hot! Linde argued that we should not regard inflation as some sort of 'patch' that is needed only to mend the holes in standard Big Bang cosmology. Instead, inflation itself is the key concept. It arises quite naturally. The standard Big Bang model that astronomers have studied all these years is an offshoot of inflation.

The key to this new way of viewing cosmology is simply to assume that the early universe contained a chaotic* jumble of all kinds and values of scalar fields. (These are not necessarily Higgs fields. The general name for such a scalar field is an *inflaton* field.) All one need do is check these various scalar fields, and see which of them leads to inflation. Domains where inflation occurs become very large; domains where inflation does not occur remain small. A domain stops inflating when the scalar field reaches the minimum of its potential energy; as it oscillates about this minimum it releases its energy

---

*The word 'chaos' is a Greek word that originally meant 'empty space'. Due to its association with creation myths, the word eventually came to refer to the confusion of matter that emerged from originally empty space. So the use of the word in the context of inflation is apt. Note that it does not refer to the chaos theory of mathematics.

in the form of elementary particles. The evolution of the universe from this point on is described by the usual theory of the Big Bang. Linde called this scenario *chaotic inflation*, referring to the complicated tangle of scalar fields that existed in the early universe. There are many possible models of inflation, but chaotic inflation (in some form) is probably the version of inflation favoured by most cosmologists.

Inflation is an active area of research, and it would be wrong to suggest that the picture is anywhere near complete. It is difficult to form a complete picture when physicists have yet to obtain experimental evidence for the existence of even one fundamental scalar field! Much theoretical work remains to be carried out, particularly in linking inflation with the current theories of elementary particles. But if the details remain unknown, what seems almost certain is that a period of inflation did take place. And every model of inflation tells us that most of the volume of the universe was carried beyond the present particle horizon. The universe is very much larger than the volume we can at present observe.

11.10   THE UNIVERSE?

Throughout this book I have written of the universe — with a lower-case 'u'. But if inflation is correct our observable universe is just a small part of a much bigger structure. Perhaps we should call this larger structure 'the Universe'? Incredibly, the Universe may be much bigger even than this.

According to the model of chaotic inflation, the spacetime region we live in sprang from a quantum fluctuation that inflation made large. But there is nothing special about 'our' quantum fluctuation. Different regions of the early universe may have undergone different amounts of inflation starting at different times. Some domains will have evolved into different vacuum states of equivalent energy, in which case they will be governed by different realisations of the laws of physics. In some domains, even fundamental properties such as the number of dimensions of space might be different.

Linde expanded the idea of chaotic inflation even further. In his model of *eternal chaotic inflation*, the major part of the volume of the Universe is in the inflationary phase. A subregion of a universe can undergo its own period of inflation, a process that generates an entirely new 'baby' universe. Once it starts, inflation never stops. (Thus when we considered earlier the 'age of the universe', we merely obtained the time back to the end of our own inflationary event.) In the model of eternal chaotic inflation the Universe — the totality of space and time — consists of a richly structured never-ending succession of universes, of which our own universe is an infinitesimal part. These universes may, in principle, for ever be unobservable. That has not stopped cosmologists from speculating about them ... but perhaps this is the point for *us* to stop.

CHAPTER SUMMARY

- We can use the Hubble law as a distance indicator: it enables us to convert redshift to distance. For large redshifts the linear redshift–distance law breaks down, but for any particular cosmological model we can calculate a relation that holds for all redshifts.

- Observational cosmologists usually measure a luminosity distance (using a standard candle) or an angular diameter distance (using a standard rod). In common parlance one often expresses distances in terms of light travel time. In general relativity one can also calculate the proper distance.

- Within the simple Einstein–de Sitter universe, the following expressions denote the relation between distance and redshift, z:

$$d_P = 2H_L\left(1 - \frac{1}{\sqrt{1+z}}\right)$$

$$d_L = 2H_L\left(1 + z - \sqrt{1+z}\right)$$

$$d_A = 2H_L\left(\frac{1}{1+z} - \frac{1}{(1+z)^{3/2}}\right)$$

$$d_M = 2H_L\left(1 - \frac{1}{\sqrt{1+z}}\right)$$

$$d_T = \tfrac{2}{3}H_L\left(1 - \frac{1}{(1+z)^{3/2}}\right)$$

where d_P, d_L, d_A, d_M and d_T are respectively the proper distance, luminosity distance, angular diameter distance, proper motion distance and light travel-time distance; the term H_L $(= c/H_o)$ is the Hubble length.

- The value of the Hubble constant H_o has been hotly debated in recent decades. Some workers favour $H_o \approx 50\,\mathrm{km\,s^{-1}\,Mpc^{-1}}$; others favour $H_o \approx 75\,\mathrm{km\,s^{-1}\,Mpc^{-1}}$. Much of the disagreement stems from the difficulty in making accurate measurements of the distances to galaxies taking part in the Hubble flow.

- At the time of writing, the most distant object is a galaxy with a redshift $z = 5.64$. The cosmic microwave background radiation consists of photons that have been redshifted by $z = 1100$. It is impossible to see farther than this because at higher redshifts the universe was opaque to photons.

- The particle horizon is the boundary of our observable universe. For an Einstein–de Sitter universe with $H_o = 50\,\mathrm{km\,s^{-1}\,Mpc^{-1}}$ the particle horizon is at a proper distance of $12\,000\,\mathrm{Mpc}$. Any region beyond the particle horizon is causally disconnected from the observable universe.

- The particle horizon recedes faster than the cosmic expansion; in the past, therefore, the horizon distance was less than the radius of the universe. This gives rise to the horizon problem: why is the present universe so isotropic when in the past it consisted of many causally disconnected regions?

- The solution to the horizon problem — and to several other problems with the standard Big Bang model — is inflation. In the inflationary model, a small region of the universe underwent a period of exponential expansion. The result is that our present universe expanded from a region that was small compared to the horizon distance; thus all parts of the universe were once in causal contact.

QUESTIONS AND PROBLEMS

11.1 Investigate the history behind some important physical constants, such as the gravitational constant, the speed of light or the mass of the electron. Has the value of any of these constants been the subject of debate in the way that astronomers have debated the value of the Hubble constant? (Make allowance for the fact that physical constants can often be measured in the laboratory.)

11.2 From the fact of your own existence, derive a rough estimate for the upper limit of H_0.

11.3 Obtain an expression for parallax distance in terms of z, in the same way that we obtained an expression for luminosity and diameter distance. Is there any likelihood of a parallax distance being observed over cosmologically interesting distances?

11.4 For each of the extragalactic distance indicators we have discussed, decide whether they provide us with a luminosity distance, an angular diameter distance or something else. For the range of applicability of these indicators, is there any practical difference between luminosity and diameter distances?

11.5 Demonstrate that the Einstein–de Sitter model we have been studying does not have an event horizon.

11.6 You see a galaxy with a redshift of 1 and another with a redshift of 2. Which was closer to us *at the time of emission*, assuming that we live in an Einstein–de Sitter universe?   *The galaxy with $z = 2$*

11.7 Obtain an expression for the distance modulus of a galaxy that is valid for small z within an Einstein–de Sitter universe. Take $H_0 = 50\,\mathrm{km\,s^{-1}\,Mpc^{-1}}$. $\mu_0 = 16.51 + 5\log cz + 0.543z$

11.8 Calculate the proper *emission* distance of a galaxy of redshift z in an Einstein–de Sitter universe.

$$d_{p(e)} = 2H_L[(1+z)^{-1} - (1+z)^{1/2}]$$

FURTHER READING

GENERAL
Barrow J D (1994) *The Origin of the Universe* (Weidenfeld and Nicolson: London)
    — An engaging and clearly written account of the latest ideas about the Big Bang.

Bondi H and Weston-Smith M (1998) *The Universe Unfolding* (Clarendon: Oxford)
    — Twenty Milne Lectures in book form, from speakers who constitute a *Who's Who* of modern astronomy. Several of the lectures refer to topics covered in this chapter.

Lederman L M and Schramm D N (1995) *From Quarks to the Cosmos* (Freeman: New York)
    — A very clear account of particle physics and inflation; one of the *Scientific American Library* series.

Overbye D (1991) *Lonely Hearts of the Cosmos* (Macmillan: London)
    — A superb popular account of what it is like to be a professional cosmologist.

http://www.astro.ucla.edu/~wright/cosmolog.html
    — Excellent tutorials on cosmology, general relativity and the distance ladder. Highly recommended.

REDSHIFT SURVEYS
Fairall A (1998) *Large-Scale Structure in the Universe* (Wiley–Praxis: Chichester)
    — A very accessible work covering all aspects of large-scale structure. Worth buying just for the beautiful 26-page atlas of nearby structures.

http://www.sdss.org/sdss.html
    — A page with news of the Sloan Digital Sky Survey.

REVIEWS OF THE HUBBLE CONSTANT DEBATE

Bonnell J T, Nemiroff R J and Goldstein J J (1996) The scale of the universe debate in 1996. *Pub. Astron. Soc. Pacific* **108** 1065–1067
— The introduction to the modern re-run of the 'Great Debate'. This journal volume contains six interesting essays on the Hubble parameter.

Mould J R, Kennicutt R. C. Jr and Freedman W L (1999) The calibration of the extragalactic distance scale: methods and problems. *Rep. Prog. Phys.* **62** to be published
— A useful overview of the *HST* Key Project methods.

Sandage A R and Tammann G A (1997) The evidence for the long distance scale with $H_0 < 65$. In: *Critical Dialogues in Cosmology* ed. N.Turok. (World Scientific: Singapore)
— A robust defence of the long distance scale.

Tammann G A (1996) The Hubble constant: a discourse. *Pub. Astron. Soc. Pacific* **108** 1083–1090
— Part of the 1996 distance scale debate (see the Bonnell reference). The author of course prefers the long distance scale: $45 < H_0 < 65 \, \mathrm{km \, s^{-1} \, Mpc^{-1}}$.

van den Bergh S (1994) The Hubble parameter revisited. *Pub. Astron. Soc. Pacific* **106** 1113–1119
— The author prefers the short distance scale: $H_0 \gtrsim 75 \, \mathrm{km \, s^{-1} \, Mpc^{-1}}$.

van den Bergh S (1996) The extragalactic distance scale. *Pub. Astron. Soc. Pacific* **108** 1091–1996
— Part of the 1996 distance scale debate (see the Bonnell reference). The author argues for the short distance scale: $H_0 = 72 \, \mathrm{km \, s^{-1} \, Mpc^{-1}}$.

van den Bergh S (1997) Early history of the distance scale problem. In: *The Extragalactic Distance Scale* ed. M. Livio, M. Donahue and N. Panagia. (Cambridge University Press: Cambridge) pp 1–5
— An interesting account, starting with Slipher's work on the redshifts of nebulae.

http://www.ipac.caltech.edu/H0kp/
— The home page of the *HST* Key Project on the extragalactic distance scale. From here you can obtain details of all the Key Project publications.

RECENT ESTIMATES OF THE HUBBLE CONSTANT

(Note: you can find more estimates in the 'further reading' section of the chapter on galaxy distances.)

Impey C D *et al.* (1998) An infrared Einstein ring in the gravitational lens PG 1115 + 080. *Astrophys. J.* **509** 551–560
— Depending upon the lens model, $H_0 = 44 \pm 4 \, \mathrm{km \, s^{-1} \, Mpc^{-1}}$ or $H_0 = 65 \pm 5 \, \mathrm{km \, s^{-1} \, Mpc^{-1}}$.

Lanoix P (1998) *Hipparcos* calibration of the peak brightness of four SNe Ia and the value of H_0. *Astron. Astrophys.* **331** 421–427
— By calibrating the Cepheid period–luminosity relation from *Hipparcos* data, the author derives a peak absolute magnitude of Type Ia supernovae of -19.65 ± 0.09 and thence $H_0 = 50 \pm 3 \, \mathrm{km \, s^{-1} \, Mpc^{-1}}$. This clearly favours the long distance scale.

Madore B F *et al.* (1998) A Cepheid distance to the Fornax cluster and the local expansion rate of the Universe. *Nature* **395** 47–50
— The authors deduce $H_0 = 70 \, \mathrm{km \, s^{-1} \, Mpc^{-1}}$ from the distance to NGC 1365 in Fornax.

Nevalainen J and Roos M (1998) Cepheid metallicity and Hubble constant. *Astron. Astrophys.* **339** 7–14
— From a combination of Cepheid and supernova data, the authors find $H_0 = 68 \pm 5 \, \mathrm{km \, s^{-1} \, Mpc^{-1}}$.

Schaeffer B E (1998) The peak brightness of SN1974G in NGC 4414 and the Hubble constant. *Astrophys. J.* **509** 80–84
— The author finds $H_0 = 55 \pm 8 \, \mathrm{km \, s^{-1} \, Mpc^{-1}}$.

Shanks T (1997) A test of Tully–Fisher distance estimates using Cepheids and SN Ia. *MNRAS* **290** L77–L83
— The author compares Tully–Fisher distance estimates to 11 spiral galaxies with Cepheid distances, and to 12 spiral galaxies with supernova distances, and shows that the Tully–Fisher distance estimates were too short by 0.43 ± 0.11 mag. The Tully–Fisher estimates of $H_0 = 84 \pm 10$ km s^{-1} Mpc^{-1} should be revised down to $H_0 = 69 \pm 8$ km s^{-1} Mpc^{-1}.

Tanvir N R *et al.* (1995) Determination of the Hubble constant from observations of Cepheid variables in the galaxy M 96. *Nature* **377** 27–31
— The authors use Cepheids to determine a distance to the Leo I group of 11.6 ± 0.8 Mpc, and thence a distance to the Coma cluster. From this they obtain $H_0 = 69 \pm 8$ km s^{-1} Mpc^{-1}.

MALMQUIST BIAS

Teerikorpi P (1997) Observational selection bias affecting the determination of the extragalactic distance scale. *Ann. Rev. Astron. Astrophys.* **35** 101–136
— A detailed review of how to recognise and correct for biases, particularly in the case of the Tully–Fisher method of distance determination.

HIGH-REDSHIFT OBJECTS

Dey A *et al.* (1998) A galaxy at $z = 5.34$. *Astrophys. J.* **498** L93–L98
— This galaxy briefly held the record for highest redshift.

Franx M *et al.* (1997) A pair of lensed galaxies at $z = 4.92$ in the field of CL1358+62. *Astrophys. J.* **486** L75–L78
— Gravitational lenses can be used to observe very distant objects.

Hu E M, Cowie L L and McMahon R C (1998) The density of Ly α emitters at very high redshift. *Astrophys. J. Lett.* **502** L99–L103
— Records a galaxy at $z = 5.64$, the most distant object known.

Schneider D P, Schmidt M and Gunn J E (1991) PC 1247 + 3406: an optically selected quasar with a redshift of 4.897. *Astron. J.* **102** 837–840
— Until recently, PC 1247 + 3406 held the record for the quasar with the highest redshift.

Weymann R J *et al.* (1998) Keck spectroscopy and NICMOS photometry of a redshift $z = 5.60$ galaxy. *Astrophys. J.* **505** L95–L98
— One of the most distant known galaxies. Near-IR imaging with NICMOS will enable the study of high-redshift galaxies out to $z \sim 10$.

INFLATION

Abbot L F and Pi S.-Y. (1986) *Inflationary Cosmology* (World Scientific: Singapore)
— A collection of 62 research papers on inflationary cosmology, including the early papers by Starobinsky, Guth, Linde, Albrecht and Steinhardt. Useful if you do not have access to the research journals.

Guth A H (1989) Starting the universe: the Big Bang and cosmic inflation. In: *Bubbles, Voids and Bumps in Time: the New Cosmology* ed. J. Cornell (Cambridge University Press: Cambridge)
— A superb non-technical explanation of inflation by one of those who devised the theory; the other reviews on different aspects of cosmology, by famous astronomers such as Lightman, Kirshner, Geller, Rubin and Gunn, are equally good.

Linde A D (1990) *Particle Physics and Inflationary Cosmology* (Harwood: Chur)
— Chapter 1 of this book is written at a relatively elementary level, and is an excellent introduction to the ideas on inflation.

Linde A D (1994) The self-reproducing inflationary universe. *Sci. American* **271**(5) 32–39
— A readable account of the latest ideas on inflation.

Narlikar J V and Padmanabhan T (1991) Inflation for astronomers. *Ann. Rev. Astron. Astrophys.* **29** 325–362
— A relatively accessible introduction to inflation.

A

The names of astronomical objects

Astronomical objects have names ranging from the imaginative (the Cartwheel galaxy, for example) to the mundane (like CL 1358+62). Often, the same object has several names. For instance, the beautifully named Lacework Nebula also goes by the less beautiful name NGC 6960. A bright star like Sirius has dozens of names. Unfortunately, there are more than 1000 naming conventions; although most of them apply only to deep-sky objects studied by professional astronomers, there is still plenty of room for confusion. In this appendix I hope to make some sense of the different naming conventions that appear throughout the book. The appendix is split into two sections.

A.1 STARS

About 1000 of the brightest naked-eye stars have *proper names*, which come from a variety of languages. Our modern-day proper names for stars generally derive from Arabic, since it was the Arabs who preserved Greek astronomy during the Dark Ages. Thus we have stars called Rigel, Aldeberan and Betelgeuse. Only a few stars, like Sirius, Castor and Pollux, have names that date back to Greek times.

The ancient astronomers grouped naked-eye stars into constellations — patterns of stars that seemed to them to resemble mythological or living beings, or inanimate objects. Thus we have constellations like Orion (a group of stars that was said to resemble the mythological hunter), Taurus (a group of stars that was said to look like a bull) and Triangulum (a group whose brightest stars really do form a noticeable triangle). The stars in a constellation are not necessarily physically associated with each other, of course. Nevertheless, the constellations helped the ancient astronomers to make sense of the starry sky, and the constellations remain with us to this day albeit in slightly altered form. (Originally there were 48 constellations. Now there are 88 constellations in total. The southernmost constellations were assigned names in relatively recent times, so we have constellations called Antlia (the Air Pump), Telescopium (the Telescope) and Horologium (the Clock) — names that would have been meaningless to the first people who saw patterns in the sky, or even to the ancient Greeks or Romans.)

There was no way of knowing which star belonged to which constellation, unless you simply memorised the information. Between 1603 and 1607, the German astronomer Johann Bayer (1572–1625) introduced some order into the naming of stars within constellations. He produced a comprehensive star catalogue, and listed the stars in each constellation in order of brightness. (Actually, Bayer was not consistent in his naming scheme, and this strict ordering-by-brightness scheme is seldom followed exactly!) The brightest star in a constellation was α, after the first letter of the Greek alphabet; the second brightest star in that constellation was β, after the second letter of the Greek alphabet; and so on. These identifiers were used with the Latin genitive of the constellation name. Consider, for instance, the constellation Orion. Under Bayer's convention, Betelgeuse is known as α Orionis ('alpha of Orion'); Rigel is known as β Orionis; and so on. (This is an example where the Bayer system is confusing. In fact, Rigel is the brightest star in Orion and Betelgeuse is the second brightest! Rigel has a visual magnitude of 0.12, whereas Betelgeuse has a visual magnitude of 0.50, so Rigel should really be α Orionis.) The Greek alphabet is as follows:

α	alpha	η	eta	ν	nu	τ	tau
β	beta	θ	theta	ξ	xi	υ	upsilon
γ	gamma	ι	iota	o	omicron	ϕ	phi
δ	delta	κ	kappa	π	pi	χ	chi
ϵ	epsilon	λ	lambda	ρ	rho	ψ	psi
ζ	zeta	μ	mu	σ	sigma	ω	omega

Astronomers still often refer to stars of the first magnitude or brighter by their proper names; for dimmer stars they are more likely to use Bayer's nomenclature. (It is safe to assume that any professional astronomer you meet will be unable to identify the star Alya, for instance. Its Bayer name, θ Serpentis, is much more common.) Stars in the southern sky that were catalogued only after Bayer's time do not have proper names, and are referred to using Bayer's nomenclature.

Bayer's system has an obvious drawback: there are only 24 Greek letters, but constellations generally have more than 24 stars. When you look at the sky through a telescope, and begin to see hundreds of thousands of stars, the need for a new naming system becomes obvious.

In 1712, Flamsteed introduced a new system whereby he ignored the brightness of the star and simply numbered each star in a constellation from west to east, in order of right ascension. Flamsteed gave each star a number irrespective of whether Bayer had assigned it a Greek letter. Thus it was that α Lyrae gained the name 3 Lyrae (though it is still better known as Vega).

The main success of Flamsteed's system was the idea of numbering stars by right ascension. When Argelander compiled the gigantic *Bonner Durchmusterung des Nordlichen Himmels* it was clear that astronomers could no longer categorise stars by constellation. There were simply too many stars for this to make sense. Instead, Argelander divided the sky into strips of declination $1°$ wide and a full 24 h in right ascension. Stars in each band were numbered in strict order of right ascension. For instance, table 4.1 on page 82 includes the star BD $+5°$1668. This means it is in the *Bonner Durchmusterung* (BD), and is the 1668th star, counting from 0^h RA, in the zone between declination $+5°$ and $+6°$. There are hundreds of thousands of stars with a BD designation. Stars with a CD designation belong to the *Cordoba Durchmusterung*, which continued the Bonn Survey into the southern half of the sky.

Several other large catalogues appeared after the *Bonner Durchmusterung*, and we have met some of them in this book: the *Henry Draper Catalogue* compiled by Cannon, for instance. Many stars are known by these HD numbers. Also common is the numbering system based on the *Smithsonian Astrophysical Observatory Star Catalogue*. About 300 000 stars have an SAO designation. Table A.1 gives the names and abbreviations of several of the major star catalogues.

TABLE A.1: Abbreviations of some of the best known star catalogues. The *Hipparcos* and *Hubble Space Telescope* catalogues are likely to become the most important star catalogues over the next few years.

Abbreviation	Catalogue
AGK *n*	*Astronomischer Gesellschaft Katalog Nummer n*
BD	*Bonner Durchmusterung*
BS	*Yale Catalogue of Bright Stars*
CD	*Cordoba Durchmusterung*
CPD	*Cape Photographic Durchmusterung*
FK *n*	*Fundamental Katalog Nummer n*
GC	*Washington General Catalogue*
GCVS	*General Catalogue of Variable Stars*
GJ	*Gliese–Jahreiss*
GL	*Gliese*
HD	*Henry Draper Catalogue*
HDE	*Henry Draper Catalogue Extension*
HR	*Harvard Revised Photometry Catalogue*
IDS	*Index Catalogue of Visual Double Stars*
L	*Luyten*
SAO	*Smithsonian Astrophysical Observatory Catalogue*

Two new catalogues may replace some of those in table A.1: the *Hubble Space Telescope Guide Star Catalogue* and the *Hipparcos* catalogue. Designations based on these catalogues will become more common. A star with a GSC designation has eight digits assigned to it: the first four refer to one of 9537 regions of sky that the *Hubble Space Telescope* covered, and the last four are a serial number within that region. A star in the *Hipparcos* catalogue is identified simply by its number in the catalogue. Barnard's star, for instance, is HIP 87937.

That takes care of 'normal' stars. What of variable stars? Bright variables that have a proper name or Bayer designation are usually referred to by name. For dimmer variables, Argelander introduced a system whereby the first variable star discovered in a particular constellation was assigned the letter R along with the Latin genitive of the constellation name. The next variable to be found was S, the next T, and so on up to Z. After Z, Argelander continued the system RR (hence RR Lyrae stars), RS, RT and so on up to RZ. After that it went SS up to SZ, and finally to ZZ. This was soon insufficient to capture all the new variable stars being discovered, so after ZZ astronomers went to AA through to AZ, BB through to BZ and so on up to QZ. (They omitted J because of potential confusion with the letter I of German script; much of the astronomical literature at that time appeared in German.) At QZ they gave up. They had 334 two-letter designations for variables within a constellation, and this was insufficient. After QZ, the next variable discovered in a constellation was called simply V335, the one after that V336, and so on.

Novae are also variable stars. They are *eruptive* variables, rather than the more common pulsating variables such as Cepheids and RR Lyrae stars. Novae were once named after the constellation in which they appeared, but since 1925 astronomers have given them variable-star designations.

Supernovae are also variable — cataclysmically so — but they are not given variable star designations. Since supernovae are much rarer than any other type of variable star, the naming system can be simpler. A supernova is named after the year in which it is discovered, the first supernova of that year being called A, the second B, and so on. Thus the first supernova of 1999 is designated SN1999A. (An amusing sign of the tremendous recent progress in this field is that, in 1998, the Supernova Cosmology Project and the High-z Supernova Search team found so many supernovae that the official letter–number designations became clumsy. The SCP members have thus begun to give supernovae nicknames. For example, SN1998eq — the most distant supernova yet discovered — has been given the nickname Albinoni, after the Venetian composer Tomaso Giovanni Albinoni (1671–1751).)

Planetary nebulae are the remains of high-mass stars. Many planetaries have proper names based upon their visual appearance, but the commonest designation consists of the letters PN followed by the appropriate galactic coordinates. Some planetaries also have catalogue names. Thus the Owl Nebula is designated PN $11^h 14^m + 55°01'$, and also M 97 and NGC 3587.

A.2   GALAXIES, GALAXY CLUSTERS, QUASARS AND NEBULAE

With just a few exceptions, extragalactic objects were invisible to the ancients. So although the best-known deep-sky objects often have 'nicknames' (usually whimsical ones, based upon their appearance), astronomers almost always refer to them by a catalogue number. (The situation is slightly different in the Local Group, where several of the smaller galaxies derive their name from the constellation in which they lie.) There are several major catalogues, and an important object like the Andromeda galaxy has several designations. Fortunately, the situation is nowhere near as bad as with stars.

The most famous catalogue is the one completed by Messier. The Messier Catalogue contains several objects that belong to our Galaxy. For example, M 1, the famous Crab Nebula, is a nearby supernova remnant. Other Messier objects include planetary nebulae, globular clusters, emission nebulae and reflection nebulae. The Messier Catalogue does contain galaxies, though; the three most prominent members of the Local Group have Messier numbers.

Perhaps the most widely used naming system for galaxies is based on the *New General Catalogue*. Thousands of galaxies have an NGC designation. The *Index Catalogue* is also common; two Local Group galaxies have IC designations.

Clusters of galaxies are often known by their number in the most famous cluster catalogue, produced by the American astronomer George Ogden Abell (1927–1983). Otherwise they have the designation CL followed by coordinates specifying their position.

Quasars often go by their designation in large radio or redshift catalogues (3C and PC are well known). Otherwise they are given the designation QSO and followed by coordinates specifying their position.

Table A.2 lists the names and abbreviations of a few of the best-known deep-sky catalogues.

TABLE A.2: Abbreviations of some of the best-known deep-sky catalogues.

Abbreviation	Catalogue
A	*Abell Catalogue*
IC	*Index Catalogue*
M	*Messier Catalogue*
PC	*Palomar Transit Grism Catalogue*
PKS	*Parkes Catalogue*
NGC	*New General Catalogue*
3C	*Third Cambridge Catalogue*

FURTHER READING

GENERAL

Arnold H J P, Doherty P and Moore P (1997) *The Photographic Atlas of the Stars* (Institute of Physics: Bristol)
— Contains the proper names of all the bright stars, in addition to a host of other beautifully presented information.

Room A (1988) *Dictionary of Astronomical Names* (Routledge: London)
— The meaning of astronomical names, from lunar craters to constellations.

B

Journal abbreviations

In the 'further reading' sections the following abbreviations are used.

Abbreviation	Journal
Ann. Harvard College Observatory	Annals of the Harvard College Observatory
Ann. Rev. Astron. Astrophys.	Annual Reviews of Astronomy and Astrophysics
Ann. Rev. Geophys.	Annual Reviews of Geophysics
Astron. Astrophys.	Astronomy and Astrophysics
Astron. J.	Astronomical Journal
Astron. Nachr.	Astronomische Nachrichten
Astrophys. J	Astrophysical Journal
Astrophys. J. Suppl.	Astrophysical Journal Supplement
Bull. Astron.	Bulletin Astronomique
Comments Astrophys. Space. Phys.	Comments in Astrophysics and Space Physics
J. Hist. Astron.	Journal of the History of Astronomy
Lick Observatory Bull.	Lick Observatory Bulletin
MNRAS	Monthly Notes of the Royal Astronomical Society
Phil. Mag.	Philosophical Magazine
Phil. Trans.	Philosophical Transactions of the Royal Society
Phys. Rev.	Physical Review
Proc. Nat. Acad. Sci.	Proceedings of the National Academy of Sciences
Pub. Astron. Soc. Japan	Publications of the Astronomical Society of Japan
Pub. Astron. Soc. Pacific	Publications of the Astronomical Society of the Pacific
Quart. J. R. Astron. Soc.	Quarterly Journal of the Royal Astronomical Society
Rep. Prog. Phys.	Reports on Progress in Physics
Rev. Geophys.	Reviews of Geophysics
Rev. Mod. Phys.	Reviews of Modern Physics
Sci. American	Scientific American
Z. Phys.	Zeitschrift für Physik

Name index

Object index

Subject index